World Sustainability Series

Series editor

Walter Leal Filho, Hamburg, Germany

More information about this series at http://www.springer.com/series/13384

Walter Leal Filho
Diana-Mihaela Pociovalisteanu
Abul Quasem Al-Amin
Editors

Sustainable Economic Development

Green Economy and Green Growth

Springer

Editors
Walter Leal Filho
HAW Hamburg
Hamburg
Germany

and

Manchester Metropolitan University
Manchester
UK

Diana-Mihaela Pociovalisteanu
Faculty of Economics and Business Administration
"Constantin Brancusi" University of Targu-Jiu
Targu-Jiu
Romania

Abul Quasem Al-Amin
International Business School (IBS)
Universiti Teknologi Malaysia (UTM)
Kuala Lumpur
Malaysia

ISSN 2199-7373 ISSN 2199-7381 (electronic)
World Sustainability Series
ISBN 978-3-319-45079-7 ISBN 978-3-319-45081-0 (eBook)
DOI 10.1007/978-3-319-45081-0

Library of Congress Control Number: 2016949567

Printed on acid-free paper

This Springer imprint is published by Springer Nature
The registered company is Springer International Publishing AG
The registered company address is: Gewerbestrasse 11, 6330 Cham, Switzerland

Preface

There are at present many attempts to seek effective ways to foster sustainable development on the one hand, and economic growth on the other. Many countries are now pursuing the so-called green growth, often described as the "circular economy" and trying to green their economies. But even though this topic is important, there is a lack of clarity around what green economy policy measures encompass, and how green growth related to sustainable economic development and poverty eradication can take place.

This book addresses this knowledge gap. It illustrates many experiences in designing, implementing and reviewing the costs and benefits of green economy policies. The emerging practices, which this book identified and disseminates, will help to provide some important insights and much-needed clarity regarding the types of green economy policy measures, and the institutional barriers, risks and implementation costs associated with them. The body of knowledge generated by this book will be very useful in offering guidance and in addressing concerns on how to foster the integration of green economic policies and green growth with national economic and social priorities and objectives.

The following elements can be found in many of the chapters of this book:

- principles and practices of green growth
- the implementation of green economy strategies
- the role of ecosystem services
- socio-economic issues
- economic growth and poverty eradication
- aspects of policy and governance

We thank the authors for their willingness to share their knowledge, know-how and experiences, as well as the many peer reviewers, who have helped us to ensure the quality of the manuscripts.

Enjoy your reading!

Hamburg, Germany
Targu-Jiu, Romania
Kuala Lumpur, Malaysia
Winter 2016/2017

Walter Leal Filho
Diana-Mihaela Pociovalisteanu
Abul Quasem Al-Amin

Contents

Industrial Symbiosis: An Innovative Tool for Promoting Green Growth

Angela Albu

Abstract

In actual context it is becoming more apparent that economic activities are inconceivably linked with the environment, a fact that is generating the need for a more complex relationship between the economy and ecology. We need to arrive at a reconciliation between the desire for economic and social development on the one hand, and environmental protection on the other. The concepts of industrial ecology and industrial symbiosis are providing viable solutions to economic growth without sacrificing environmental quality. The implementation of industrial symbiosis in the economy represents an innovative method to promote green economy and to create a new culture of economic growth. The chapter will present the concepts of industrial ecology and industrial symbiosis in the context of eco-innovation, the advantages and limitations in their implementation, and some significant case studies with positive results in promoting green growth through industrial symbiosis.

Keywords

Industrial ecology · Industrial symbiosis · Eco-design · Green growth

A. Albu (✉)
Faculty of Economics and Public Administration, University "Stefan cel Mare" Suceava, Suceava, Romania
e-mail: angelaa@seap.usv.ro

W. Leal Filho et al. (eds.), *Sustainable Economic Development*,
World Sustainability Series, DOI 10.1007/978-3-319-45081-0_1

1 Introduction

From the first stages of the development of human society, all human activities were developed in the environment and used environmental resources. For a long period of time, the equilibrium between the impact of human actions on the environment and its ability to maintain its natural features was kept. However, the intensification of development, the new industries, products, by-products, pollutants, the huge increase in transportation, demand of resources and many other reasons have led to the deterioration of the environment to the point where it can no longer be restored. Human society has to evolve and the future generations have the same rights as past and actual generations regarding access to a clean environment and to natural resources. This is the starting point for a re-thinking of the whole process of production of goods and the providing the services in society.

Many scientists and scholars contributed to the new concepts used as a basis for the development of a new economy known as the "green economy", an economy in which all the decisions have two pillars: the economic motivation and the forecasted environmental impact. Bearing in mind this dual manner of thinking, the environmental aspects are not a final problem that needs a solution, but are part of the "whole picture". The reality shows that it is possible to have economic growth with less impact on the environment or, even better, with a positive environmental impact. According to one of the first scientists in the field of environmental aspects of the anthropogenic activities (Frosch 1992), it is possible to maximise the use of materials, including waste, with notable positive influences on the environment if the design of the product also involves the design of waste, if the waste costs are internalised and if the producers of goods and service providers are totally responsible of their actions.

In the process of re-thinking and re-designing the new economic activities, we don't have to blame the old processes—they represented big progress, big developments at that moment, the answer to society's needs. But, in the actual context, these old processes are no longer responding to the economic and ecologic requirements specific to our period. Some of the effects of those processes on the environment could not be forecasted and assessed 50–60 years ago; for example, the ozone hole is the result of releasing into the atmosphere chlorofluorocarbons, substances used as refrigerants due to their 'beneficial' properties (nontoxic and nonflammable products). The replacement of nitrogen with chlorofluorocarbons was a big improvement in the refrigeration process, but no assessment was undertaken at the time on the effects of chlorofluorocarbons on the atmospheric ozone.

The interaction between the economy and the environment is complex and we need to have a clear understanding of it. The actions associated with the industrial ecology aim to minimise the negative impact of the processes on the environment, not only "at the end of the pipe", but especially to create a new manner of thinking, understanding and designing the human activities in accordance with the natural process and natural equilibria. From this point of view, the changes are not only technological but also sociological and behavioural. From a technological point of view, the changes involve the design of industrial processes, products and services from a dual perspective of product competitiveness and environmental concerns. The word industrial is not only addressed to the industry sector, but it involves also agriculture, transportation, production and consumption, all levels of industrialisation from a national economy (Socolow et al. 1994). From a sociological and behavioural point of view, it is about human culture and education, the role of societal institutions, the quality of life, the relationships between people and between people and the environment.

Due to the pollution problems at the present time, very often the environmental requirements are seen as barriers that hinder the development of existing business or future initiatives. The new manner of treating the link between economic activities and their environmental impact can transform the constraints in business opportunities. Table 1 presents some examples.

The classical way of doing business has nothing to do with taking into account the environmental aspects, as the main focus is earning money. Our society is more concerned with ecological and social issues connected with its own activity, but not all companies are acting properly. The big challenge for the new concepts, including industrial ecology, is to demonstrate that eco-economical (ecological and economical) investment and attitude represent an asset for the company, and not a pressure. In this scientific and practical approach, we can use the example of numerous ecosystems which are very effective in recycling resources and can be taken as examples for loop closing in industry (Lifset 1999).

We have to understand that we are still living in a largely non-sustainable system with limited resources, which is not big enough "to swallow" all the pollutants that result from our activities without suffering any changes. Now, we need to find solutions for problems caused by past generations while, at the same time, not to create fresh problems for future generations. The dynamics of the science and research, with their new concepts and knowledge, will allow us to make appropriate decisions for assuring the continuity of economic growth and to have a cleaner environment for future generations.

Table 1 From pressure to business opportunities

Pressures	Business opportunities	Required changes
Stricter environmental legislation	Developing "green" products and services	Designing new products and services
Intensifying of the competition	The development of new markets for ecological products and services	Diversification of the range of products and services
Rising costs	Reduction of the quantity of waste	Intensification of the environmental control, adoption of new technologies
Higher requirements and standards asked by beneficiaries	Using "clean" technologies (eco-technologies)	Adoption of new technologies

2 The Concept of Industrial Ecology

Every industrial activity is connected with hundreds other activities, all of them with different levels of impact on the environment's components. The term of industrial ecology (IE) is an approach that describes the industry–environment interactions—and offers solutions to assess and minimise those impacts. From this general definition, we can understand how difficult and complicated the process of identification of the environmental impacts is, then ranking them according to their level of risk and offering the appropriate solution in order to minimise every impact, taking into account the connections between them.

A broader definition of the industrial concept was given by Graedel and Allenby (2002) in their book *Industrial ecology*:

> Industrial ecology is the means by which humanity can deliberately and rationally approach and maintain sustainability, given continued economic, cultural, and technological evolution. The concept requires that an industrial system be viewed not in isolation from its surrounding systems, but in concert with them. It is a system view in which one seeks to optimize the total material cycle from virgin materials, to finished materials, to component, to product, to obsolete product, and to ultimate disposal. Factors to be optimized include resources, energy, and capital.

The content of industrial ecology was formulated by way of an analogy with a biological system: in this system, energy and materials are consumed or used by the organisms and the wastes resulting from one organism are used or transformed by another one in its own processes. Similarly, the industrial systems must be viewed as part of a larger whole, with connections and interrelations with other systems. This picture is the opposite of the linear approach, based on the concept of unlimited resources and unlimited waste disposed in the environment (Fig. 1).

The linear concept of industrial processes is not a viable one, as the natural resources are not unlimited and the environment is not capable of processing all the wastes and turning them into non-dangerous substances. Analysing the biological system appears to be the logical idea for cyclical evolution inside the components of the system and/or between the components. This description characterised a system with lower consumption of material and energy and lower quantities of waste, which use the products and waste from a component in another one. The similarity of an industrial process with a biological system is not perfect, but has the advantage that it introduces a new manner of analysing and designing for industrial

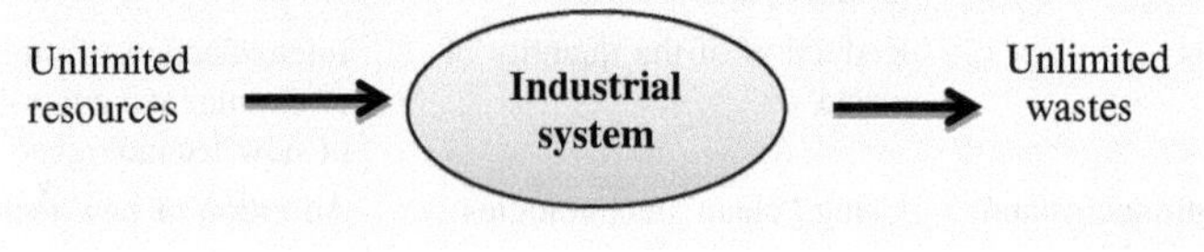

Fig. 1 Description of a linear system. *Source* Own elaboration

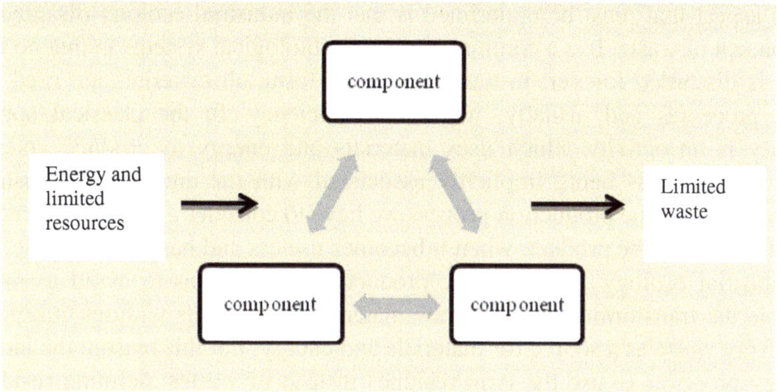

Fig. 2 Quasi-cyclical system. *Source* Adapted after Jelinski et al. (1992)

processes, with the fundamental goal of minimising the consumptions of materials and energy and the emissions of all type of wastes (Fig. 2).

The concept of Industrial ecology can be interpreted very broadly: even the name includes the word ‘industrial’. The concept is not limited to the domain of industry, as it includes all the impacts produced by human actions and presence: mining, manufacturing, energy production and utilisation, transportation, construction, agriculture, forestry, services, and waste disposal. The industrial ecology can be oriented toward two different directions of study:

(a) To study individual products and their environmental impacts at different stages of their life cycle. In this approach, taken into consideration as inputs are raw materials, other processed materials or finished products used in the production process, and energy; outputs are considered as, first, the finished product, then the waste and pollutants emitted during the production process and the energy residues (usually as heat and noise). The goal of using industrial ecology is to find the solutions for reusing and recycling wastes and energy residues as much as is possible. For this, it is necessary to elaborate a total material balance (for all material fluxes) and an energy balance and to find innovative ways to use the waste into or out of the system. This approach has both economic and environmental aspects and the difficulty consists in finding the optimum level which satisfies, at the same time, the economic and ecologic requirements.
(b) To study the facilities used in the production process. In this case, we are dealing with more technological aspects because every facility, apparatus, and technical device is analysed from the design point of view and then from its capabilities to fulfill the projected tasks.

An aspect that must be underlined is that the industrial ecology disagrees with the concept of waste. It is a common point with biological systems in that no type of waste is discarded forever; in nature, in some form, all materials are used in different processes and, usually, with great efficiency. In the classical approach, industry is an activity which uses materials and energy to produce goods, the production process being implicitly associated with the emission of wastes and pollutants. After the production process we have to consider also the moment of the end of the life of the product, when it becomes useless and has to be disposed of. In an industrial ecology approach, the production process has to avoid as much as possible the transformation of the raw materials in products without utility and to treat every waste as a source for materials and energy. For this reason, the industrial ecology proposes to use the term residues instead of wastes, defining residues as materials that our economy has not yet learnt to use efficiently. Some authors (Graedel and Allenby 2002) are going further and propose the term *experienced resources* for all the residues and obsolete products in order to gain the attention of specialists, but also society, about the value represented by them.

Industrial ecology is a concept which acts both at microeconomic and macroeconomic levels. It can be applied starting with one single entity and developed at regional/global levels, as shown in Fig. 3.

Companies play an important role in implementing industrial ecology, firstly because the company is where improvements and innovations take place, where the ideas are put into practice. Secondly, because of the managerial process, more and more organisations are considering the command and control process as inefficient and even counterproductive. Industrial ecology is seen as a flexible and more cooperative tool for managing the process and is focused on accomplishing simultaneously economic and environmental goals (Lifset and Graedel 2002).

According to Fig. 3, the industrial ecology can be implemented in specific, particular ways, depending on the type of activity and the aims of the organisation.

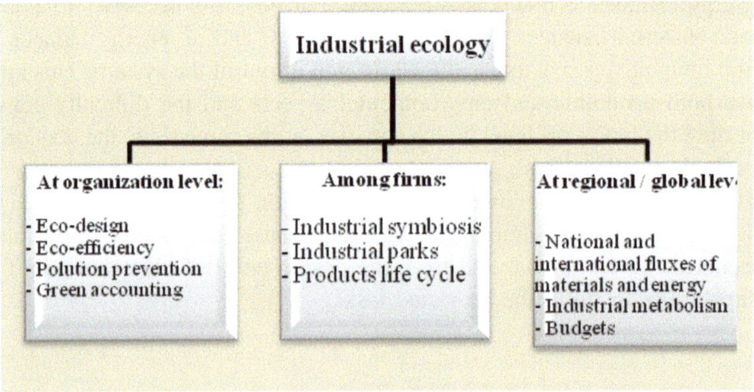

Fig. 3 Spheres of activity for industrial ecology. *Source* Adapted after Lifset (1998)

2.1 Industrial Ecology at Organisation Level

Eco-design is an oriented design in which the final result (product, service, process) must accomplish several environmental requirements. A designer has to understand the relation between the final result of the designing process and the environment and to project a proper set of characteristics. For this, it is necessary to have a database with information regarding the possible or assessed impacts of various processes and products on the environment. For simple products or services, the volume of information handled by the designer is important and it grows significantly for an industrial process. To fulfill the requirements of industrial ecology, during the eco-design process the so-called "life cycle thinking" is used—a reasoning which takes into consideration economic and ecologic aspects related to the designed final result. Given the subject of this chapter, the further discussion will be focused on the ecological aspects. The eco-design must not design "green products" but "environmentally sound product life cycles". It means not to think only of the impacts of the product during its use, but to all the impacts that may occur during its whole life cycle, including the disposal phase. The inputs can be divided in two groups:

- *environmental inputs*: raw materials and energy
- *economic inputs*: other products or semi-finished products, energy, which are outputs from other processes.

Similarly, there are two types of outputs:

- *environmental outputs*: represented by emissions
- *economic outputs*: final products, semi-finished products, energy.

All this information is organised in a MET matrix (M: materials, E: energy, T: toxicity) with three levels, according to the main stages of a product during its life cycle (http://www.pre.nl/ecodesign/ecodesign.htm) (Fig. 4):

The industrial ecology pays great attention to the energy sources, production and consumption. Accordingly, the eco-design process has to evaluate all types of energy involved during the life cycle of the product, taking into consideration several details such as the source of energy, the consumption and possibility to recover energy. Geographical and national aspects can change drastically the final picture of energy involved in the eco-design of a similar product in different

	Materials	Energy	Toxicity
Production	…	…	…
Use	…	…	…
Disposal	…	…	…

Fig. 4 Example of MET matrix

locations. Often in the design process, the energy consumption is underestimated, so the eco-design has to consider this detail.

The eco-design process has the possibility to increase the product life span, which means especially an optimisation of the material consumption and a means for environmental education of the population. Through eco-design, the product is more durable from a technical point of view and, if it is possible, upgradable. Finally, but not without importance, the eco-design is looking to use a minimum quantity of material for products, but not forgetting the quality and use characteristics of the product. This minimisation of the quantity of material is important not only for raw material consumption, but also for energy aspects: less weight means less fuel consumption.

The eco-design for a product is not simple and it becomes more complicated when it is applied to an industrial process. Here, the economic-ecological decision is fundamental for fulfilling the desired economic efficiency and environmental impacts at the same time. The traditional goals for an industrial process are presented in Table 2.

Through the eco-design process developed at the organisational level, there are indirectly fulfilled other requirements associated with industrial ecology, namely eco-efficiency and pollution prevention.

Eco-efficiency is a concept that can be traced back to the 1970s. It was first used in the form of "environmental efficiency" and introduced for the assessment of environmental impacts of the economic activities. Eco-efficiency deals with the use of resources, the volume of products produced and the environmental impacts caused by the production—the conception chain. The World Business Council for Sustainable Development (WBCSD) gave a definition for eco-efficiency: "*Eco-efficiency is reached by the delivery of competitively priced goods and services that satisfy human needs and bring quality of life, while progressively reducing ecological impacts and resource intensity…*" (United Nations Conference on Trade and Development 2004). The definition emphasises clearly that eco-efficiency is addressing economic activities with their main components: production and consumption.

Table 2 Economic and environmental goals for eco-designing an industrial process

Economic goals	Environmental goals
Accomplish the desired technological results	Prevent pollution
Achieve high precision and high efficiency in the manufacturing process	Reduce risks to the environment
Design the process for high reliability over a long period of time	Perform process design from a life cycle perspective
Design the process to be upgradable and modular (if it is possible)	
Design the process for minimum operating costs	
Make the process safe for the workers	

Source Adapted from Graedel and Allenby (2002)

In the production field, eco-efficiency aims to produce equal or greater outputs with fewer resources and, at the same time, with less pollutants and waste. To achieve this goal, it is necessary to involve technology and innovation; the technology represents the base of the production process, which assures the transformation of raw materials into goods. Eco-efficiency requires technologies that are capable of assuring very high levels of transformation of the raw materials (high yields) with low emissions. These technologies are designed using the eco-design process, targeting economic and ecologic goals (Table 2) adapted to specific requirements of each industry. Innovation is a powerful tool which supports eco-efficiency; through innovation, it is possible to improve the processes, to find new ways for production, to reduce consumptions and emissions, and to find methods of use for waste.

For the production sector, the eco-efficiency concept has the potential to discover a win-win solution for different problems/situations. Improving eco-efficiency entails costs reduction and less environmental impact. We can nominate several instruments used to implement eco-efficiency in the production sector: (a) research, development and innovation, which are the sources of eco-efficient technologies, both for carrying out and their dissemination; (b) Environmental Management Systems (EMS) and environmental standards, which help firms to be more aware of their environmental impact and sustain the control and monitoring processes; (c) economic incentives, which aim to internalise the external costs, like eco-tax or tradable permits (Moll and Gee 1999).

The consumption is viewed in two forms: resource consumption and final consumption (final demand). The resource consumption is integrated into the production sector with all its aspects. The final consumption is linked with the production, but has an opposite trend. While an eco-efficient production means more goods are produced with fewer resources, an eco-efficient consumption promotes the decrease in goods consumption. These two opposite trends can be harmonised, i.e. the production can offer products with higher value which fulfill the requirements for a sustainable consumption. The eco-efficient consumption is based on some simple concepts: extend the life of the products (use phase); repair, reuse the goods (a shift from products to services); education for consumption of more eco-efficient goods and services.

The instruments used for implementing the eco-efficient consumption are: (a) information instruments such as eco-label, Fair Trade label, FSC label, environmental statement, awareness campaigns for the environment and eco-efficient consumption; (b) economic incentives and programmes to stimulate eco-efficient consumption and behaviour.

Implementation of the eco-efficiency concept is not easy due to several barriers: financial problems, which do not allow the replacement of old technologies and production systems, inappropriate public policies, low public awareness, lack of education.

Adopting eco-efficiency means one must not only understand the concept and introduce improvements, but also find appropriate methods to measure the results. Therefor, there were proposed *eco-efficiency indicators* to assess the different aspects of economic and ecologic efficiency of the production and consumption processes. This means they are useful also for the assessment of the industrial ecology progress at organisational level. Without appropriate measuring methods, the concepts of eco-efficiency and green growth would remain at a declarative stage. According to the United Nations Economic and Social Commission for Asia and the Pacific (2009), the eco-efficiency indicators are useful both at organisational and national level for:

- measurement of eco-efficiency for an organisation or for a sector of economy;
- identification trends in eco-efficiency at both levels mentioned above;
- comparison of the eco-efficiency of different organisations in the same area of activity or different countries;
- as support for decisions and policies in economic and environmental fields.

The basic principle of the elaboration of eco-efficiency indicators is a determination of the ratio between the value of the product or service and its environmental impact.

$$Eco\text{-}efficiency = \frac{Value\ of\ a\ product}{Environmental\ impact\ of\ a\ product}$$

There are two ways to enhance the eco-efficiency: to raise the value of the product (for example, improving the quality) or to reduce its environmental impact. The concept of eco-efficiency leads to the identification and utilisation of a large number of eco-efficiency indicators, which can be divided into two groups (Table 3).

The reverse ratio—environmental impact reported to the value of the product, called eco-intensity—is used for assessing the environmental impact of economic activities (Ehrenfeld 2005). Generally, eco-intensity is used to measure the welfare and expresses how much "nature" is necessary to produce a certain level of welfare (Klein et al. 1999).

Table 3 Eco-efficiency indicators for organisation level

Scope indicators	Subject indicators
Indicators for assessing the eco-efficiency at micro and macro level	Indicators for resource-use intensity: water, energy, material and land use intensity
Sector-specific indicators: industry, agriculture, transport, services	Indicators for environmental impact intensity: refers to emissions and waste

Source Adapted from United Nations Economic and Social Commission for Asia and the Pacific (2009)

Eco-efficiency and eco-intensity represent basic elements for decision-making in the process of adoption of industrial ecology. The industrial ecology itself considers eco-efficiency as one of its main features, and focuses on the minimisation of the environmental impact of the product.

The concept of eco-efficiency and the set of indicators can be used for integrated analysis and assessment of industrial ecology, e.g. to assess the efficiency of changes introduced by industrial ecology to a production process.

Pollution prevention (*P2*) is an approach which deals with the protection of the environment and more efficient use of the resources. Both aspects are connected directly with industrial ecology, especially at the level of an organisation. The main concept of pollution prevention is the reduction or elimination of the pollutants and waste at their sources. It is obvious that less wastes and emissions will cause fewer problems and will lead to less action needed for the management of pollution of the environment. Pollution prevention is a participative approach; it involves not only emission control, but it aims to collaborate with other areas like eco-design for achieving its goals—a cleaner environment and a better use of the resources. Pollution prevention can have different aspects: re-designing the products to have less emissions, pollutants and waste during their entire life cycle (extraction, production, use and final disposal), better practices in the production processes to avoid/minimise leaks and fugitive releases, actions for the reduction of energy consumption or new production processes with better environmental performances. The concept of pollution prevention addresses mainly to the industry, but all other areas can contribute with their efforts: transportation (through minimisation of air emission and optimisation of the routes), agriculture (using fewer quantities of fertilisers and other chemicals), offices (with a better use of paper and printing cartridges), and every household and every person.

Pollution prevention is a medium and long-term approach with significant environmental, economic and social benefits; however, these benefits differ from one country to another, with many different variables. According to United States Environmental Protection Agency (1998), the main factors that will determine the success of pollution prevention actions are presented in Table 4.

It was mentioned previously that industrial ecology disagrees with the concept of waste; it means that the pollution prevention process is integrated in industrial ecology and represents a particular method for adopting industrial ecology concepts

Table 4 Success factors for pollution prevention

Availability of resources	Technical, financial, scientific and engineering resources available to develop and implement pollution prevention
Stage of development	Includes the specific degrees and forms of pollution for each area, the capabilities and the willingness of the responsible parties, or society in general, to solve the environmental problems
Cultural and educational issues	The cultural acceptance of the environmental aspects by the population and the ecologic education and behaviour

Source The author with information from United States Environmental Protection Agency (1998)

in an organisation. The reduction of waste and pollutants represents the main goal, both for pollution prevention and for industrial ecology, and there has been a great deal of effort made in this direction at firm level (Salhofer et al. 2008). Industrial ecology is focused on source reduction, which prevents the generation of wastes and environmental releases and conserves natural resources. This is in perfect accordance with EU policies, which are promoting mainly the prevention of waste production, i.e. a complex of measures taken before a product, substance or material has become waste (Albu and Chasovschi 2014). The pollution prevention approach involves several methods to reduce waste at the source (Phipps 1995):

- material substitution—the replacement of hazardous material in a product composition with other materials that are less toxic and with equal performances;
- improvements in process efficiency—improving or designing new production systems and making them more effective in terms of production yields and resource conservation;
- preventive maintenance—refers to a set of activities that prevent equipment malfunctions and environmental emissions;
- in-process recycling—can be applied to some industries and means the reusing of some materials back into the process before they become waste (waste is not generated). These materials include scraps (low-quality final products that don't fulfill the qualitative requirements), metal pieces, some products in food industry or pulp and paper industry;
- inventory control—a management issue and deals with the reduction of product losses due to over-stocking and product expiration.

Pollution prevention is now generally accepted as part of sustainable development and its principles and methods are applied in all fields of activity. The results are goods that are improving with every positive experience gained by the organisations. However, it is necessary to identify also the barriers and restrictions of pollution prevention. One of these can be the organisational culture and norms; if the business leaders are not convinced with the importance of environmental issues, they will not make decisions to support the pollution prevention process and other actions in this direction. Very often, the better ideas for improvements are originating from the workers that are dealing every day with different problems. Breaking down hierarchical barriers and offering the possibility to communicate and to offer their opinion is fundamental to the success of the pollution prevention process.

Costs can also represent a restriction to the successful implementation of pollution prevention actions. The final goal is, of course, the improvement of the environmental performance of the organisation; but, besides this goal, pollution prevention is a means of saving money. For a manager, it is important to understand that, first, there are necessary investments for the reduction of emissions, for the improvement of energy efficiency, replacing hazardous materials, re-designing the products and for other actions in the frame of pollution prevention. For SMMs enterprises in particular, these costs can be more than the organisation can afford. This will not stop the development of the industrial ecology approach at the organisation level or pollution prevention actions; it is only a problem of priorities and resource allocation. The authorities have the task of adopting legislation and financial measures which support the economic entities in their efforts for a sustainable development.

Green accounting, or environmental accounting, has developed as a tool for assessing the environmental performance of organisations, together with a set of eco-efficiency indicators. As environmental preoccupation has risen during the last few years, a new type of accounting has developed to assess environment-related economic activities. According to Huiguan et al. (2014), green accounting studies and discloses artificial assets, natural assets and conversions between these assets. Environmental accounting is viewed in two different ways: (i) in a simplistic approach, green accounting plays a role of "book noting" like general accounting; (ii) in a larger approach, according to which environmental accounting has a managerial role, to confirm, distribute and analyse material flows and related cash flows through the use of the environmental accounting management system. For sustainable development, the latter description of green accounting is the most suitable because it offers a broad picture of the connections between the activities developed in an organisation and its impacts on the environment, all in financial terms.

Green accounting operates with the notion of environmental costs; they are defined as 'additional expenditure caused by enterprises' attempt to reduce negative effect of producing activities on environment' (Wei and Jinglu 2015). This definition shows that all the activities involved in the company's efforts to improve its environmental performance are resulting in extra costs. This is only the formal part of the green accounting, because there are also non-formal aspects, which are not always measurable, regarding the benefits (financial and non-financial) brought by the environmental improvements. The environmental costs fall into the following categories, as presented in Table 5.

For industrial ecology, green accounting represents a very useful tool for the assessment and improvement of the results at organisation level. Green accounting takes into consideration the environmental costs and calculates the following (Patrut et al. 2008):

Table 5 Classification of environmental costs

Criterion	Description of costs
Implications of environmental costs	*Environmental costs in narrow sense*—include organisation costs for recovery and improvement of environmental conditions
	Environmental costs in broad sense—include the costs for natural resources consumed and the environmental pollution costs
Level of occurrence	*Internal costs*
	External costs

Source The author with information from Wei and Jinglu (2015)

- conventional calculations with the aim to obtain a detailed report about environmental protection costs and pollutants costs;
- calculation of costs connected with losses;
- cash flow accounting, based on all environmental costs.

Green accounting represents a useful tool for the management of decision making. With information provided by green accounting, the manager sees confirmation of environmental costs, can control these costs and is able to make decisions with positive economic and environmental impacts.

2.2 Industrial Ecology Among Firms

The second level where industrial ecology is utilised is among organisations which have the possibility to cooperate for mutual advantages. The concept of industrial ecology itself emphasises the necessity of collaboration between different entities to reduce the environmental impact and to improve the resource use, which means interaction and interdependence. Because the industrial symbiosis is treated separately in this chapter, here we will focus only on industrial parks and products life cycle as manifestations of industrial ecology among firms.

Industrial parks are modern patterns of organisation within the economic environment, where the proximity of different organisations creates the prerequisites of business development. If the operating of an industrial park is aimed at environmental improvement, it uses the term eco-industrial park to emphasise this aspect. The eco-industrial parks are the most visible form of industrial ecology; several companies are located near one another and have the opportunity to change fluxes of materials and energy. The eco-industrial parks are designed from the beginning to search for the best possibilities for collaboration regarding the environmental impacts associated with business. This collaboration includes energy management, water resources and waste water management, material resources and wastes resulting from technological processes. The principles that underlie the operation of an eco-industrial park respect the concept and the goals of industrial ecology (http://makinglewes.org/2014/02/25/eco-industrial-parks-and-industrial-ecology/):

- industrial processes from the eco-industrial park are connected systematically to reduce consumption of raw materials, energy and water;
- in an eco-industrial park, the largest proportion of wastes (or all wastes, if it possible) become raw materials for different activities located in the same eco-industrial park;
- co-location of organisations together in the same space allow the reduction of transport activities, transport costs and logistic simplification;
- best practices and expertise can be shared among the organisation situated in an eco-industrial park.

The eco-industrial parks are very good examples of business integration: here are integrated the principles of industrial ecology with principles of pollution prevention as well as sustainable development and eco-design. The co-located firms are acting symbiotically, the overall result being more than the results obtained, but of each organisation separately. Although the discussion about eco-industrial parks is focused on the environmental benefits, all the scholars involved in this area recognise the economic benefits for all the companies situated in the park. According to Tudor et al. (2007), the main economic benefits fall in two categories, as presented in Fig. 5.

The analysis of the economic aspects summarised below in Fig. 5 shows that the companies involved in eco-industrial parks have a competitive advantage resulting from a better use of the resources, a minimisation of energy consumption, a better use of land and an added value to their products or services. Besides these very important aspects, the eco-industrial parks involve some indirect, but equally important, benefits: the collaboration between organisations lead to innovation, eco-innovation and knowledge sharing. Taking into consideration the EU policy to support the innovation, technology transfer and continuous development, we can conclude that eco-industrial parks are effective patterns of organisation, capable of fulfilling the specific requirements of sustainable development.

The industrial ecology and its application, the eco-industrial parks, are innovative concepts with positive results confirming the possibility of developing "cleaner business" and economic efficiency. However, we cannot underestimate the limitations put in front of the development of eco-industrial parks, of which three are considered as being major: (i) The system is considered fragile because it strongly depends on the organisations being located in the park; if one of them leaves or closes the activity, the entire chain is affected; (ii) The mechanism of results control, and evaluation for an eco-industrial park, is complex and complicated; it is possible to experience a lack of clear understanding as to what an eco-industrial park is exactly, the best way to manage such a system, what parameters are used for evaluating the efficiency of the eco-industrial park; (iii) The types of activities and wastes categories will determine the possibility of establishing an eco-industrial

Economic benefits of eco-industrial parks	
Benefits of sustainable production processes	**Benefits for the location**
- Exchange fluxes of raw materials, energy and water	- More intensive use of space
- Joint use of utilities	- Efficient use of public utilities
- Collective gathering and management of wastes	- Joint commercial firm facilities
- Combining transport of goods and people	- High-quality and efficient public transport

Fig. 5 Two categories of economic benefits of eco-industrial parks. *Source* The author with information from Tudor et al. (2007)

park and its efficiency. The reality shows that there will always be wastes that are impossible to recycle/recover; plus, even with every effort made, it isn't possible to create appropriate links between companies according to industrial ecology principles.

As a final remark, we can affirm that eco-industrial parks play a significant role in achieving sustainable development, both for organisations and organisation networks. If the specific factors are taken into consideration, it is possible to design and to manage a successful eco-industrial park.

Products life-cycle is a concept that has a major impact on resource consumption, pollution prevention and, generally, on sustainable development. The life cycle assessment (LCA) of products aims to conserve non-renewable resources and ecological systems, to minimise the magnitude of pollution and to promote appropriate methods for pollution prevention, to maximise the recycling of materials and wastes and to develop and use cleaner technologies. LCA liaises between companies and links them in a common effort toward a development based on eco-economic decisions. For each organisation, as well as for a group of organisations, there are several opportunities to reduce waste output and to optimise the consumption of resources. All these opportunities can be analysed from different perspectives with the final goal—to produce goods with minimum impact on the environment. Figure 6 represents a simplified diagram for a product's life cycle, which shows the fluxes and the connections between different stages during the life of the product.

In the frame of industrial ecology, the role of an LCA is to explain the materials and energy fluxes associated with a product manufacturer, use and disposal, the emissions released during the entire life of the product and how it is possible to

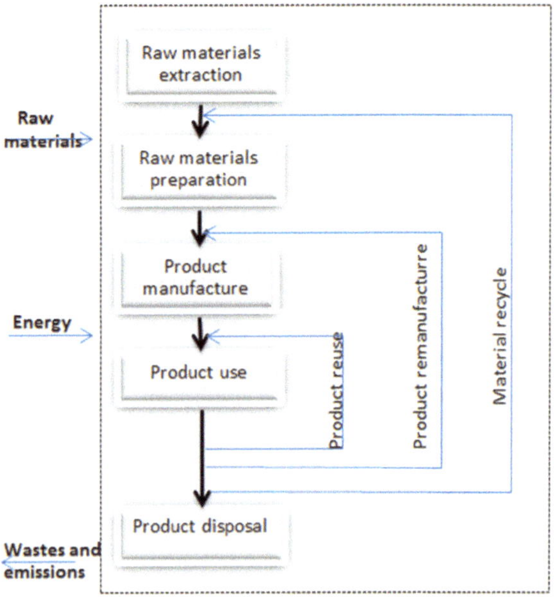

Fig. 6 Main stages from a product life cycle. *Source* Adapted from http://www.utexas.edu/research/ceer/esm282/dfe/LCAoverview.PDF

minimise all of these negative impacts. We can say that LCA is analysing the environmental impact from the product's perspective. Besides this role, the LCA represents a very important tool for comparing and improving the products. The method reveals the energy and material fluxes associated with each phase during the product life, which enables us to act where it is necessary to minimise any negative effects.

2.3 Industrial Ecology at Regional/Global Level

The third level of action for industrial ecology is the macro-level, which involves the study of materials and energy fluxes between regions or countries, environmental policies and financial support promoted by the governments. The concept of industrial ecology is consistent with sustainable development and offers solutions to the current major problems of mankind: generalised pollution, depletion of non-renewable resources, and population growth accompanied by increasing consumption. Being a new concept, industrial ecology needs innovation at institutional and managerial levels and to be able to promote appropriate measures for the creation of a competitive environment. At the same time, in order for it to contribute to pollution reduction and optimisation of resource consumption, it needs an innovative approach. This is why the success of the application of industrial ecology depends on the policies and economic measures promoted by authorities. Taxation policy may help or, to the contrary, act against utilisation of industrial ecology as support for sustainable development. For example, if the taxation policy favours import-export activities, this approach will hamper the interest and the diffusion of industrial ecology among organisations. Access to the production resources (raw materials and energy) will determine firms' orientation toward recycling, reuse, re-design, eco-design or other tolls specific to sustainable development and industrial ecology.

Another aspect controlled by the authorities, which affects companies' decisions for adopting industrial ecology, is the government system of regulations (Jelinski et al. 1992). This system can make reuse/recycling so complicated and expensive that firms will be discouraged and will not take into consideration the environmental aspects in the decision-making process. Conversely, the regulations can promote the environmental initiatives and collaborations between organisations with better economic and ecologic benefits. The price system represents another macro-economic element which will impact the adoption and diffusion of industrial ecology; if the price includes the externalities associated with its environmental performance, the company will pay more attention to the possibilities to reduce its negative impact, using all types of methods and tools, including industrial ecology.

Related to the issue of the national/global extent of industrial ecology, it is necessary to bring into the discussion the relation between industrial ecology and the standard of living of consumers and level of education. The standard of living can encourage two types of behaviours: a long-term use of the products or a short-term use, followed by early disposal of the products. A medium level of education, including environmental issues, can guide the population through a sustainable behaviour in all stages of a person—whether as an employee or as a customer.

At a regional/global level, industrial ecology is part of the process of globalisation; we are witnesses not only to the pollution generalisation, but also the joint efforts of emissions reduction, for a correct management of the resources, for restoration of polluted sites or for the adoption among companies of a responsible attitude. At regional level, industrial ecology provides guidelines for the improvement of environmental performance by promoting a 'win-win' collaboration, eco-efficiency, reducing consumption and cost saving.

We can affirm that implementation of industrial ecology will lead to local, regional and national development; new business and new jobs can be created, organisations can save money and improve their economic efficiency, while ensuring a minimal environmental impact. Besides these aspects, industrial ecology has an educative potential. The employees (which are also citizens) become more careful with environmental aspects in their jobs and in society, which represents added value for the communities.

Unfortunately, industrial ecology is moving slowly in terms of application and results; even for the organisations that are using this tool, the results came after long periods of efforts and sometimes failures. This situation is not normal considering the good results obtained with industrial ecology. The main problem is represented by the behaviour of the organisations: if not all, or almost all, organisations act according to the principles of industrial ecology, it is hard to improve economic and ecological efficiency at the global/national levels and to see notable positive environmental effects (Ehrenfeld 2006).

Industrial ecology brings a new, innovative manner of thinking and decision-making for managers. In the traditional economic thinking, the competitive advantage was obtained with low-cost inputs; now, the competitive advantage is not only about lowering the costs, but finding innovative ways to increase product values, have a lower environmental impact, gain money from waste, to use less energy and many others.

Industrial ecology brings many benefits, as presented briefly previously, but we have to be conscious of its limits. According to Esty and Porter (1998), there are two aspects when considering the limits of industrial ecology:

(a) The situation when costs exceed benefits. There are many cases when using eco-design, recycling, reusing, recovering and other processes connected with industrial ecology are too expensive and do not represent a viable strategy for the organisation. For this reason, many scholars are suggesting that industrial ecology should focus on the substantial zone that overlaps the competitive strategy for the organisation; in this way, more managers will be attracted to industrial ecology in their business.
(b) Imperfect environmental regulations. In our current regulation there exists the permission to emit different substances to specified levels of concentrations. In these circumstances, the organisations will not make any efforts to find

solutions for these pollutants as long as the emissions are under the legal limit. If the company is still interested in using industrial ecology in order to avoid pollution when its emissions are below the required limits, it will face pollution costs that their competitors don't have. The social benefits resulting from applying industrial ecology do not outweigh the costs.

We can conclude that industrial ecology can promote innovation and help managers find opportunities both inside and outside the organisation to add value to their products. However, industrial ecology cannot, by itself, be an independent tool for all the problems within a company.

3 Industrial Symbiosis—Or How We Can Close the Loop

Natural symbiosis occurs between at least two organisms that are changing materials or energy for mutual benefit. This was the model for the so-called 'industrial symbiosis', an innovative process included in industrial ecology, which focuses on the inter-changes among firms for by-products, energy, materials, and water for mutual benefit. The key to industrial symbiosis is the location—the entities which are changing fluxes have to be settled in the same location or at a short distance one from the other. Being part of industrial ecology, industrial symbiosis is oriented through the fluxes which can have negative effects on the environment, e.g. residual energy, emissions, by-products, water vapours, wastes, and others. In practice, industrial symbiosis seeks to 'close the loop', giving new life to the materials and products at their end of life moment, or recover the energy for optimum consumption of the resources. Sometimes, the loop can be closed inside the company if it is able to reintegrate the by-products and residual energy; however, more often, the cycle includes other entities more appropriate for the use of residual fluxes.

As part of industrial ecology, industrial symbiosis proposes another meaning for the wastes: it is not a useless, polluting and expensive element for a company, but a new form of resource that can be re-integrated in the economic cycle. This approach encourages economic entities to look beyond traditional markets for business, without affecting the environment (Albu 2011).

The process of industrial symbiosis has some very important strong points, both from an economic and ecological point of view. First, industrial symbiosis is based on the collective efforts and actions; working together in synergy, entities can obtain better results than the sum of individual benefits that could be achieved by acting alone. Second, industrial symbiosis encourages firms to look for options and ways for re-using and recycling their own by-products, wastes and residual energy with an overall goal—to reduce material and energy costs. This will involve a sustainable consumption of the resources and have less environmental impact. Third, industrial symbiosis will lead to social benefits for communities because the process involves not only the companies, but also institutions, universities, research centres, and NGOs. All these participants are striving for common interests which offer the possibility to find appropriate solutions for the community and strengthen the relation between them. The main benefits of industrial symbiosis are:

Exchange fluxes of materials and energy—co-products, waste or residues of a company may have practical uses (replacement of raw materials and fuels) in production processes or providing the services of another company.

Common infrastructure—the companies are using and managing in common the usual industrial infrastructure (water, energy, water vapours, sewerage) for the optimisation of the consumption.

Common utilisation of the services—the entities involved in industrial symbiosis can use common services such as firefighting, transportation and catering, for a better use of resources and to cut costs.

Costs minimisation—the industrial symbiosis allows a reduction of costs for supplying, production and waste management.

New jobs, cleaner environment, better relation between participants.

Industrial symbiosis can appear natural (such as the classical example of Kalundborg), or be created; a modern management and decision-making process can have as a strategy identification of the intern and extern cycles from which it is possible to use the residual fluxes.

Approaching things from an ecological perspective, a big problem worldwide is the over-exploitation of natural resources. Current models of production and consumption lead to negative environmental effects which, most often, have an inequitable global impact. Industrial symbiosis focuses on increasing the eco-efficiency of production and consumption processes, leading to better economic performance of firms taking part in symbiotic activity.

Industrial symbiosis is closely linked with economic efficiency; it uses familiar practices such as recycling, reuse, remanufacturing and energy recovery, which obviously determine a better use of materials and energy—a better economic performance. However, it is not the only aspect connected with economic efficiency; industrial symbiosis creates a link between firms which contributes to an increasing performance for all the partners. One firm, analysed alone both from economic and ecologic points of view, can have an unsatisfactory result. The connection with other partners will contribute to the improvement of its performances with mutual benefits for everybody.

This kind of relation between companies is often called a 'cascade link' for illustrating the involvement of several entities in the symbiotic process (Ehrenfeld and Gertler 1997). The cascading use of materials in industrial symbiosis means the utilisation of by-products resulting from one process as feedstock for another process. This creates economic and ecological advantages due to the replacement of the raw materials with by-products and to the reduction of the emissions in the environment. The cascading use of energy is more complex because it involves not only companies, but also other institutions. The residual energy of liquids, steam or gases is used in industrial processes or as a source for heating and hot water for populations and institutions. The group of firms that participate in the symbiotic process, including all the links between them, form an industrial ecosystem. This structure resembles a cluster which has many elements in common.

The administrative organisation of an industrial ecosystem often takes the form of an eco-industrial park—a location that hosts different economic entities with links between them, which develop collaborative strategies such as: waste-to-feed exchanges, cascading use of energy, common logistics and shipping and receiving facilities, green technology purchasing, green building system, shared sewerage and treating water facilities and local education and resource centres.

Because the companies and other entities that participate in the industrial symbiosis process have to be located near to one another, this collaborative approach addresses especially local and regional economies. Closing the loop at local level gives the opportunity to involve several production and service provider companies with benefits for the community. The implementation of industrial symbiosis is part of industrial ecology; however, in a more practical approach, there are many examples of projects implemented or in the implementation stage that deal with this modern approach. By analysing the goals of these projects or actions and the fluxes between participants, Chertow (2000) proposes a taxonomy of industrial symbiosis types:

Type 1: Characterised by one single material circuit—the waste flux. This situation is not exactly an industrial symbiosis because waste fluxes exist also in traditional business. There are entities which collect certain types of wastes for processing, reusing or selling. Even local authorities, with their waste management programmes, are part of the chain, but the waste flux is only one-way, oriented to the disposal at the end of the life cycle.

Type 2: The industrial symbiosis established inside a company. The situation is possible for large entities formed by several parts which can change fluxes of materials and energy between them. Outside, the company will exit only the materials and energy that cannot be recycled/recovered at the internal level. Significant results can be achieved if the company uses eco-design for its products, the life cycle assessment, the cradle-to-grave approach or the cradle-to-cradle approach.

Type 3: Industrial symbiosis among firms collocated in a defined area (possibly an eco-industrial park). This is the model of an industrial symbiosis projected for this special purpose—to create from the beginning appropriate links among firms in order to maximise the use of by-products and residual energy.

Type 4: Industrial symbiosis among firms which are not collocated. This model characterises the situation where firms exist in a certain area; they act according to their strategies, but they realise it is possible to establish connections and change material and energy fluxes for their mutual benefit. For sure, the location contributes to the development of business relations between firms, but industrial symbiosis opens opportunities for new business. The proximity of the companies reduces transport costs and enhances the collaboration.

Type 5: The model of industrial symbiosis developed on a broader region. At first sight, this model seems complicated and risky. Some costs become higher due to the distances among firms and the fluxes scheme can become very complicated. However, Type 5 industrial symbiosis gains value through the number of economic entities involved and through the contribution to the local/regional development. Of course, the fluxes of material and energy have to be optimised, reaching the best partner for a certain by-product, waste or energy. Regarding the contribution of the local/regional community, the Type 5 model of industrial symbiosis can involve SMMs, NGOs, local farmers, and small family businesses in the process of closing the loop. This will be materialised in new businesses and the development of existing ones, new jobs, better lifestyle conditions, cleaner environment, more confidence and collaboration among firms and with the local community.

A very good tool that helps the initiation of an industrial symbiosis is the materials and energy map. For the entities involved in the industrial symbiosis, or those that want to participate, it is necessary to draw a map with all materials and energy fluxes. From the map, it is possible to identify the partner and the proper flux for each partner.

Similar to industrial ecology, industrial symbiosis is not an attractive alternative for a large number of managers. According to Ehrenfeld and Gertler (1997), there were identified several reasons for not adopting industrial symbiosis:

- the process is interesting only if the company has big fluxes of energy, wastes or by-products. Otherwise, changing the usual patterns, managerial changes and possible supplementary expenses are serious reasons for managers not to adopt industrial symbiosis;
- the fact that the firms don't know, and are not conscious of, the savings that can be obtained by changing materials and energy. The situation can be improved with accounting tools like ante calculation, cost comparing, managerial accounting and others that can prove the benefits of industrial symbiosis;
- regulatory context and local/regional/national policies can support or break the initiatives for adopting industrial symbiosis. In the states where the decision bodies and policy makers have a constant dialogue with the economic sector, companies are stimulated to find innovative ways of improving their environmental performance, while being profitable at the same time. Denmark, where such a symbiotic system was developed for the first time at Kalundborg, has such a consultative system of regulation.

The current vision of closing the loop using industrial symbiosis is an innovative one and it is gaining more and more followers in all economic branches. The new currents of industrial ecology and industrial symbiosis changed the way of thinking not only for scholars, but for managers, too. The waste has had a long history during which it were ignored and abandoned (Erkman 1997). Now, it is difficult to change this vision and to see wastes as sources of materials, energy and money, but the large number of eco-industrial parks and projects in the industrial symbiosis field proves that the new concepts are correct. Actually, we are witnessing a cultural change that involves the production chain, the business management and decision-making process, the authorities' attitude and the behaviour of every member of a community.

4 Application of Industrial Symbiosis—Possibilities and Barriers

In the scientific literature are examples of implementation of industrial symbiosis all over the world, some of which are considered 'classic' and are used as a case study for explaining how the concept functions. In this work, we decided not to introduce these examples or other projects in the implementation phase, but to identify the possibility of building from the beginning a set of symbiotic relations among some existing firms that had never previously considered this type of collaboration.

According to the concepts discussed previously, we initiated a research in two counties in the north-eastern part of Romania—Botosani and Neamt—in locations where several companies are functioning with different domains of activities. As a model for industrial symbiosis, we selected the Type 4 model, in which the firms exist in a certain area and act individually, with no symbiotic connections.

In Romania, the initiatives for promoting industrial symbiosis are almost nonexistent; even sustainable growth represents a subject of interest for both authorities and the business environment. In this situation, the research started from the hypothesis that there are no circuits, links, materials and energy flux exchanges specific to industrial symbiosis among firms situated in a defined area. The research aimed to propose symbiotic circuits and links for a certain combination of companies and was developed according to the following steps:

- setting the boundary of the research, i.e. setting the location (town, small area, big area);
- identification of all economic entities situated in the established location;
- defining the partners that may be involved in a process of exchanging such as industrial symbiosis;
- defining the waste material and energy flows resulting from economic activities concerned;
- discussing the potential of using these waste streams in other productive activities, services or other fields;
- asking what could be the expected results of such exchange

The research methodology consists of two stages. The first stage was designed according to the information and study of the activities developed in the two counties, following the identification of: the main companies acting in each county, location and area of activity, the products and services provided by each company, the residual fluxes of materials and energy specific to each activity/company.

The second stage of the research consists of selecting a sample from each county consisting of large firms and/or firms with negative impact on the environment. The board, or the manager, of each company was contacted and asked to participate in the study by participating in an interview. All the persons agreed and the interviews were scheduled direct (for the companies from Botosani and Piatra Neamt towns) or over the telephone (for the companies situated in other localities of each county).

Because the area of activity differs greatly, a questionnaire or template was not used. Instead, the discussions were structured as follows:

- if the company has identified its residual fluxes of materials and energy;
- if the company has a monitoring system for these fluxes;
- if the company falls within legal limits regarding the environmental pollution;
- what is the manager's vision regarding the wastes and residual fluxes, and does he/she have knowledge regarding the possibility of finding practical utilisation for these wastes;
- if the company has, or has had, initiatives regarding the waste recycling, reuse, selling wastes or using residual fluxes;
- if this type of initiative exists at company level, who are the partners (stakeholders) and the results;
- if the company is interested in participating in future projects/activities focused on pollution minimisation through utilisation of residual fluxes and wastes.

As was specified previously, the industrial symbiosis can be an interesting alternative when the residual fluxes are substantial. As a result, we eliminate from the study the small entities with small streams of waste, energy or by-products and

Table 6 Enterprises selected for the research

No.	Economic entity	Domain of activity	Residual fluxes
Neamt county			
1.	ARCELORMITTAL TUBULAR PRODUCTS, ROMAN SA	Manufacture of tubes, pipes and hollow profiles and fittings from steel	Metal scrap Slag Hot gas
2.	T. C. E. 3 BRAZI SRL, PIATRA NEAMT	Growing of cereals (except rice), leguminous crops and oil seeds plant	Vegetal wastes
3.	RIFIL SA, SAVINESTI	Preparation and spinning of textile fibres	Synthetic fibre and yarn waste
4.	CERSANIT ROMANIA SA, ROMAN	Manufacture of ceramic sanitary fixtures	Ceramic wastes Hot gas
5.	DINAMIC 92 DISTRIBUTION SRL, PIATRA NEAMT	Wholesale automotive parts and accessories	
Botosani county			
1.	TEHNIC ASIST SRL, BOTOSANI	Construction of residential and nonresidential buildings	Construction wastes
2.	MODERN CALOR SA, BOTOSANI	Supply of steam and air conditioning	Ash Slag

(continued)

Table 6 (continued)

No.	Economic entity	Domain of activity	Residual fluxes
3.	BIG CONF SRL, BOTOSANI	Manufacture of garments	Textile wastes
4.	GRIGIOVERDE COMPANY SRL, BOTOSANI	Manufacture of garments	Textile wastes
5.	ITALPANT SRL, BOTOSANI	Manufacture of garments	Textile wastes
6.	LUCA DAMILANO SRL, BOTOSANI	Manufacture of furniture	Wood wastes
7.	COMCEREAL SA, BOTOSANI	Growing of cereals (except rice), leguminous crops and oil seeds plant	Vegetal wastes
8.	SERCONF SA, BOTOSANI	Manufacture of garments	Textile wastes
9.	NOVA APASERV SA, BOTOSANI	Collection, purification and distribution of water	Non-toxic sludge
10.	NORDIC ROMAR SRL, DOROHOI	Retail sale of audio/video equipment in specialised stores	Wood wastes

focus instead on the large enterprises. Table 6 shows the selected economic entities with their domain of activity and residual fluxes.

5 Results and Discussion

The study developed in the counties of Botosani and Neamt from the north-east part of Romania revealed some interesting aspects, not all positive, for placement on a sustainable basis the development in future. The interview was an effective method as it offered the opportunity to discuss different aspects of the company's activities without remaining in a rigid structure (as in an ordinary questionnaire). Table 7 contains a synthesis of the answers provided by the managers-participants in the study.

For Neamt County, even the activities of the selected companies cover various fields, however, they are incompatible with the initiation of a process of industrial symbiosis. Taking into account only the waste stream energy results from two enterprises in Roman town (Arcelormital and Cersanit), we propose their use as a source of heating and hot water for institutions and small businesses in the vicinity (tourist pensions, offices, kindergartens, etc.). This simple industrial symbiosis would lead to saving conventional energy resources and reduce costs with utilities for the receiving entities. Also, we can talk about a positive social impact and a good image for the two enterprises.

Table 7 Synthesis of interviews

Neamt county	Botosani county
The wastes are identified, in some companies by weight The residual fluxes of energy are identified, but are not monitored with a system and the managers can't appreciate what is the quantity of lost energy All the companies fall within legal limits regarding environmental pollution There were few initiatives to recycle waste, especially metal scrap; the managers affirmed that there is a lack of a recycling network at regional level There were no initiatives to use residual fluxes of energy The companies are interested to participate in projects in the environmental field, including the establishment of symbiotic relationships	The wastes are identified, for wood wastes there is a reuse cycle For companies from Botosani county the residual energy fluxes are not significant All the companies fall within legal limits regarding the environmental pollution Textile wastes represent the major residual flux for the companies in Botosani county; for this type of waste, there isn't a recycling network or other possibilities of recovery There were no collaboration initiatives between companies from the same location/county in the frame of environmental issues The companies are interested in participating in projects in the environmental field, including the establishment of symbiotic relationships

Botosani County, represented by 10 companies in our study, also has a low potential for the use of industrial symbiosis as a tool to promote green development. The main reason that led us to this conclusion is the fact that, in this county, there are several companies in the same field (e.g. in the manufacture of textiles or furniture) from which result the same types of waste material flows, but there no areas that can be targeted by these streams. For this situation, we propose an organisation on the model of eco-industrial parks, in which the companies can use in common the infrastructure for utilities, can optimise the supply process or can join into a cluster and be competitive on the textile products market.

The result of the research shows how difficult it is to establish symbiotic connections among enterprises which are already situated in a certain area and are less compatible for industrial symbiosis collaboration. The international examples, like Kalundborg, were lucky to have, in the same location, companies with the potential for residual flows exchanges.

Analysing the results of the research from the point of view of industrial symbiosis, we conclude that the Type 4 model is not the best model and that perhaps it is better to extend the area of interest for implementation of the symbiotic relation, thus adopting Type 5 model. The distances between counties are not big (with a maximum of 300 km), with the possibility to involve companies from different counties in the symbiosis. Also, the study can involve two other neighbouring counties with firms in other domains of activity in order to discover further possibilities for establishing symbiotic relations.

6 Conclusion

The economic crisis of recent years, the high level of pollution, the scarcity of natural resources and the necessity for continuing economic growth for future generations brought into discussion, and led to, an increased interest in innovative techniques for gaining economic and environmental advantages. Industrial ecology and industrial symbiosis are methods to fulfill these two sets of goals for every company. In spite of multiple advantages brought about by the implementation of these concepts, there are still many barriers of manner of thinking the development and growth for an enterprise, economic and regulatory limits, complex environmental aspects which brakes their the application on a large scale.

The preliminary research presented above shows the practical difficulties faced within the process of shifting the economic development to a 'green' path. These difficulties must not cause us to give up; instead, we must push the scientific research and the practical application to find better economic and environmental solutions.

References

Albu A (2011) Business innovation using industrial ecology. Metalurgia International 11 (XVI):28–31

Albu A, Chaşovschi C (2014) Landfilling or incineration of waste? Practices for choosing the appropriate solution for waste management, supplement of "Quality—access to success". Journal 15(S1):189–197. ISSN 1582–2559

Chertow MR (2000) Industrial symbiosis: literature and taxonomy. Annu Rev Energy Environ 25:313–337. http://www.researchgate.net/profile/Marian_Chertow/publication/249558396_INDUSTRIAL_SYMBIOSIS_Literature_and_Taxonomy/links/0deec53a0ff2f1d91b000000.pdf

Ehrenfeld J (2005) Eco-efficiency philosophy, theory and tools. J Ind Ecol 9(4):6–8

Ehrenfeld J (2006) Environmental and economic boon. file:///C:/Articole/Green%20economy%20and%20green%20growth/industrialecology.pdf

Ehrenfeld J, Gertler N (1997) Industrial ecology in practice. The evolution of the interdependence at Kalundborg. J Ind Ecol 1 (1):67–79. http://planet.botany.uwc.ac.za/nisl/ESS/Documents/Industrial_ecology_Kalundborg_in_practice.pdf

Erkman S (1997) Industrial ecology: an historical view. J Cleaner Prod 5(1–2):1–10. http://www.cob.sjsu.edu/osland_a/Substantiability%20Articles/Industrial%20ecology%20An%20historical%20view%20.pdf

Esty DC, Porter ME (1998) Industrial ecology and competitiveness. Faculty scholarship series. Paper 444. http://digitalcommons.law.yale.edu/fss_papers/444

Frosch R (1992) Industrial ecology: a philosophical introduction. In: Proceedings of national science USA, vol 89, pp 800–803. Colloquium paper. http://planet.botany.uwc.ac.za/nisl/ESS/Documents/Industrial_ecology_a_philosophical_approach.pdf

Graedel TE, Allenby BR (2002) Industrial ecology, 2nd edn. Prentice Hall, New Jersey, p 18

http://makinglewes.org/2014/02/25/eco-industrial-parks-and-industrial-ecology/

http://www.pre.nl/ecodesign/ecodesign.htm

http://www.utexas.edu/research/ceer/esm282/dfe/LCAoverview.PDF

Huiguan D, Shiyou L, Tingting L, Boyang L (2014) A literature survey of research on environmental accounting: theoretical review and prospect. Int J Financ Econ 2(1):13–25

Jelinski LW, Graedel TE, Laudise RA, McCall DW, Patel KN (1992) Industrial ecology: concepts and approaches. In: Proceedings of national academic science USA vol 89, pp 798–797. Colloquium paper. http://www.pnas.org/content/89/3/793.full.pdf

Klein R, Bringezu S, Fischer-Kowalski M, Palm V (eds) (1999) Ecologizing societal metabolism: designing scenarios for sustainable materials management, ConAccount workshop, Centre of Environmental Science (CML) Leiden University. http://www.leidenuniv.nl/cml/ssp/publications/proc-con98.pdf#page=45

Lifset R (1998) Ecologizing societal metabolism in the US materials flow accounting and industrial ecology, third conaccount meeting: ecologizing societal metabolism. http://www.leidenuniv.nl/cml/ssp/publications/proc-con98.pdf

Lifset R (1999) The industrial ecology of renewable resources. http://archive.unu.edu/zef/publications_e/ZEF_EN_1999_02_D.pdf

Lifset R, Graedel TE (2002) Industrial ecology: goals and definitions in a handbook of industrial ecology. http://planet.uwc.ac.za/NISL/ESS/Documents/Industrial_Ecology_Overview.pdf

Moll S, Gee D (eds) (1999) Making sustainability accountable: eco-efficiency, resource productivity and innovation. European Environmental Agency, topic report no. 11. http://www.eea.europa.eu/publications/Topic_report_No_111999/download

Patrut V, Ciuraru C, Luca M (2008) Green Accounting—a Challenge for the Accountant Specialist. http://steconomiceuoradea.ro/anale/volume/2008/v3-finances-banks-accountancy/252.pdf

Phipps E (1995) Pollution prevention concepts and principles, in introductory pollution prevention materials, National Pollution Prevention Center for Higher Education, University of Michigan. http://www.umich.edu/~nppcpub/resources/GENp2.pdf

Salhofer S, Obersteiner G, Schneider F, Lebersorger S (2008) Potentials for the prevention of municipal solid waste. Waste Manag 28(2):245–259. http://www.aseanenvironment.info/Abstract/41016648.pdf

Socolow R, Andrews C, Berkhout F, Thomas V (1994) Industrial ecology and global change. https://www.princeton.edu/mae/people/faculty/socolow/industrial_ecology_overview.pdf?__toolbar=1

Tudor T, Adam E, Bates M (2007) Drivers and limitations for the successful development and functioning of EIPs (eco-industrial parks): a literature review. Ecol Econ 61:199–207. http://www.researchgate.net/profile/Terry_Tudor/publication/4842084_Drivers_and_limitations_for_the_successful_development_and_functioning_of_EIPs_(eco-industrial_parks)_A_literature_review/links/5409e1bf0cf2df04e7491b92.pdf

United Nations Conference on Trade and Development (2004) A manual for the preparers and users of eco-efficiency indicator, New York and Geneva. http://unctad.org/en/docs/iteipc20037_en.pdf

United Nations Economic and Social Commission for Asia and the Pacific (2009) Eco-efficiency indicators: measuring resource-use efficiency and the impact of economic activities on the environment. United Nations publication, ST/ESCAP/2561. http://sustainabledevelopment.un.org/content/documents/785eco.pdf

United States Environmental Protection Agency (1998) "Principles of pollution prevention and cleaner production. An international training course". http://infohouse.p2ric.org/ref/02/01993.pdf

Wei L, Jinglu J (2015) Discussion about enterprises environmental cost management and control based on the concept of ecological civilization. In: Proceedings of the AASRI international conference on industrial electronics and applications. Atlantis press, pp 544–547. ISBN (on-line): 978-94-62520-65-3. doi:10.2991/iea-15.2015.135

Author Biography

Angela Albu Ph.D is associate professor at University "Stefan cel Mare" Suceava, Romania, Faculty of Economics and Public Administration, where she teaches Environmental Management, Eco-technology and Integrated Quality, Environmental and Occupational Health Management. She is a graduate of Chemical Engineering at the Polytechnic University from Bucharest, where she also obtained her Ph.D title and holds a master degree diploma in Business Administration. She participated in several national and international projects focused on environment, eco-tourism and innovation and has published papers in international scientific journals and in the proceedings of international conferences.

Socio Economy Impact in Relation to Waste Prevention

Antonis A. Zorpas, Irene Voukkali and Pantelitsa Loizia

Abstract

Globally waste is driven by population and economic growth, and urbanization. One important instrument to meet this increasing demand and to decrease the pressure on food production is to minimize food loss, food waste, green waste, recycling waste and other wastes etc. It has been estimated that globally one third of the edible parts of food destined for human consumption is lost or wasted each year. Much of the waste that comes from high-income countries has been attributed to poor marketing practices and consumer behaviour, with consumers being identified as a bigger contributor than food manufacturing, distribution, grocery retail and the hospitality sectors. There are environmental as well as socio-economic, financial and legal reasons for being concerned about waste prevention which are: (i) Environmental reasons: Municipal waste is merely the final evidence of consumption which entails environmental footprints which are rarely sustainable; (ii) Socio-economic reasons: Production of waste appears to be the result of wasting natural resources and these resources are both limited

A.A. Zorpas (✉)
Faculty of Pure and Applied Sciences, Environmental Conservation and Management,
Laboratory of Chemical Engineering and Engineering Sustainability,
Cyprus Open University, Gianou Kranidioti, 33, Latsia, Nicosia 2220, Cyprus
e-mail: antonis.zorpas@ouc.ac.cy; antoniszorpas@yahoo.com

I. Voukkali · P. Loizia
Department of Research and Development, I.E.S.T - EnviTech Ltd,
(Institute of Environmental Technology and Sustainable Development),
Paralimni 34073, 5309, Cyprus
e-mail: irenevoukkali@envitech.org; voukkei@yahoo.com
URL: http://www.envitech.org

P. Loizia
e-mail: pantelitsaloizia@envitech.org
URL: http://www.envitech.org

W. Leal Filho et al. (eds.), *Sustainable Economic Development*,
World Sustainability Series, DOI 10.1007/978-3-319-45081-0_2

and unequally distributed; (iii) Financial reasons: Reducing the quantities of waste produced means it should be possible to reduce the budget required for the collection, transportation and treatment of waste products; (iv) Legal reasons: The European framework directive on waste requires national waste prevention programmes to be drawn up. Waste prevention is performed in the privacy of our own home and is personal as it is driven by deeply held beliefs and attitudes rather than social norms.

Keywords
Waste Prevention • Food waste • Waste management • Waste generation • Strategic planning • Waste composition analysis • Waste technologies • Social impact • Economy impact • Technologies costs

1 Waste Production

With a rising level of prosperity in industrialized countries, an increasing number of products and services are being produced and consumed. The amounts of municipal solid waste (MSW) have been increasing for years in many countries. This development is reflected in the amount of waste generated. According to the latest official Eurostat statistics (Eurostat 2011), the total waste generation in the EU-27 was more than 2.62 billion t. The statistics indicated that the total amount of municipal solid waste is continuously rising (Zorpas et al. 2015b, c; Zorpas and Lasaridi 2013; Salhofer et al. 2008). From those a total amount up to 98 million t or 3.7 % were categorized as hazardous waste. Typical, during 2008 each European citizen produced more or less 5.2 t/year of waste, of which 196 kg were hazardous (Eurostat 2011). According to the OECD (2002), the generation of municipal solid waste increased about 54 % in the Denmark, Switzerland, Netherland, Portugal, and Greece between 1980 and 2000. In the Organisation for Economic Co-operation and Development (OECD) countries, an increase of 18 % has been reported between 1995 and 2007 (OECD 2010). Studies of the relationship between quantity of waste generated and economic growth indicators have proposed that a decoupling is needed in order to reduce the increasing burden from waste management (Zorpas et al. 2014a; OECD 2002; Jacobsen and Kristoffersen 2002). According to Zorpas et al. (2014b) the municipal waste per capita has slightly decreased in the EU-27 since 2000 (−4 %), although the GDP (Gross Domestic Product) increased by 33 % between 2000 and 2013. However, waste generation in new Member States has remained relatively stable by weight since the 1990s. This may be due to a reduced incidence of heavy mining and construction waste and increased lighter paper and packaging waste. Decoupling economic growth from the environmental impacts associated with waste generation is a key objective of the EU's revised Waste Framework Directive (Mazzanti and Zoboli 2008). To

Table 1 Waste generation projections for 2025 by region

Region	Available data			Projections for 2025			
	Total urban population millions	Urban waste generation		Projected population		Projected urban waste	
		Per capital (kg/capital/day)	Total (t/d)	Total population millions	Urban population millions	Per capital (kg/capital/day)	Total (t/d)
AFR	260	0.65	169,119	1152	518	0.85	441,840
EAP	777	0.95	738,958	2124	1229	1.5	1,865,379
ECA	227	1.1	254,389	339	239	1.5	354,810
LCR	399	1.1	437,545	681	466	1.6	728,392
MENA	162	1.1	173,545	379	257	1.43	369,320
OECD	729	2.2	1,566,286	1031	842	2.1	1,742,417
SAR	426	0.45	192,410	1938	734	0.77	567,545
TOTAL	2980	1.2	3,532,252	7644	4285	1.4	6,069,703

AFR Africa Region, *EAP* East Asia and Pacific, *ECA* Europe and Central Asia region, *LCR* Latin America and the Caribbean region, *MENA* Middle East and North Africa region, *OECD* Organisation for Economic Co-operation and Development, *SAR* South Asia region

Source World Bank (2012)

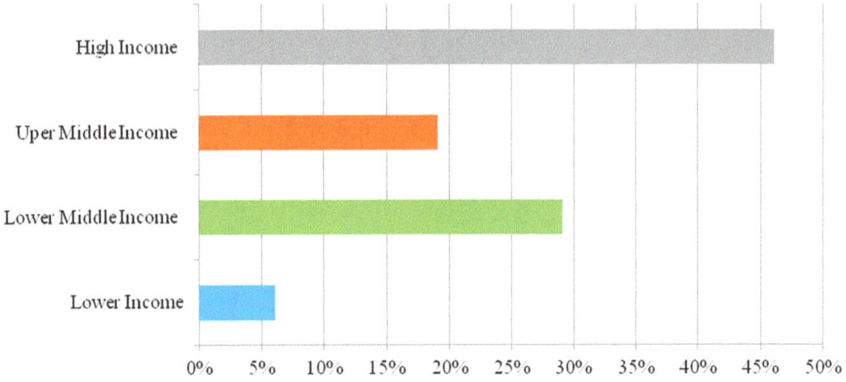

Fig. 1 Waste generation in relation with incomes

control waste generation is no longer enough, waste growth in the EU must now reverse (Zorpas and Lasaridi 2013).

World Bank Report (2012) focuses on waste generation (projection for 2025) indicated that in all Regions we will have a continual waste amounts (Table 1). The per capital waste production varies from 0.77 kg/day for South Asia Region (SAR) to 2.1 kg/day for OECD region. OECD region include 29 countries. Among them are Andorra, Australia, Austria, Belgium, Canada, Czech Republic, Denmark, Finland, France, Germany, Greece, Hungary, Iceland, Ireland, Italy, Japan, South Korea, Luxembourg, Monaco, Netherlands, New Zealand, Norway, Portugal, Slovak Republic, Spain, Sweden, Switzerland, UK, USA. SAR region include the countries of Bangladesh, Bhutan, India, Maldives, Nepal, Pakistan and Sri Lanka. At the same time the SAR population on 2025 is estimated to be 734 millions (426 million on 2012) with the urban waste generation on 2025 to be 0.77 kg/day

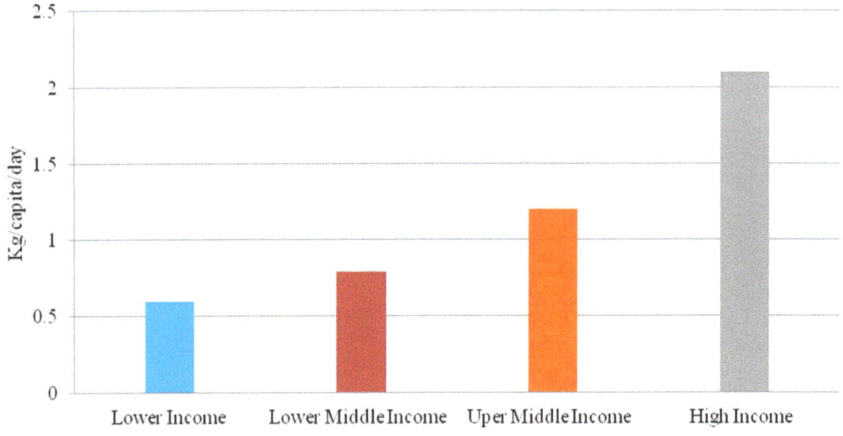

Fig. 2 Waste generation per capita in relation with incomes

than 0.45 kg/day on 2012. On the other hand, OECD region produced 2.2 kg/day in 2012 with total population to 729 million while in 2025 the population is estimated to be up to 842 million.

High-income countries produce (Figs. 1, 2) the most waste per capita, while low income countries produce lowest waste per capita. While the total waste generation for lower middle income countries is higher than that of upper middle income countries, likely skewed as a result of China's inclusion in the lower middle income group, the average per capita waste generation amounts for the various income groups reflect the income level of the countries (Fig. 1). The high, upper-middle, lower-middle, and low income descriptions are fairly incorrect as these classifications are country-wide, and in several countries average national affluence can be very different from average affluence of the urban populations. Only the affluence of urban residents is important in projecting Municipal Solid Waste (MSW) rates. If we look on India, and more specific China, have income ensure which high generation of urban waste rates per capita in relation to the overall economic status due to the present of high relatively poor rural populations that tend to dilute and influent national figures. The per capita waste generation according to income level (Fig. 2) the average rate is decreasing.

2 Waste Composition

Waste compositional analysis is a technique used to evaluate and estimate in detail the nature, scale and origin of any kind of waste and more specific for house hold waste (Zorpas et al. 2015b, c). Through this household and local attitudes, social behaviour as well as socio-demographics are determined (WRAP 2008; Zorpas and Lasaridi 2013). The establishment of waste composition analysis helps to confirm and assess several data focuses on MSW generation, treatment and disposal (IPCC 2006). Waste compositions, as well as the classifications used to collect data on waste composition in MSW vary widely in different regions and countries (IPCC 2006; Zorpas and Lasaridi 2013), as well as they are influenced by many factors, such as level of economic development, cultural norms, geographical location, energy sources, and climate (Lebersorger and Beigl 2011).

Composition of waste mainly is influenced by several factors. Among them (World Bank 2012; Zorpas et al. 2015b, c) are the development of one area, urban planning, economic variability, energy requirements, ethical issues, culture, existing waste management systems, strategic planning, zero waste approach, prevention activities, treatment technologies. Generally, MSW classified into two main streams: inorganic and organic. The main categories of waste include: organics, plastic, papers, glass, metals, and others. Organics mainly include food scraps, yard waste (leaves, grass, brush), wood, process residues; plastic includes bottles, packaging, containers, bags, lids, cups; papers contain paper scraps, cardboard, newspapers, magazines, bags, boxes, wrapping paper, telephone books, shredded paper, paper beverage cups; glass comprises glass bottles, broken glassware, light

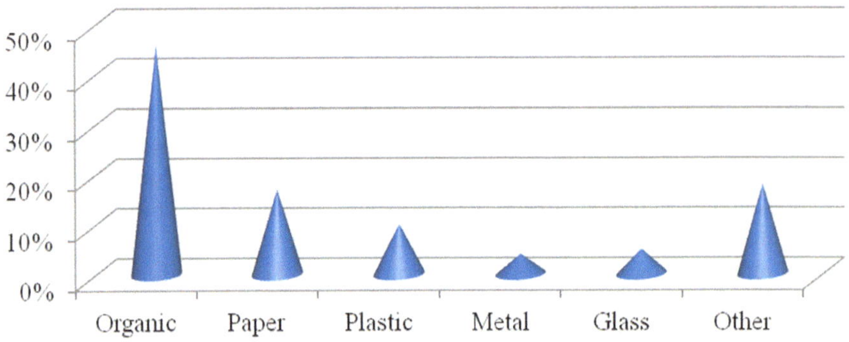

Fig. 3 Global solid waste composition

bulbs, colored glass; metals mainly include cans, tins, non-hazardous aerosol cans, and other mainly cover textiles, leather, inert materials etc. According to the World Bank (2012), the global solid waste composition analysis (Fig. 3) consist of 46 % of organics followed others waste 18 % and papers 17 %. All those waste are mainly produced in urban area due to their metabolism. The concept of the urban metabolism (mentioned for first time from Wolman 1965), is essential and important to developing sustainable cities and communities or general urban or peri-urban area. Urban metabolism according to Kennedy et al. (2007, 2010) is defined as "the sum total of the technical and socio-economic processes that occur in cities, resulting in growth, production of energy, and elimination of waste". This typically means that urban metabolism mainly includes quantification of the inputs, outputs and storage of energy, water, nutrients, materials and wastes for an urban region. Factors such as urban structure, form, climate, quality and age of building stock, urban vegetation and transportation technology can influence the rate of a city's metabolism (Zorpas and Voukkali 2015).

As a country urbanizes and populations become wealthier, consumption of inorganic materials (such as plastics, paper, and aluminum) increases, while the relative organic fraction decreases. Generally, low and middle-income countries have a high percentage of organic matter in the urban waste stream, ranging from 40 to 85 % of the total. Paper, plastic, glass, and metal fractions increase in the waste stream of middle- and high-income countries.

The total amount of organic waste in China is up to 65 % followed by plastics which was 13 % while 1 % is consist of metals (Fig. 4; this composition is for the citizens who used gas for the energy needs). On the other hand, Cyprus which is a very small country with 800,000 citizens according to the last inventory report (Zorpas et al. 2015a) and belongs to Europe and Central Asia Region region (ECA) (World Bank 2012) has % 42 organics, 17 % plastics and 23 % papers (Fig. 5).

The annual per capita production of waste in Cyprus is estimated at 468 kg for residential areas and 670 kg for tourist areas, while the total waste is estimated at the end of 2012 up to 630,000 t (Zorpas et al. 2015b, c). 75 % of the population of

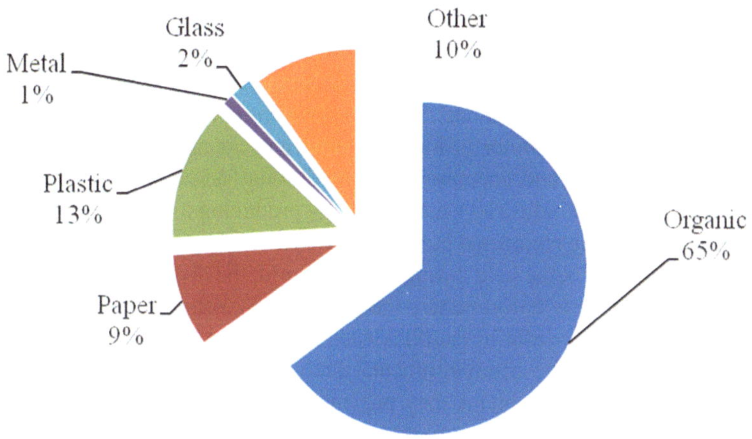

Fig. 4 China's municipal solid waste composition

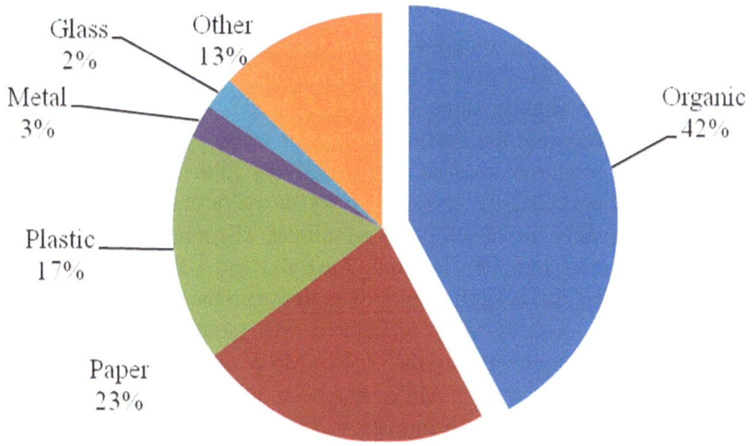

Fig. 5 Cyprus waste composition analysis

Cyprus is served by 5 landfills while the remain 25 % disposed their waste uncontrolled (Ministry of Environment 2013; ENVITECH 2013).

3 Waste Disposal

MSW directors are charged with an enormous task: get the waste out from underfoot and do so in the most economically, socially, and environmentally optimal manner possible. Solid waste management is mainly local government's responsibility. Moreover, governments mainly in developing countries offer their efforts and high budget to solve the problem. Mainly in the developing countries, urban employments count 1–5 % and according to the report of World Bank (2012) solid waste workers tend to be Youngers. Furthermore, solid waste remains one of the most pernicious local pollutants. It is well known that uncollected solid waste is usually the leading contributor to local flooding and air and water pollution.

Landfills (controlled and uncontrolled) receive today a combination of municipal, commercial and mixed industrial waste, typically producing a wide range of pollutant compounds influencing nature and human health in many ways. Despite the promotion of waste management being held throughout the world, the disposal of untreated waste into final landfill sites is unfortunately the most widely solution these days. Uncontrolled management of waste in landfills leads to negative environmental issues and there is a need to undertake an environmental analysis of existing facilities and services in order to analyses the problems they present and take the necessary measures for reducing adverse effects in order to propose the most valuable restoration solution especially in insular communities which they have limited available space.

According to Fig. 6, municipal solid waste landfilling reduced by almost 40 million t, whereas incineration increased by 15 million t and recycling grew by 29 million t. Observing the EU-27 only, landfilling decreased by 41 million t, incineration by 15 million t and composting with recycling with increased at the same time by 28 million t. Moreover, Fig. 6 shows that the total amount of MSW recycled has declined slightly since 2008 (EEA 2013).

MSW landfills represent the leading choice for waste disposal in all around the world mainly for economic reasons. The relatively high costs of treatment and disposal of several options are considered to be the major reason for the reliance on MSW landfills, mainly in developing economies (Brunner and Fellner 2007). Nevertheless, countries like UK, Finland, Australia and USA (the most industrial countries) mainly depend on landfilling. For example, in the US, 54 % of the 250 Tg (1 Tg = 106 metric tons) of MSW generated was landfilled during 2008, with composting and recycling being around 33 % of MSW management (USEPA 2009). In Australia during 2002, 70 % of MSW, has been dumped to landfills without any pretreatment stage (Productivity Commission 2006). The direct disposal of MSW in

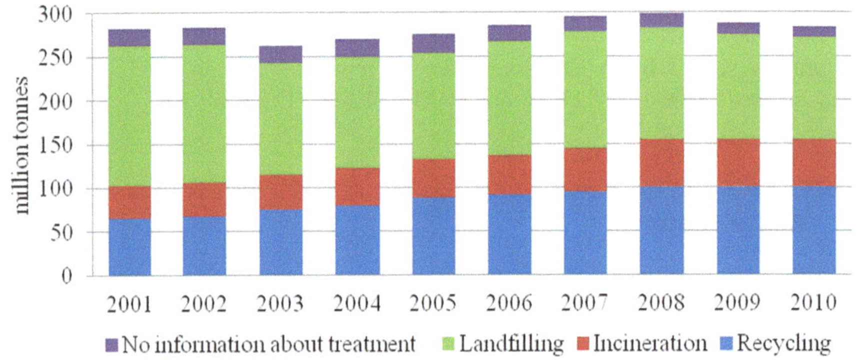

Fig. 6 Development of municipal waste management in 32 European countries, 2001–2010

Japan was less than 30 % of MSW generation in 2000 with high incineration rates during the last decades due to the historic scarcity of land (Tanaka et al. 2005). Finland, UK, and Greece are the main EU countries where landfilling is the priority solution. According to Eurostat (2010), MSW landfilled in 2008 was for Greece 77 %, for UK 55 %, and for Finland 51 %. On the other hand, other EU countries like Netherland, Germany, Austria, Denmark, Sweden this option was less than 5 %.

Engineers and Scientist around the world agree that any measurable waste disposal data are extremely difficult to be collecting (ISWA 2013). As mentioned in the study "What a Waste" (World Bank 2012; ISWA 2013), thermal treatment of waste and landfilling is considered to be the most common methods of municipal solid waste disposal mainly in high-income countries. Even though quantitative data is not direct available, most low- and lower middle-income countries dispose of their municipal solid waste in open dumps. This consists one of the main socio problems as degrade the nearby environment. Some middle-income countries have poorly operated landfills, which should likely be classified as controlled dumping. According to ISWA (2013) report the "Waste Atlas" defines waste which is disposed or burnt in controlled and uncontrolled dumpsites as "unsound disposal". 96 countries around the World from North to South and from East to West indicated that 38 % of the total MSW generated is disposed in such a way. As less developed countries cannot provide data for this indicator, it can be supposed that the actual rate of unsound disposal is reasonably higher. The study of 22 reference cities "Solid Waste Management in the World's Cities" concludes that 48 % of the total waste generated is not disposed of under controlled way. The existing data indicates a clear correlation between a high GDP and sound disposal.

Tadesse et al. (2008) indicate the difficulties to understand (in a successfully way) and manage the produced waste if management does not consider and understand waste generation and the main waste producers. In order to maintain a good waste management system, it is necessary not only accurate data on waste generation but also to collect data and information on the social behaviour, demographic data, and attitude of citizens in respect to waste and their daily habits as well as their ethical issues. This is because waste is the final product of human activities and behaviour (Bisson 2002). A main, household waste attitudes which is «NIMBY» (not in my back yard), will affect success and acceptability in such waste management practices (like prevention, reused, recycling) as the siting of solid waste containers and landfill sites in quarters of the city (Schubeler 1996; Tadesse et al. 2008). However, a new household attitude is «INMBY» which means IN MY BACK YARD. Typically this means that after several awareness campaigns (Zorpas et al. 2012, 2014a; Guidelines on waste prevention program 2015) citizens change their behaviour and treat their waste in their yards. For example, they participated in home composting (Zorpas et al. 2014a, 2015a; Chroni et al. 2015) or reuse several wastes like wooden pallets (making fence) or other recycling materials using the concept of «up-cycling» than recycling. Up-cycling as a definition the process of converting old or discarded or unused items into something useful or mainly visible «beautiful». For example, interior designers may use newspapers as wall papers, or they will make small counters using wooden pallets as well as tires for flowerpots etc.

4 Waste Treatment Technologies

The impact on the environment from the uncontrolled disposal of waste has been already mentioned from many researchers. On the other hand, any proposed treatment technology has indirectly affected the economics of a community as citizens are enforced to cover mainly the operation cost of the proposed management systems. For example, for Cyprus according to Zorpas et al. (2012, 2014a, 2015b, b), the current waste management plan consists of the collection of the waste (door by door) twice a week mainly in all Municipalities and Communities. Some regions like Famagusta and Larnaka have the potential to transfer their waste to the Koshis Municipal Waste Treatment Plant (KMWTP). The KWMTP was constructed as a Build Operation and Transferred (BOT) plan and covers mechanical sorting, recover of recycling material and composting. According to the Ministry of Interior Affairs KWMTP, for a period of 10 years, will be run under a private commercial company. KMWTP charged 54.8 €/t for the mix waste, 46.8 €/t for the green waste, 80.80 €/t for the recyclable waste and 100.80 €/t for the residual waste until 2012. Those fees were paid to the KMWTP directly from each Municipality that is participated in the system and the Municipalities charge directly their citizens. According to the Clauses 84Z, of the Cyprus National Law N(I) 111/1985, each Local Community is approved to charge waste fees according to the follow main principle: (i) for houses no more than 171 €/year, (ii) for shops, stores, coffee shops and similar no more than 855 €/year, (iii) for restaurants, bars, tourist apartments less than 6848 €/year, (iv) for Hotels up to 17,100 €/year and (v) for industries, private hospitals or other manors not included above fees no more than 13,680 €/year.

Mainly any proposed treatment technology (Golomeova et al. 2013) is basic requirement in each strategic planning in centralized or non-centralized waste management systems (Tsilemou and Panagiotakopoulos 2006). Available cost data are often reflecting and cover specific cases with limited or incomplete information's regarding costs breakdown, specific local conditions, operating practices, system performance, responsibilities, returned of investments as mainly related cost covers initial planning (Tsilemou and Panagiotakopoulos 2006). The main proposed solid waste treatment technologies include the following methods: (i) waste to energy (Tsilemou and Panagiotakopoulos 2006), (ii) composting (Zorpas et al. 2000; Zorpas 2012, 2014), (iii) anaerobic digestion (Mata-Alvarez et al. 2000; Tsilemou and Panagiotakopoulos 2006) and (iv) landfilling (Tsilemou and Panagiotakopoulos 2006; Zorpas et al. 2015b, c). However, new ones are under development, to digest, compost, sort, convert, select, destroy, and store waste in final sinks (World Bank 2012). Yet the spread of these is limited by public opposition in many countries, or by the lack of financial instruments to attract and lure investors. To design, develop and build up a complete waste management

Table 2 Relate cost regarding waste treatment technologies

Country	Capacity (10^3 t/y)	Type of energy recovery	Initial capital investment (10^6 €)	Operating cost (€/t)	Annual total cost (€/t)
Waste to energy plans					
France	18.7	Electricity	11.8–13.3	74–79	129–141
Greece	36.5	Electricity	23.57–27.03	182.03–188.08	
Sweeten	40	Heat	17.2–21.51	59–66	105–120
Denmark	40		26	48.8	
Belgium	140	Electricity	59.49	38.79	82.57
Italy	584		200		
Germany	200	Electricity	121.93	57.66	105
Composting units (closed systems)					
Finland	1.3		1.22	40.77	189
Greece	6		1.1		
France	7		1.35	48.71	76
Spain	16		5.40		
Italy	20		5.72	13.90	41
Portugal	45		5.40		
Composting systems (open systems)					
Finland	10		3.35	37.37	77
Greece	13		1.30		
France	6		1.38–1.86	28–37.5	58–78
Italy	12		2.95	25.11	
Sweden	24		3.75	15	
Anaerobic digestion					
Germany	2.5		0.51	60	79.5
UK	20		8.45	25	
Europe	5		2.90–3.10	24	
France	72		12.97	53	57
Landfilling					
Greece	18.25		2–2.3	5.20–5.80	
France	20		0.1–0.18	36–45	72–94
Luxemburg	32.20		32.05	48.45	147
Italy	125		34.57	13.10	52.48
Germany	300		21.72	11.29	20

infrastructure, financing, especially in developing countries, must become an international priority. The World Bank (2012) mentions that to cover waste collection estimates that the gap will be up to 40 billion US dollar per year, which will treble in the next 15 years. It is increasingly and highly important that cities are fast growing and metabolized and mainly need to be proactive in long term planning with good expertise and enough capacity to build infrastructure, instead of simply trying to overcome short term emergencies such as epidemics.

The cost of a Waste to Energy plan, and composting (mainly to produce electricity) depends mainly from on its capacity. For example, a Waste to Energy plan includes initial capital investment, operation costs and annual cost. Table 2 presents some of those data according to the report of Tsilemou and Panagiotakopoulos (2006).

5 Waste Prevention

The concept of waste prevention, set out by the OECD (2002), includes: source reduction, strict avoidance (not generating waste in the first place), reuse of products (in its original form) and finally the avoidances of hazardous waste in landfill sides. Moreover, the above definition ignores all forms related with recycling (like food collection or industrial composting as well as re-manufacturing).

According to Regarding the Council Directive 75/442/EEC (published on 15 July 1975) nine Member States support appropriate actions for reducing the quantities of certain wastes. Based on this directive, the first Community Strategy for Waste Management (SEC (89) 934 Final 1989) established the hierarchical system for waste management, under which waste prevention and minimization were given the highest priority, followed by recycling and disposal. This development was continued in the Community Strategy for Waste Management of 1996 (COM (96) 399) and in the Thematic Strategy on the prevention and recycling of waste (COM (2003) 301 final, COM (2005) 666 final). Regarding to the Waste Framework Directive (WFD-2008/98/EC, in Article 3 Clause 12 and 13) (The European Parliament and the Council of the European Union 2008), waste prevention is "the measures taken before a substance, material or product has become waste, that reduce the quantity of waste, the adverse impacts of the generated waste on environmental and human health or the content of harmful substances". Preventing waste means reducing the amount of waste generated, reducing the hazardous content of that waste and reducing its impact on the environment.

According to Zorpas and Lasaridi (2013), waste prevention comprises strict avoidance of the generation of waste, quantitative and qualitative reduction at source and the reuse of products. It is clearly mentioned that recycling is not included in the prevention activities or separation of waste (Pre-waste 2012). The last decade, in national and/or international level, emphasis is given to define waste prevention and minimisation, proposing and established specific measurable targets per waste stream (Salhofer et al. 2008).

Moreover, according to WFD (The European Parliament and the Council of the European Union 2008) waste prevention focuses on the quantity of waste, including through the re-use of products or the extension of the life span of products, the adverse impacts of the generated waste on the environment and human health or the content of harmful substances in materials and products and that "'re-use' means any operation by which products or components that are not waste are used again for the same purpose for which they were conceived". WFD lays down the five-step hierarchy of waste management options, with waste prevention as the preferred option, and then in descending order, reuse, recycling, recovery (including energy recovery) and safe disposal.

UK, has funded a large research program which focuses on waste prevention. This project included among others a review of evidence studying the behavioural chances and barriers in household waste prevention, related with the effectiveness of various policy measures (Cox et al. 2010; Zorpas and Lasaridi 2013). The impact of waste prevention campaigns (Sharp et al. 2010a), and methods to monitor, measure (Zorpas and Lasaridi 2013) as well as evaluate waste prevention through mass reduction as well as behavioural and communication studies (Sharp et al. 2010b), were also developed.

Current decoupling indicators of waste prevention were reviewed (Fell et al. 2010). Among them, according to Zorpas et al. (2015c) and Inglezakis and Zorpas (2015), are included (a) Environmental indicators (level of material reuse, prevention actions, participation in public awareness activities, evaluation of home composting, reduce specific waste streams, exchange material etc.), (b) Economic indicator (cost of actions taken, cost of infrastructure, cost of waste avoidances etc.) and (c) Social Indicators (job development, social acceptances, changes of attitudes etc.)

This unprecedented government led research initiative has demonstrated the key issues of waste prevention from a behavioural perspective. It is, however, quite critical also to assess the environmental significance of waste prevention as this can reinforce evidence for policy development. In the United States, the US EPA has undertaken a program to support local authorities and waste management organisations to quantitatively assess the consequences of waste prevention on Global Warming Potential (GWP) reduction with the WAste Reduction Model (WARM) but does not include other environmental impact categories (USEPA 2006). Some countries, such as Taiwan (Young et al. 2010), South Australia (Zero Waste South Australia 2007) and New Zealand (Ministry for the Environment 2007) have adopted the target of "zero waste" as a form of strategic waste prevention. Regrettably, zero waste is usually interpreted as zero waste to uncontrolled disposal or landfill, mostly includes recycling and generally excludes environmental assessment.

6 Food Waste

According to Parfitt et al. (2010) food waste in food supply chains occurs at several points covering the cultivation, production, manufacturing, storage, transportation, leftovers, end of life etc. It is estimated that in a yearly base the food that are discharge to landfills in a European level was up to 100 million t for the 2012. It is estimated that by the end of 2020 the total food waste in EU level will be more than 120 million t (http://ec.europa.eu/food/safety/food_waste/index_en.htm). Wasting food is not only an ethical and economic issue but it also depletes the environment of limited natural resources. All the involvement stakeholders have a significant role to do starting from prevention and reduce of the food waste. All those actions must cover the producers (including farmers, food manufacturers and processors) to those who are responsible to make foods safe and available for consumption (mainly hospitality industry) as well as consumers themselves. According to EU (Food and Agricultural Organizations), approximately 1.3 billion t/y food are wasted. In developing countries more than 40 % of food are waste after harvest and during the processing while at the same time in the industrialized countries over 40 % occurs at consumer level and retail points. Generally, in world level millions of € or $ are spent every year advertising and promoting new foods.

Schneider (2011) mentioned that food has been wasted throughout civilisation and one of the main modifications with the modern situation is that the reasons for the wastage could often not be controlled by human beings. Food which was no longer suitable for consumption (Schneider 2013) by people beings at a specific level of society was given to other people with a lower social position. This typically reduced food waste. However, due to ethical issues (Schneider 2013; Papargyropoulou et al. 2014) in our days those practices could not be implemented. However, during the thirteenth century, there are evidences that there was a lack of wheat in Genoa and a corresponding surplus in Sardinia. The people responsible for the nutrition of the inhabitants of the City of Genoa decided to decrease the population and transported foreigners and poor people from Genoa to Sardinia (Schneider 2013). In the 14th century, Peter IV of Aragon separate musty/hard and mouldy bread, low quality wine (high presentation of acetic), spoiled cheese, vegetables, fruits and other food for donations to people in need (Schneider 2013). Informal materials from the late of 18th century indicated that the left-overs from the Viennese Imperial Court were first offered to the employees, then to "special restaurants" where the citizens could eat the dishes free and finally the rest was given to mendicants and hospitals in Vienna (Bauer 2009). According to Schneider (2013) issues such as hygiene or other contaminations were not very important in those days. Moreover, nowadays due to economic crisis especially in Greece, in Cyprus and other EU countries several individual targets groups collect foods (mainly dry foods or foods that their life will end in 7 or 30 days) and they offer them to family's that can't afford the cost of living. Citizens fill very proud and usually they cannot accept to receive help from other peoples if the total population of one area knows that they face economic problems. That's why several targets group remain unnamed. Also, in our days, Schneider (2013), the European Federation of Food Banks which established in 1984 and operating in 21 European countries collect more than 401,000 t of food. Those were distributed to 31,000 social welfare organisations in 2011. It is estimated that 5.2 million citizens were supported from those foods preventing thousands tonnes of food to be wasted. Almost 55 % of the food products distributed by the European Food Banks are donated by the European Union or by member states in a percentage of 4 %. Some of those have been subject to the withdrawal or intervention approach which is used to fixed the market prices. In the other of Atlantic Ocean in USA are exist 50 states (including Puerto Rico and Columbia) that participate in the America Food Bank. Those (61,000 organizations participating in America Food Bank) help with food supplies 37 million citizens in US. According to Feeding America (2011), 3 billion bounds of food were collected and distributed to people that needed them during 2009.

7 Conclusion

It is well known that any waste prevention programs, or zero waste strategy plans mainly provide emphasis on resource savings, on waste reduction in order to avoid wastes to discharge to landfill. Any strategic plan focuses on waste management or plan has as first target to protect human health and environment as well as to reduce any negative (mainly) impact on society, on biodiversity, on environment. The second target is to support and promote sustainability through economic development. From many researches it is clear and obvious that, nations are not able to "understand something that is not there" (Zorpas and Lasaridi 2013) and mainly prevention activities as they cannot realize when they participate in any waste management plan (prevention activities, recycling, reuse, etc.) how and what they actually offer as they can't see the direct results. Citizens all over the world, regardless of what category their countries are (low or high incomes), need motivation (mainly less taxes or to receive extra money somehow) to react to anything and especially to participate in environmental performances actions. Mainly to have success plan in all levels a continual public awareness event is necessary.

References

Bauer C (2009) Hofkóche und Hofkeller (kitchen and cellar). Cultural magazine of Viennese guides 2009

Bisson K (2002) Attitudes to waste. In: Bisson K, Proops J (eds) Waste in ecological economics, 2000. Edward Elgar, Cheltenham, pp 56–97

Brunner PH, Fellner J (2007) Setting priorities for waste management strategies in developing countries. Waste Manag Res 25(3):234–240

Chroni C, Lasaridi K, Kyriakou A, Zorpas AA, Abeliotis K (2015) Comparative evaluation, of compost stability measurement methods, CEST 2015, 14th international conference on environmental science and technology, 3–5 September, Rodes, Greece. Oral Presentation

Cox J, Giorgi S, Sharp V, Strange K, Wilson DC, Blakey N (2010) Household waste prevention—a review of evidence. Waste Manag Res 28:193–219

EEA (2013) Managing municipal solid waste, a review of achievements in 32 European countries, European environmental agence. file:///C:/Users/user/Downloads/Managing%20municipal%20solid%20waste%20(1).pdf Accesses on April 2015

ENVITECH (2013) Development of the strategic plan for waste prevention in Cyprus. Funded from the Department of Environment, Cyprus

Eurostat (2010) Municipal waste generated, 1000 tonnes, 1995–2008 (update 11 March 2010). Environmental data centre on waste [WWW Document]. URL http://epp.eurostat.ec.europa.eu/portal/page/portal/waste/documents/Municipal_%20waste_%20generated_1000t_%20update_%20website180112.mht

Eurostat (2011) http://epp.eurostat.ec.europa.eu/statistics_explained/index.php/Waste_statistics. Accessed on 12 May 2015

Feeding America (2011) Food banks: Hunger's new staple. Research brief. A report on visitation patterns and characteristics of food pantry clients in the United States in 2009. Chicago. http://feedingamerica.org/hunger-in-america/hunger-studies/_/media/Files/research/HungerNewStapleFullReport_FINAL.ashx?.pdf. Accessed on 20 June 2015

Fell D, Cox J, Wilson DC (2010) Future waste growth, modelling and decoupling. Waste Manag Res 28:281

Golomeova S, Srebrenkoska V, Krsteva S, Spasova S (2013) Solid waste treatment technologies. Mach Technol Mater 9:59–61. ISSN 1313-0226
Guide Lines on Waste Prevention Program. http://ec.europa.eu/environment/waste/prevention/pdf/Waste%20Prevention_Handbook.pdf. Access on March 2015
Inglezakis JV, Zorpas AA (2015) Environmental indicators. In: AA Zorpas (ed) Sustainability behind sustainability. Nova Science Publisher, New York, pp 87–107
IPCC (2006) Guidelines for national greenhouse gas inventories, waste generation, composition and management data. http://www.ipcc-nggip.iges.or.jp/public/2006gl/pdf/5_Volume5/V5_2_Ch2_Waste_Data.pdf. Accessed on 20 June 2015
ISWA (2013) International solid waste association, promoting sustainable waste management worldwide. www.iswa.org. Access on 15 June 2015
Jacobsen H, Kristoffersen M (2002) Case studies on waste minimisation practices in Europe. European Topic Centre on Waste, Copenhagen, Denmark
Kennedy CA, Cuddihy J, Engel Yan J (2007) The changing metabolism of cities. J Ind Ecol 11:43–59
Kennedy C, Pincetl S, Bunje P (2010) The study of urban metabolism and its applications to urban planning and design. Environmental pollution, pp 1–9
Lebersorger S, Beigl P (2011) Municipal solid waste generation in municipalities: quantifying impacts of household structure, commercial waste and domestic fuel. Waste Manag 31:1907–1915
Mata-Alvarez J, Mace S, Llabres P (2000) Anaerobic digestion of organic solid wastes. An overview of research achievements and prospectives. Bioresour Technol 74(1):3–16
Mazzanti M, Zoboli R (2008) Waste generation, waste disposal and policy ffectiveness: evidence on decoupling from the European Union. Resour Conserv Recycl 52:1221–1234
Ministry for the environment (2007) Targets in the New Zealand waste strategy 2006: review of progress. New Zealand Ministry for the Environment, Wellington
Ministry of Environment (2013) www.moa.gov.cy. Access on 30 Oct 2015
OECD (2002) OECD workshop on waste prevention: toward performance indicators (No. ENV/EPOC/WGWPR/SE(2002)1/FINAL). Paris, France
OECD (2010) Total amount generated of municipal waste. In: OECD factbook 2010: economic, environmental and social statistics. Environment—air and land municipal waste. OECD Publishing, Paris
Papargyropoulou E, Lozano R, Steinberger KJ, Wright N (2014) The food waste hierarchy as a framework for the management of food surplus and food waste. J Clean Prod 76:106–115
Parfitt J, Barthel M, Macnaughton S (2010) Food waste with in food supply chains: quantification and potential for change to 2050. Philos Trans R Soc B 365:3065–3081. doi:10.1098/rstb.2010.0126
Pre-waste (2012) http://www.prewaste.eu/prevention-policies/item/55-what-iswaste-prevention?.html. Accessed on 18 June 2015
Productivity Commission (2006) Waste management—productivity commission inquiry report. Commonwealth of Australia, Canberra, p 562
Salhofer S, Obersteiner G, Schneider F, Lebersorger S (2008) Potentials for the prevention of municipal solid waste. Waste Manag 28:245–259
Schneider F (2011) The history of food wastage. In: 3rd international conference "Waste—The Social Context", May 08–11, 2011, Edmonton, Alberta. Proceedings online: www.ewmce.com/library/2011-conference-proceedings
Schneider F (2013) The evolution of food donation with respect to waste prevention. Waste Manag 33:755–763
Schubeler P (1996) Urban management and infrastructure: conceptual framework for municipal solid waste management in low-income countries. United Nations Development Program, UMP working paper no. 9. St. Gallen, Switzerland, SKAT
Sharp V, Giorgi S, Wilson DC (2010a) Delivery and impact of household waste prevention intervention campaigns (at the local level). Waste Manag Res 28:256–268

Sharp V, Giorgi S, Wilson DC (2010b) Methods to monitor and evaluate household waste prevention. Waste Manag Res 28:269–280

Tadesse T, Ruijs A, Hagos F (2008) Household waste disposal in Mekelle city, Northern Ethiopia. Waste Manag 28:2003–2012

Tanaka N, Tojo Y, Matsuto T (2005) Past, present, and future of MSW landfills in Japan. J Mater Cyc Waste Manag 7(2):104–111

The European Parliament and the Council of the European Union (2008) Directive 2008/98/EC of the European parliament and of the council of 19 November 2008 on waste and repealing certain directives. Off J 312:3–30

Tsilemou K, Panagiotakopoulos D (2006) Approximate cost functions for solid waste treatment facilities. Waste Manag Res 24:310–322

USEPA (2006) Solid waste management and greenhouse gases: a lifecycle assessment of emissions and sinks. Washington, DC

USEPA (2009) Municipal solid waste generation, recycling, and disposal in the United States detailed tables and figures for 2008. ASM Press, Washington, DC

Wolman A (1965) The metabolism of cities. Sci Am 213(3):179–190

World Bank (2012) Urban development series—knowledge papers. What a waste. A global review of solid waste management. In: Hoornweg D, Bhada-Tata P (eds) Urban development and local government. World Bank, Washington

WRAP (2008) Material change for a better environment. Food waste report: the food we waste. http://wrap.s3.amazonaws.com/the-food-we-waste.pdf. Accessed on 26 Jan 2014

Young CY, Ni SP, Fan KS (2010) Working towards a zero waste environment in Taiwan. Waste Manag Res 28:236

Zero Waste South Australia (2007) South Australia's waste strategy 2005–2010. Benefit cost assessment. Volume 1: summary report. Melbourne, Australia

Zorpas AA (2012) Sewage sludge compost evaluation and utilization. In: Zorpas AA (ed) Sewage sludge management; from the past to our century. NovaScience Publisher, New York, pp 173–216

Zorpas AA (2014) Recycle and reuse of natural zeolites from composting process: a 7 years project. Desalin Water Treat 52:6847–6857

Zorpas AA, Lasaridi K (2013) Measuring waste prevention. Waste Manag 33:1047–1056

Zorpas AA, Voukkali I (2015) Tourist area metabolism and its potential to change through a holistic strategic plan in the framework of sustainable development» 5th international conference on environmental management, engineering, planning and economics SECOTOX (society of ecotoxicology and environmental chemistry) conference on June 14–18, on Mykonos island, Greece

Zorpas AA, Kapetanios E, Vlyssides A, Zorpas AG, Karlis KP, Haralambous I, Loizidou M (2000) Compost produced from organic fraction of municipal solid waste, primary stabilized sewage sludge and natural zeolite. J Hazard Mater B 77:149–159

Zorpas AA, Lasaridi K, Voukkali I, Chroni C, Loizia P, Foitiri L (2012) Waste minimization and prevention activities in insular communities. The case studies of a municipality in Cyprus· 3rd international conference on industrial and hazardous waste management "CRETE 2012" September 12th–14th, 2012, Chania, Oral Presentation

Zorpas AA, Lasaridi K, Abeliotis K, Voukkali I, Loizia P, Fitiri L, Chroni C, Bikaki N (2014) Waste prevention campaign regarding the Waste framework directive. Fresenious Environ Bull 23(11a):2876–2883

Zorpas AA, Lassaridi K, Voukkali I, Loizia P, Chroni C, Georgiou A, Fanou C (2014b) Social measuring in the framework of sustainable waste prevention activities. In: Kalogerakis V, Manio Th (eds) International phytotechnologies conference, heraklion, Crete, Greece, Sept 30–Oct 3, p 370

Zorpas AA, Voukkali I, Loizia P (2015a) The impact of tourist sector in the waste management plans. Desalin water Treat 56(5):1141–1149

Zorpas AA, Lasaridi K, Voukkali I, Loizia P, Chroni C (2015b) Household waste compositional analysis variation from insular communities in the framewok of waste prevention strategy plans. Waste Manag 38:3–11

Zorpas AA, Lasaridi K, Voukkali I, Loizia P, Chroni C (2015c) Promoting sustainable waste prevention activities and plan in relation to the waste framework directive in insular communities, environmental processes

Author Biographies

Antonis A. Zorpas is a Lecturer Prof. in Cyprus Open University and in the Department of Pure and Applied Sciences, Department of Environmental Conservation and Management. He is a Chemical Engineer and he holds a PhD in the Section of Environmental Engineer/Management. Until now, he is the editor in 3 two scientific books: "Sludge Management; From the Past to Our Century", "Natural Zeolites" and "Sustainability Behind Sustainability". He has more than of 75 publications in scientific journals and with impact factor from 0.751 to 6.072 and more than 200 papers in International and National Conferences. He is research back-round includes: (a) Implementation of Natural minerals in the treatment of solid and liquid waste; (b) Implementation of Composting methods for the treatment of biosolids; (c) Implementation of chemical oxidation, electrochemical oxidation for the treatment of heavy polluted waste; (d) Waste water treatment plans and energy; (e) Waste minimization monitoring and evaluation, EMAS, ISO 14001, strategic planning, zero waste approach, easte prevention and the metabolism of the cities; (f) Sustainable development programs (environmental, social culture and economics); (g) Management of Liquid waste, Management of solid waste; (h) Environmental Impact and Environmental/Risk Assessment Analysis/multi criteria analysis models/LCA.

Irene Voukkali holds a Master Degree in Environmental Engineering and she's the Quality Control Manager of the Institute of Environmental Technology and Sustainable Development. Until now she has published 5 chapters on behalf of Scientific Books; more than 15 research papers in Scientific Journals and more than 30 papers in International Conferences. She's research mainly focus on Sustainable Development, EMAS, Quality Control, Environmental Management Systems, Strategic Development, Liquid waste treatment, cities metabolism.Mrs Voukkali is a PhD student in Cyprus Open University

Pantelitsa Loizia studied Chemical Engineer and she's now the European Affaire's on behalf of the Institute of Environmental Technology and Sustainable Development. Until now she has published 2 chapters on behalf of Scientific Books, more than 10 research papers in Scientific Journals and more than 20 papers in International Conferences. She's research mainly focus on Sustainable Development, Environmental Management Systems, Strategic Development, Waste Prevention, waste water treatment plans etc.

Ways of Fostering Green Economy and Green Growth

Begum Sertyesilisik and Egemen Sertyesilisik

Abstract

There is need for international collaboration among developing and developed countries for fostering green economy and sustainability. Despite of high awareness on climate change, even the developed countries are underperforming with respect to the environmental footprint of their activities. Furthermore, countries having cheap labour tend to attract global investment. Countries have been classified based on their virtual carbon trade, which requires calculation of carbon emission deficit or surplus of the countries through their imports and exports (Slay in Virtual carbon trade, transition and developing economies: Sinners, or sinned against? UNDP (United Nations Development Programme), Regional Bureau for Europe and CIS 2011). This chapter focuses on fostering green economies and green growth from political, social and economic perspectives so that sustainable development can be achieved, and poverty as well as environmental footprints can be reduced. The chapter covers the following topics: relationship between green economies and green growth; the ways for recruiting change agents for sustainability (i.e. the roles of media, higher education, policies, and companies in recruiting change agents for sustainability); ways for enhancing green economies (i.e. carbon taxes, international protocols, virtual water, virtual carbon, eco-industrial parks, green supply chain); as well as introduction of new perspective towards carbon tax and carbon-motivated border tax adjustments concepts. This chapter reveals that political, social and economical aspects are interrelated and that they support

B. Sertyesilisik (✉) · E. Sertyesilisik
Faculty of Architecture, Department of Architecture,
Istanbul Technical University, Istanbul, Turkey
e-mail: bsertyesilisik@itu.edu.tr

E. Sertyesilisik
e-mail: egemens@alumni.bilkent.edu.tr

W. Leal Filho et al. (eds.), *Sustainable Economic Development*,
World Sustainability Series, DOI 10.1007/978-3-319-45081-0_3

each other. For this reason, all three aspects should be considered together so that green economy and green growth are fostered.

Keywords
Virtual carbon trade · Green economy · Change agents · Green growth

1 Introduction

The world's habitat is being deteriorated (i.e. water depletion, loss of biodiversity) (Tukker 2013: 274). The growing industrialization and increase in the scale of economic activities have transformed the world's resources into wealth causing adverse effects on ecosystems and resources (Linnenluecke and Griffiths 2013: 382). Nature is under the combined pressure of human population growth and the growth in the wealth per capita (Tukker and Butter 2007: 102). Throughout the twentieth century 142 billion tons petroleum, 265 billion coal, 38 billion iron 760 million aluminum, 480 million cooper have been consumed (Xiaowei et al. 2011: 1368). The 'economy is crashing against the earth' (Tukker 2013: 274). The world economy is expected to grow by 3 % per year until 2030 and more than 9 billion humans are expected to live on earth by 2050 (Rohn et al. 2014: 32). The economic growth puts pressure on the environment as quoted by Tukker and Butter (2007: 102):

> … if the environmental pressure was to remain at the same level as now, the fulfillment of needs would have to be a factor of 10 more effective (Weizsacker et al. 1997; Factor 10, 1997)." (Tukker and Butter 2007: 102).

The economy needs to be sustainable. An economy is sustainable only if it simultaneously caters all human needs and accepts the limitations imposed by the need to sustain the environment's ability to meet present and future needs (Lorek and Spangenberg 2014: 33).

Despite of rising prices for natural resources, there is an increase in the global consumption of natural resources (Rohn et al. 2014: 32). The Earth, however, has its limitations (Ayres and Kneese 1969; Daly 1991; Meadows et al. 1972; Tukker 2013: 272). As stated by "The Club of Rome's (1972) Limits to Growth, economic growth cannot continue indefinitely due to the limits of the capacity of the global environment (Jung et al. 2013). Essential needs are, however, not substitutable (Lorek and Spangenberg 2014: 33).

As the environmental degradation continues to occur in an accelerated way, time is of the essence for taking effective precautions. The report from the Intergovernmental Panel on Climate Change (IPCC) called for immediate action (EC website 2014). There is need for abandoning the 20th century's traditional growth model, and for a shift towards low carbon economic model and lifestyle, and for achieving SD (sustainable development) (Xiaowei et al. 2011: 1368), which has economic, human, environmental and technological dimensions (WRI 1992: 2; Mehmet 1995: 11) and which can be supported by the green economy (Diyar et al. 2014: 696). "... the green economy aims at improved human well-being and social equity, while significantly reducing environmental risks and ecological scarcities" (UNEP 2011; Rueff et al. 2014). Green economy is based on the understanding that economy depends on natural environment (Diyar et al. 2014: 695) and on coexistence of people and nature (Xiaowei et al. 2011: 1367). SD supported by the green economy can result in green growth which can enable increase in resource productivity and in efficiency through environmental as well as resource saving processes and products (Janicke 2012: 19). Furthermore, green growth can lead to (Mehmet 1995: 14): greening of consumer values through market; employment creation; sustainable urban development increasing employment rate (i.e. the Singapore River cleanup project which generated thousands of productive jobs while promoting SD) (Ling 1992).

Both SD and green economy have been referred to in the title of the Agenda 21, adopted by the UN General Assembly as one of the main outcomes of the Rio Conference (the United Nations Conference on Environment and Development UNCED 1992; Zaccai 2012: 80) as well as of the Rio + 20 declaration (UN 2012; Allen and Clouth 2012; Sutton et al. 2014: 77; Rueff et al. 2014).

The fact that the world's habitat is being deteriorated despite of the precautions (e.g. Kyoto protocol, the Resource Efficiency and Cleaner Production Program, the 10 year Framework of Programs on Sustainable Consumption and Production, the UN Resource Panel, and the Green Economy Initiative, Rio + 20, green building certificates, etc.) taken, reveals the need for acting strategically (e.g. green growth, the integration of the green economy and green growth, enhancing green economy and recruiting the change agents for sustainability) for the survival of humanity. For this reason, this chapter focuses on fostering green economies and green growth from political, social and economic perspectives reducing poverty and environmental footprint. The chapter covers the following topics: relationship between green economies and green growth; ways for recruiting change agents for sustainability; ways for enhancing green economies (i.e. carbon taxes, international protocols, virtual water, virtual carbon, eco-industrial parks, green supply chain) as well as introduction of new perspectives towards existing concepts.

2 Fostering Green Economy and Green Growth

Ways of fostering green economies and green growth have been analysed with respect to political, social and economic aspects.

2.1 Political Aspects

Policies play the key role in improving local and global sustainability (Editorial Journal of Cleaner Production 2005: 967–969). Governments and corporations are increasingly being held responsible for their sustainability performance (Roy and Goll 2014: 849). Environmental policies started to cover SD topic especially since the Brudtland Report (1987) (Zaccai 2012: 79). The politicians need to enhance the citizens' interest in protecting the environment and in encouraging companies for "greening" their production processes as well as their products. "*Local governments should invest in a SD policy to satisfy citizens and benefit companies and act with companies as partners to increase resilience and sustainability*" (Musson 2012: 75). Politicians can encourage green economy and green growth mainly through: their commitment; green GDP; EID (eco-industrial development) policies; green technology policies fostering innovation in green technology; international collaboration in combating climate change; policy instruments (i.e. carbon taxes; carbon-motivated border tax adjustments; virtual water) as explained in the following paragraphs.

Politicians commitment for green economy and green growth: The politicians can establish green growth strategies, and support the green growth as well as green economy with the help of their commitment. For example, Kazakhstan carries out the shift towards green economy and integrates green growth and green economy with the help of the National Strategy for SD, the regional management strategies, and programs as well as with the help of the State initiatives (i.e. transition to the Green Economy, formulation of the National Strategy for SD, the Green Bridge Program and holding the International Specialized Exhibition EXPO—2017) (Diyar et al. 2014: 695–697). Similarly, China and South Korea have 5-year green growth plans, and strategies supported by industrial policies encouraging green industries based on green technologies (Mathews 2012: 762). Another example is the UK, where low carbon economy is proposed in energy white book entitled as "Our Energy future: Create Low Carbon Economy" which has been published by the British government in 2003 (Xiaowei et al. 2011: 1367).

Green GDP: Politicians can encourage green development with the help of "green GDP" which "*is an attempt by economists to measure the growth of an economy compared to the harm production does to the environment. This is done by subtracting the costs of environmental and ecological damage done in a specific period of time from the gross domestic product from that some time.*" (WiseGEEK 2014). In this way, the consumers can have a tendency for consuming the products which are less harmful to the nature.

EID policies: Politicians can encourage policies for EID which is a framework for the development of the industry reducing its adverse effects to the environment (Cohen-Rosenthal 2003). EID aims to improve business and environmental performances mainly through: resource efficiency (Babu and Meyer 2012); cleaner product; industrial ecology; industrial symbiosis; environmental management systems; and design for the environment.

Green technology policies fostering innovation in green technology: Politicians can support innovation in green technology with the help of national green technology policy [i.e. Malaysia (Chua and Oh 2011: 2850), subsidy systems or patent rights (Encaoua et al. 2006; Li et al. 2013: 1529)]. As low carbon economy is based on low energy consumption, and low pollution, it requires innovation in energy technology and changes in consumers' consumption habits (Xiaowei et al. 2011: 1367). Policies can support innovation which enhances sustainability performance of the production process as well as of the product as "*...technological improvements... must be combined with and integrated into structural change and sufficiency policy initiatives...*" (Lorek and Spangenberg 2014: 36).

International collaboration in combating climate change: Politicians can encourage international collaboration in combating climate change. There is need for international collaboration among both developing and developed countries for fostering green economy and sustainability as "*no nation can secure its future alone, but together we can–in a global partnership for SD...* (UBCED 1992)..." (Mehmet 1995: 11). Despite of high level of awareness on climate change, even the developed countries are underperforming with respect to the environmental footprint of their activities. 15 % of the world's population from the industrial developed countries consumed 56 % of the world's petroleum, 60 % of the world's gas, and 50 % of the major mineral resources (Xiaowei et al. 2011: 1368). The UNEP launched "*Global new green policy and green economic plan*" in 2008 and called all countries for the establishment of 'green' system so that the global crises can be eliminated and worldwide global industry revolution can be fostered (Xiaowei et al. 2011: 1369–1370). Similarly, the "*Global Green New Policy Synopsis*" report launched by the UNEP in 2009 called all countries' leaders for investing 1 % of their GDP into regeneration energy (Xiaowei et al. 2011: 1369–1370). If all countries invest in 1 % of the GDP per year, 5–20 % of the GDP loss in the future is expected to be avoided (Stern Report 2006). Rueff et al. (2014) emphasized that the green economy agenda of developed and developing countries differ due to the fact that developed countries mainly focus on economic growth and on addressing climate change whereas developing countries mainly focus on poverty alleviation, equity improvement, and food security (Messerli et al. 2012; UNCSD 2012; UNEP 2011).

Policy instruments: Politicians can use policy instruments to enhance green economy and green growth. These policy instruments include: carbon taxes, CBTAs (Carbon-motivated border tax adjustments), and virtual carbon trade.

- *Carbon tax*: To stay under the 2 °C global warming target, total emissions between 2000 and 2050 should be less than 1000 Gt CO_2 (Meinshausen et al. 2009). Reduction in the CO_2 emissions is within the agenda in the EU policies, and targets (i.e. the 25 year Strategic Research Agenda for the European Construction Sector (ECTP 2015); GHG Protocol (2004) by the World Business Council for SD, the World Resources Institute, and the Carbon Disclosure Project; the EC's 20-20-20 targets; the new EU framework on climate and energy for 2030; the European Commission's Roadmap for moving to a low-carbon economy in 2050). Carbon tax is an effective tool in reducing the CO_2 emission level. A carbon tax imposes a cost on carbon emissions resulting in polluters to internalize the cost of the externality (Mathur and Morris 2014: 326) and in optimization of the resource allocation by internalizing the environmental externality caused by anthropogenic CO_2 emissions (Jiang and Shao 2014: 269) (i.e. reduction in energy intensity and electricity use in manufacturing plants in the UK) (Martin et al. 2014: 1). Just after the Kyoto Protocol, carbon tax policy was firstly introduced in Finland in 1990 (Liu et al. 2015: 1), and then applied in various countries (i.e. Norway, Sweden, Denmark, Switzerland, India, Ireland and Costa Rica) (Chen et al. 2015: 45). The practices of carbon taxes in Europe contributed to the reduction in carbon emissions (Liu et al. 2015: 1). Revenues earned from carbon taxes can support promotion of the utilization of renewable energy (Jiang and Shao 2014: 269–270) and finance cuts in ordinary income taxes (Duan et al. 2014: 345). Carbon taxes can also encourage: companies to adopt less emission intensive production technologies (Allan et al. 2014: 41) and invest in research and development activities for carbon-free technologies (Duan et al. 2014: 345). Dissou and Siddiqui (2014: 88) emphasized the importance of the carbon tax level as carbon taxes have been identified as being regressive in Canada (Hamilton and Cameron 1994); Denmark (Wier et al. 2005); Netherlands (Dinan and Rogers 2002), and the US (Kerkhof et al. 2008; and Shammin and Bullard 2009). A policy is called as being regressive in case it "… *burdens lower-income households relatively more than higher-income households as a share of household income…*" (Mathur and Morris 2014: 326).

- *CBTAs*: "*CBTA is a kind of import tax which requires imported goods to be taxed according to their carbon content incurred in the production process. It could be levied according to the carbon content of exports or the carbon content of imports.*" (Li et al. 2013: 927–928). CBTAs can encourage companies to invest in new technologies (Almutairi and Elhedhli 2014: 89). CBTAs can result in relocation of output from the target countries to CBTA users (Li et al. 2013: 927).
- *Introduction of new concepts*: The scope of the carbon taxes and CBTA concepts can be extended from product level to country level as CO_2 emission rates per capita and per country differentiate. China emits higher amount of CO_2 compared to the US and Canada (Rogers and Evans 2011) whereas the per capita CO_2 emission levels are 18 tones in the USA, 6 tones in China, 1.38 tones in India and 152 tones in Gibraltar (Rogers and Evans 2011). According to Davis and Calderia (2010), in 2004, 6.2 gigatones of CO_2 have been traded globally. In countries like Sweden, Switzerland, Britain and France over 30 % of consumption based emissions accrue from imports (Ketterer et al. 2011). Some countries (i.e. China, Russia, India, South Africa, Ukraine, Venezuela, Australia etc.) are net carbon exporters whereas other countries (i.e. USA, Japan, Singapore, Korea, Switzerland, Hong Kong, Mexico, Norway and Turkey) are net carbon importers (Peters et al. 2011; Slay 2011). In order to reduce the carbon emission in our planet, the countries can be ranked by weighted average of their carbon emission. This weighted average can be calculated by taking the countries' virtual carbon trade rate and carbon emission per capita. Accordingly, a country having both high carbon export rate and high carbon emission per capita

can be marked as performing poorly and it can face high taxation rates to its products. Accordingly, new concepts introduced within the scope of this chapter include: "global CBTA"; "global CBTA ranking"; and "global CBTA tax rate".

- "global CBTA" is the extended scope of the CBTA at the country level.
- "global CBTA ranking" concept is based on the countries' ranking with respect to their weighted average of their carbon emission. This weighted average can be calculated by taking the countries' carbon emission per capita and per country. Accordingly, a country having both high carbon export rate and high carbon emission per capita can be marked as performing poorly.
- "global CBTA tax rate" is based on the countries' "global CBTA ranking". Based on the country's "global CBTA ranking", the carbon tax rate level to be applied to the products of that country's exports can be determined. This carbon tax rate level applied to the products of a specific country's exported goods is called as "global CBTA tax rate".

- *Virtual water*: Virtual water is "*...the volume of water used to produce a certain commodity*" (Fracasso 2014: 215). In other words, it is the water embodied in the commodity. Accordingly, virtual water trade (VWT), proposed by Allan (1997, 1998), is based on the idea that "*as commodities are internationally traded, one can depict a network of fluxes of the water that is somehow 'embodied' in the goods exchanged across countries*" (Fracasso 2014: 215). Countries who are poor in water resources can import water intensive products so that they can make efficient use of their water resources.

Various countries have carried out researches for successful transition towards green economy. Reports for the EU, Egypt and Washington are among these researches. Brief summary for these examples have been provided as follows:

- *EU*: the Policy Studies Institute (PSI), Institute for European Environmental Policy (IEEP), BIO and Ecologic Institute have accomplished a research to provide an initial scoping assessment of potential priorities and policy options to support the transition to a circular economy in the EU (2014).
- *Egypt*: the CEDARE (the Centre for Environment and Development for the Arab Region and Europe)'s report (2014) for Egypt. In Egypt, green economy and greening key economic sectors (agriculture, waste, energy, water, tourism, industry and transportation) have been perceived as effective tools in coping with Egypt's problems (i.e. poverty, unemployment, water, energy and food shortages) and in achieving long-term prosperity and human welfare (CEDARE—the Centre for Environment and Development for the Arab Region and Europe, 2014: 62). The CEDARE's report revealed the importance of (CEDARE 2014): green economy policy measures; changing consumption and production patterns in greening the economy; good governance; inter-ministerial coordination and the inter-generational equity principle.

- *Washington*: The Strategic Framework of the Washington State's Green Economy (CTED—Community, Trade and Economic Development, 2009: 7) revealed that "the state's green policies must be strong, incontrovertible and signal a long-term, deep commitment to change" and that state commitment as well as supportive policies and incentives are important for achieving the green economy.

2.2 Social Aspects

Change agents for sustainability can foster green economy and green growth. Public awareness can influence political and economic aspects as well as their success. According to Wangel (2011: 873), the term 'agency' refers to 'the social' part of the socio-technical society, which consists of formal institutions, (i.e. policies, taxes, and organisations), and informal institutions (i.e. norms, values, and social practices). Individuals and organisations having the capacity to act can act as agents (Wangel 2011: 873). Change agents, on the other hand, are opinion leaders and driving forces in change processes (Hesselbarth and Schaltegger 2014: 26). Sustainability leaders are the change agents who play the key role for the successful transformation towards sustainability as well as for green economy and green growth. Change agents for sustainability can be individuals at all levels internal or external of an organization (Hesselbarth and Schaltegger 2014: 26). The Aarhus Convention Strategic plan (2008) emphasized the importance of change agents as follows (Zilahy and Huisingh 2009: 1057):

> the serious environmental, social and economic challenges faced by societies worldwide cannot be addressed by public authorities alone without the involvement and support of a wide range of stakeholders, including individual citizens and civil society organizations.

The roles of education, and media in recruiting change agents for sustainability, SD, green economy and green growth have been briefly summarized in the following paragraphs (Sertyesilisik and Sertyesilisik 2015):

- *Education*: As the future generations are future's potential change agents, schools play the key role for raising next generation's consciousness level on the importance of SD and on acting proactively. In addition to the formal education, social learning can also contribute to the consumers' consciousness on SD and their demand for sustainable products. Furthermore, as social learning enables people to learn from and with each other, it can encourage the communities to adopt sustainability principles, ethics and values (Schmitt 2011).
- *Media*: Media can have economic and social impacts supporting demand for sustainable production and consumption (Syed 2014). Media can act as a facilitator affecting the consumers' consciousness and demand for sustainable consumption as well as their willingness to act as change agents. Both mass and social media can act as facilitators as the mass media can provide people

information on sustainability related topics, whereas the social media enables people to have active interaction.

2.3 Political Aspects

Economic aspects which can foster green economy and green growth include: innovation in green technology; entrepreneurship; eco labelling; EIPs and green supply chains.

Innovation in green technology and its impacts on the fostering the green economy: Innovation can decouple growth from natural capital depletion and contribute to the establishment of new markets and to the creation of new jobs (OECD 2011a). Green innovation can "… *encourage business solutions and entrepreneurial ideas for tackling environmental challenges*." (OECD 2009). "*Innovation in green technology … can lead to a 'technological revolution*' (Tran 2012: 132) *which can be a solution to global warming*." (Li et al. 2013: 1529). Green economy can be achieved through innovation in green technology as "*green economy is related to the low carbon economy which is a low energy-consumption, low-pollution, low-emission economic mode*" (Xiaowei et al. 2011: 1367). Similarly, green growth requires environmentally friendly production processes/products and supports 'low-carbon economy' (Janicke 2012: 13–14) "… '*green growth*' … *focuses on the contribution of environmental technologies to a growing economy* (OECD 2011b). (Sutton et al. 2014: 77) "…*radical growth in environmental and resource-saving technologies leads to radical 'de-growth' in products and processes*…" (Janicke 2012: 13) as well as to dematerialisation which can contribute to the alleviation of environmental problems (Vringer and Blok 2000: 713) (i.e. dematerialization can lead to decarbonization). Dematerialization is the decline over time in weight of the materials used in industrial end products as well as the decline in the 'embedded energy' in industrial products (Herman et al. 1990: 333). Innovation in green technologies can be enhanced by policy instruments and universities. Patent rights, as a policy instrument, can foster green technology innovation through incentives provided for research and development (Encaoua et al. 2006; Li et al. 2013: 1529). Universities have vital role in enhancing innovation in green technologies as they are knowledge-producing institutions acting as engines of economic growth (Yusuf and Nabeshima 2007; OECD 2007; Trencher et al. 2014: 155). Universities' capacities and impact on innovation in green technologies can be enhanced through cross-sector university partnerships and collaboration between the university and societal stakeholders (Trencher et al. 2014: 154–155). Trencher et al. (2014: 154) summarized universities' role in innovation as follows:

> … Universities generate scientific, technological and social innovation and educate next generation leaders (Cortese 2009; M'Gonigle and Starke 2006), link differing sectors of expertise and mediate across these networks (Arbo and Benneworth 2007; Cash et al. 2003; Sedlacek 2013)…" (Trencher et al. 2014: 154)

Furthermore, green innovation can be encouraged through (OECD 2011a): clear and stable market signals; international cooperation in research and developments for green innovation; interventions to overcome specific market failures (i.e. dominance of existing technologies); policies focusing on the creation and supply of new technologies and innovations, as well as on the diffusion and take-up of green innovations in the market place. Government action can encourage innovation through empowering people to innovate; unleashing innovation in firms; creating and applying knowledge; applying innovation to address global and social challenges; and improving the governance of policies for innovation (OECD 2010).

Ecolabelling: As eco-labels provide the consumers information about the environmental impacts of products (Reczkova et al. 2013: 498), they have the potential for influencing consumers' purchasing decision towards environmentally friendly products (Reczkova et al. 2013: 498). They influence the individual consumers' demand for the end product, as well as the demand of the companies in the supply chain for sustainable/environmentally friendly input materials or byproducts.

Entrepreneurship fostering green economy and green development: Entrepreneurship can foster innovative sustainability technologies (Ustaoglu and Yildiz 2012: 142) and affect the clean technologies market which consists of a wide range of goods, services, processes that ensure high productivity with efficient use of natural resources (Diyar et al. 2014: 696). Entrepreneurship can act like "*a catalyst for solutions to sustainability problems* (York and Venkataraman 2010) *and it can play an important role in the development of an ecologically as well as socially sustainable economy* (Pacheco et al. 2010)" (Muñoz and Dimov 2015).

EIPs (Eco-industrial park)s reducing environmental footprint of the production: EIP can be defined as "... *a community of manufacturing and service businesses located together on a common property. Member businesses seek enhanced environmental, economic, and social performance through collaboration in managing environmental and resource issues.*" (Veiga and Magrini 2009: 653). EIPs are perceived as a new industrial model to address the three dimensions of sustainability, namely: social, economic and environmental dimensions (Veiga and Magrini 2009: 653). EIPs are based upon industrial ecology principles which suggest that industrial systems can operate like natural ecological systems (Jung et al. 2013: 50). EIP is based on the idea of the industrial symbiosis which aims to engage separated industries in a collective approach so that their economic performance is improved whereas their environmental footprints are reduced (Chertow 2000; Fang et al. 2007; Oh et al. 2005: 271; Yu et al. 2014). EIPs lead to the connection of different waste-producing processes, plants, or industries as well as consumers into an operating web (Fang et al. 2007: 319). The EIPs enable both tangible exchanges [i.e. the physical exchange of materials, energy, water, and by-product (Chertow 2007)] and intangible exchanges of knowledge and human or technical resources (Mirata and Emtairah 2005; Panyathanakun et al. 2013: 71). The exchanges of resources and collaboration among collaborative companies in the production process enable synergy which is greater than the sum of the individual benefits each company could get if it optimised its own performance only (Cote and Cohen-Rosenthal 1998; Lowe 1997; Panyathanakun et al. 2013: 71; Romero and Ruiz 2014: 394).

Green supply chain reducing environmental footprint of the production: Companies can be major contributors to SD as they are perceived as major contributors to ecological problems (Clifton and Amran 2011; Roy and Goll 2014: 851–852). Companies can enhance their overall sustainability performance and reduce their environmental footprint with the help of green supply chain as the focal companies are held responsible for environmental and social problems caused by their suppliers and as an increasing share of value is created at the supplier level (Koplin et al. 2007: 1053). A company "*is no more sustainable than its supply chain*" (Krause et al. 2009: 18). For this reason, different industries work on greening their supply chain (De Brito et al. 2008; Caniels et al. 2013; Foerstl 2010; Goger 2013; Koplin et al. 2007; Zhu et al. 2007; Zhu and Sarkis 2004).

3 Conclusions

Green economy and green growth need to be fostered as both of them are needed for a SD. Political, social and economic aspects of sustainability need to be considered for achieving green growth and green economy. This chapter focused on fostering green economies and green growth from political, social and economic perspectives reducing poverty and environmental footprint.

Political aspects: Political aspects fostering green economy and green growth include establishment of green growth strategies, policies for EID, green GDP, national sustainability development plans, and national green technology policy to enhance innovation in green technology. The politicians can support international collaboration in combating climate change. The politicians can establish green growth strategies, and support the green growth as well as green economy with the help of their commitment. The politicians play important role in encouraging companies for 'greening' their production processes as well as their products. They can support green growth and green economy with the help of policy instruments (i.e. carbon taxes, CBTAs, and virtual carbon trade) so that companies are encouraged for having environmentally friendly production processes in order to avoid the costs of their footprints. New concepts introduced within the scope of this chapter include: "global CBTA"; "global CBTA ranking"; and "global CBTA tax rate". The scope of the concepts carbon taxes and CBTA can be extended from product level to country level as CO_2 emission rates of the countries differ in per capita and per country values. This extended scope of the CBTA introduced in this chapter is called as "global CBTA". Accordingly, the "global CBTA ranking" concept introduced within the scope of this chapter is based on the countries' ranking with respect to their weighted average of their carbon emission. This weighted average can be calculated by taking the countries' carbon emission per capita and per country into consideration. Accordingly, a country having both high carbon export rate and high carbon emission per capita can be marked as performing poorly. Based on the country's "global CBTA ranking", the carbon tax rate level to be applied to the products of that country's exports can be determined. This carbon tax rate level applied to the products of a specific country's exported goods is called as "global CBTA tax rate" and it is based on the countries' "global CBTA ranking".

Social aspects: Green economy and green growth can be achieved in case they are supported and embraced by the individuals. Individuals play key role in fostering the change towards sustainability as they can affect the production processes and products as well as policies and institutions through their demand and consciousness. Individuals and organizations, as sustainability leaders, can act like change agents for sustainability and can foster green economy and green growth. The emergence of these change agents can be supported mainly through formal and informal education as well as through media (Sertyesilisik and Sertyesilisik 2015).

Economic aspects: Green economy and green growth can be fostered mainly through innovation in green technology; entrepreneurship; EIPs and green supply chains. Green growth and green economy can be achieved through innovations which can result in greening the production processes and products. Innovation in green technologies can be enhanced by policy instruments (i.e. patent rights Encaoua et al. 2006), universities as knowledge-producing institutions acting as engines of economic growth (Yusuf and Nabeshima 2007; OECD 2007; Trencher et al. 2014: 155), cross-sector university partnerships and collaboration between the university and societal stakeholders (Trencher et al. 2014: 154–155). Entrepreneurship can foster innovative sustainability technologies (Ustaoglu and Yildiz 2012: 142) and the clean technologies market (Diyar et al. 2014: 696). Entrepreneurship can play an important role in the development of a sustainable economy (Pacheco et al. 2010; Muñoz and Dimov 2014). Green technologies' market and green products' market can be supported by eco labelling as it can influence consumers' demand for such products (Reczkova et al. 2013: 498). Not only the products but also production processes need to be environmentally friendly so that green economy and green growth can be achieved. EIPs and green supply chains can enhance the overall sustainability performance of the production process. Green production process can result in decrease in embodied environmental footprints of the products.

Political, social and economic aspects are interrelated and they support each other. For example, political and economic aspects are interrelated as policy instruments can affect the companies' decisions to invest in environmentally friendly production processes. For this reason, all three aspects should be considered together so that green economy and green growth are fostered.

References

Arbo P, Benneworth P (2007) Understanding the regional contribution of higher education institutions: a literature review. OECD education working papers, No. 9. OECD Publishing, Paris

Allan J (1997) Virtual water: a long term solution for water short Middle Eastern economies? Technical report no. 3. Occasional paper. Water Issues Study Group, School of Oriental and African Studies, University of London

Allan J (1998) Virtual water: a strategic resource global solutions to regional deficits. Ground Water 36:545–546

Allen C, Clouth S (2012) A guidebook to the green economy. Issue 1: green economy, green growth, and low-carbon development—history, definitions and a guide to recent publications, UN Division for Sustainable Development

Allan G, Lecca P, McGregor P, Swales K (2014) The economic and environmental impact of a carbon tax for Scotland: a computable general equilibrium analysis. Ecol Econ 100:40–50

Almutairi H, Elhedhli S (2014) Modeling, analysis, and evaluation of a carbon tax policy based on the emission factor. Comput Ind Eng 77:88–102

Ayres RU, Kneese AV (1969) Production, consumption, and externalities. Am Econ Rev 59:282–297

Babu NR, Meyer AK (2012) Pathway to eco industrial development in India—concepts and cases. Deutsche Gesellschaft für Internationale Zusammenarbeit (GIZ) GmbH. http://www.igep.in/live/hrdpmp/hrdpmaster/igep/content/e48745/e50194/e50195/121004_Pathway_EID_ISOX3uncoated-1.pdf

Brudtland Report (1987) Report of the world commission on environment and development: our common future, Oslo. www.un-documents.net/our-common-future.pdf

Cash D, Clark W, Alcock F, Dickson N, Eckley N, Guston D, Jager J, Mitchell R (2003) Knowledge systems for sustainable development. Proc Natl Acad Sci USA 100(14):8086–8091

Caniels MCJ, Gehrsizt MH, Semeijn J (2013) Participation of suppliers in greening supply chains: an empirical analysis of German automotive suppliers. J Purch Supply Manag 19:134–143

CEDARE (the Centre for Environment and Development for the Arab Region and Europe) (2014) Green economy scoping study Egypt. Job Number DTI/1854/GE

Chen PY, Chen BY, Tsai PH, Chen CC (2015) Evaluating the impacts of a carbon tax on imported forest products—evidence from Taiwan. For Policy Econ 50:45–52

Chertow MR (2000) Industrial symbiosis: literature and taxonomy. Ann Rev Energy Environ 25:313–337

Chertow MR (2007) Uncovering industrial symbiosis. J Ind Ecol 11(1):11–30

Chua SC, Oh TH (2011) Green progress and prospect in Malaysia. Renew Sustain Energy Rev 15:2850–2861

Clifton D, Amran A (2011) The stakeholder approach: a sustainability perspective. J Bus Ethics 98:121–136

Cohen-Rosenthal E (2003) What is eco-industrial development? Greenleaf publishing. http://greenleaf-publishing.com/content/pdfs/eich1.pdf

Cortese A (2009) Foreword. In: Bardaglio P, Putman B (eds) Boldly sustainable: hope and opportunity for higher education in the age of climate change. NACUBO, Washington, DC, pp 1–9

Cote RP, Cohen-Rosenthal E (1998) Designing eco-industrial parks: a synthesis of some experiences. J Clean Prod 6:181–188

CTED (Community, Trade and Economic Development) (2009) Washington state's green economy: a strategic framework discussion draft January 2009 Olympia, WA, USA

Daly H (1991) Elements of environmental macro-economics. In: Costanza R (ed) Ecological economics: the science and management of sustainability. Columbia University Press, New York, p 525

Davis S, Calderia K (2010) Consumption based accounting of CO_2 emissions. PNAS 107:5687–5692

de Brito M, Carbone V, Blanquart C (2008) 'Towards a sustainable fashion retail supply chain in Europe: Organisation and performance'. Int J Prod Econ 114:534–553

Dinan T, Rogers LD (2002) Distributional effects of carbon allowance trading: how government decisions determine winners and losers. Natl Tax J LV 55(2):199–211

Dissou Y, Siddiqui MS (2014) Can carbon taxes be progressive? Energy Econ 42:88–100

Diyar S, Akparova A, Toktabayev A, Tyutunnikova M (2014) Green economy—innovation-based development of kazakhstan. Procedia Soc Behav Sci 140:695–699

Duan HB, Zhu L, Fan Y (2014) Optimal carbon taxes in carbon-constrained China: a logistic-induced energy economic hybrid model. Energy 69:345–356

Editorial (2005) Management and policy for sustainable consumption and production. J Clean Prod 13:967–969

Encaoua D, Guellec D, Martínez C (2006) Patent systems for encouraging innovation: lessons from economic analysis. Res Policy 35(9):1423–1440

ECTP (European Construction Technology Platform) (2015) Strategic research agenda for the European construction sector achieving a sustainable and competitive construction sector by 2030. www.ectp.org/documentation/ECTP-SRA-2005_12_23.pdf

European Union (2014) Scoping study to identify potential circular economy actions, priority sectors, material flows and value chains. Publications Office of the European Union, Luxembourg. ISBN 978-92-79-40166-4

Fang Y, Cote R, Qin R (2007) Industrial sustainability in China: practice and prospects for eco-industrial development. J Environ Manage 83:315–328

Foerstl K, Reuter C, Hartmann E, Blome C (2010) Managing supplier sustainability risks in a dynamically changing environment—Sustainable supplier management in the chemical industry, J Purchasing Supply Manage 16:118–130

Fracasso A (2014) A gravity model of virtual water trade. Ecol Econ 108:215–228

Goger A (2013) The making of a 'business case' for environmental upgrading: Sri Lanka's eco-factories. Geoforum 47:73–83

Hamilton K, Cameron G (1994) Simulating the distributional effects of a Canadian carbon tax. Can Public Policy 20(4):385–399

Herman R, Ardekani SA, Ausubel JH (1990) Dematerialization. Technol Forecast Soc Chang 38:333–347

Hesselbarth C, Schaltegger S (2014) Educating change agents for sustainability—learnings from the first sustainability management master of business administration. J Clean Prod 62:24–36

Janicke M (2012) "Green growth": from a growing eco-industry to economic sustainability. Energy Policy 48:13–21

Jiang Z, Shao S (2014) Distributional effects of a carbontax on Chinese households: a case of Shanghai. Energy Policy 73:269–277

Jung S, Dodbiba G, Chae SH, Fujita T (2013) A novel approach for evaluating the performance of eco-industrial park pilot projects. J Clean Prod 39:50–59

Kerkhof CA, Moll CH, Drissen H, Wilting CH (2008) Taxation of multiple greenhouse gases and the effects on income distribution: a case study of the Netherlands. Ecol Econ 67:318–326

Ketterer J, Lippelt J, Ruta G (2011) "Trade in Virtual Carbon", CESifo forum 2/2011. https://www.cesifo-group.de/portal/page/portal/A7C85DE89A1D246FE04400144FAFB1DA

Koplin J, Seuring S, Mesterharm M (2007) Incorporating sustainability into supply management in automotive industry—the case of the Volkswagen AG. J Clean Prod 15:1053–1062

Krause DR, Vachon S, Klassen RD (2009) Special topic forum on sustainable supply chain management: Introduction and reflections on the role of purchasing management. J Supply Chain Manage 45(4):18–24

Li A, Zhang A, Cai H, Li X, Peng S (2013) How large are the impacts of carbon-motivated border tax adjustments on China and how to mitigate them? Energy Policy 63:927–934

Linnenluecke MK, Griffiths A (2013) Firms and sustainability: mapping the intellectual origins and structure of the corporate sustainability field. Glob Environ Change 23:382–391

Ling OG (1992) Harmonizing environment and development: ideological and other considerations. Southeast Asian J Soc Sci 20:107–112

Liu X, Wang C, Niu D, Suk S, Bao C (2015) An analysis of company choice preference to carbon tax policy in China. J Clean Product xxx:1–8

Lorek S, Spangenberg JH (2014) Sustainable consumption within a sustainable economy—beyond green growth and green economies. J Clean Prod 63:33–44

Lowe EA (1997) Creating by-product resource exchanges: strategies for eco-industrial parks. J Clean Prod 5:57–65

Martin R, de Preux LB, Wagner UJ (2014) The impact of a carbon tax on manufacturing: evidence from microdata. J Public Econ 117:1–14

Mathews JA (2012) Green growth strategies—Korean initiatives. Futures 44:761–769
Mathur A, Morris AC (2014) Distributional effects of a carbon tax in broader US fiscal reform. Energy Policy 66:326–334
Meadows DH, Meadows DL, Randers J, Behrens WW III (1972) The limits to growth. Universe Books, New York
Mehmet O (1995) Employment creation and green development strategy. Ecol Econ 15:11–19
Meinshausen M, Meinshausen N, Hare W, Raper CB, Frieler K, Knutti R, Frame DJ, Allen MR (2009) Greenhouse gas emission targets for limiting global warming to 2 C, Nature 458:1158–1162. doi:10.1038/nature08017
Messerli P, Ludi E, Breu T (2012) 'Green economy': development opportunity or buzzword? Evidence for policy series, global edition no. 6. NCCR North–South, Bern
M'Gonigle M, Starke J (2006) Planet U Sustaining the world, reinventing the university. Gabriola Island. New Society Publishers, BC, Canada
Mirata M, Emtairah T (2005) Industrial symbiosis networks and the contribution to environmental innovation: the case of the Landskrona industrial symbiosis programme. J Clean Prod 13:993–1002
Muñoz P, Dimov D (2015) The call of the whole in understanding the development of sustainable ventures. J Bus Venturing 30:632–654
Musson A (2012) The build-up of local sustainable development politics: A case study of company leaders in France. Ecol Econ 82:75–87
OECD (2011a) Fostering innovation for green growth. http://www.oecd.org/sti/inno/48663228.pdf
OECD (2011b) Towards green growth. Organization for Economic Cooperation and Development, Paris
OECD (2010) Delivering a new approach to innovation. http://www.oecd.org/site/innovationstrategy/deliveringanewapproachtoinnovation.htm
OECD (2009) Eco-innovation in industry: enabling green growth. http://www.oecd.org/sti/ind/44424374.pdf
OECD (2007) Higher education and regions: globally competitive, locally engaged. OECD Publishing, Paris
Oh DS, Kim KB, Jeong SY (2005) Eco-industrial park design: a Daedeok technovalley case study. Habitat Int 29:269–284
Pacheco D, Dean T, Payne D (2010) Escaping the green prison: entrepreneurship and the creation of opportunities for sustainable development. J Bus Ventur 25(5):464–480
Panyathanakun V, Tantayanon S, Tingsabhat C, Charmondusit K (2013) Development of eco-industrial estates in Thailand: initiatives in the northern region community-based eco-industrial estate. J Clean Prod 51:71–79
Peters et al (2011) Growth in emission transfers via international trade from 1990 to 2008. In: Proceedings of the National Academy of Sciences, 2011. http://www.pnas.org/content/108/21/8903.full.pdf+html
Reczkova L, Sulaiman J, Bahari Z (2013) Some issues of consumer preferences for eco-labeled fish to promote sustainable marine capture fisheries in peninsular Malaysia. PSU-USM international conference on humanities and social sciences. Procedia Soc Behav Sci 91:497–504
Rogers S, Evans L (2011) World carbon dioxide emissions data by country: China speeds ahead of the rest. The Guardian, 31 Jan 2011. http://www.theguardian.com/news/datablog/2011/jan/31/world-carbon-dioxide-emissions-country-data-co2
Rohn H, Pastewski N, Lettenmeier M, Wiesen K, Bienge K (2014) Resource efficiency potential of selected technologies, products and strategies. Sci Total Environ 473–474:32–35
Romero E, Ruiz MC (2014) Proposal of an agent-based analytical model to convert industrial areas in industrial eco-systems. Sci Total Environ 468–469:394–405
Roy A, Goll I (2014) Predictors of various facets of sustainability of nations: the role of cultural and economic factors. Int Bus Rev 23:849–861

Rueff H, Kohler T, Mahat TJ, Ariza C (2014) Can the green economy enhance sustainable mountain development? The potential role of awareness building. Environmental science and policy

Schmitt K (2011) Going big with big matters. The key points approach to sustainable consumption. GAIA 20/4:232–235. In: GAIA Ecol Perspect Sci Soc 21(2):91–94

Sedlacek S (2013) The role of universities in fostering sustainable development at the regional level. J Clean Prod 48:74–84

Sertyesilisik B, Sertyesilisik E (2015) Sustainability leaders for sustainable cities. In: Opoku A, Ahmed V (eds) Leadership and sustainability in the built environment. Spon Research, Routhledge

Shammin RM, Bullard WC (2009) Impact of cap-and-trade policies for reducing greenhouse gas emissions on US households. Ecol Econ 68:2432–2438

Slay B (2011) Virtual carbon trade, transition and developing economies: Sinners, or sinned against? UNDP (United Nations Development Programme), Regional Bureau for Europe and CIS. http://europeandcis.undp.org/uploads/public1/files/vulnerability/Senior%20Economist%20Web%20site/Virtual%20carbon%20trade%20and%20transition%20economies.pdf

Sutton MA, Skiba UM, van Grinsven HJM, Oenema O, Watson CJ, Williams J, Hellums DT, Maas R, Gyldenkaerne S, Pathak H, Winiwarter W (2014) Green economy thinking and the control of nitrous oxide emissions. Environ Dev 9:76–85

Syed S (2014) Media as a change agent, mass media. http://saeedasyed.blogspot.com.tr/p/mass-media.html. Accessed 28 March 2014

Tran S (2012) Expediting innovation. Harv Environ Law Rev 36:123–155

Trencher G, Bai X, Evans J, McCormick K, Yarime M (2014) University partnerships for co-designing and co-producing urban sustainability. Glob Environ Change 28:153–165

Tukker A, Butter M (2007) Governance of sustainable transitions: about the 4(0) ways to change the world. J Clean Prod 15:94–103

Tukker A (2013) Knowledge collaboration and learning by aligning global sustainability programs: reflections in the context of Rio + 20. J Clean Prod 48:272–279

UBCED (1992) A guide to agenda 21, a global partnership. UBCED, Geneva

UNCED (1992) A guide to agenda 21, a global partnership. UBCED, Geneva

United Nations Conference on Environment and Development: The Earth Summit (UNCED) (1992) June 3–14 1992. Rio de Janeiro, United Nations

United Nations (2012) The future we want. Resolution adopted by the general assembly on 27 July 2012, A/RES/66/288

UNCSD [United Nations Conference on Sustainable Development] (2012) The future we want. UNCSD, New York

UNEP [United Nations Environment Programme] (2011) Towards a green economy. Pathways to sustainable development and poverty eradication. UNEP, Nairobi

Ustaoglu M, Yildiz B (2012) Innovative green technology in Turkey: electric vehicles' future and forecasting market share, international conference on leadership, technology, and innovation management. Procedia Soc Behav Sci 41:139–146

Veiga LB, Magrini A (2009) Eco-industrial park development in Rio de Janeiro, Brazil: a tool for sustainable development. J Clean Prod 17:653–661

von Weizsacker EU, Lovins A, Lovins LH (1997) Factor four. Doubling wealth, halving resource use. UK: Earthscan Publishers

Vringer K, Blok K (2000) Long-term trends in direct and indirect household energy intensities: a factor in dematerialisation? Energy Policy 28:713–727

Wangel J (2011) Exploring social structures and agency in backcasting studies for sustainable development. Technol Forecast Soc Chang 78:872–882

Wier M, Birr-Pedersen K, Jacobsen KH, Klok J (2005) Are CO_2 taxes regressive? Evidence from the Danish experience. Ecol Econ 52:239–251

WiseGEEK (Clear answers for common questions) (2014) What is green GDP? http://www.wisegeek.com/what-is-green-gdp.htm

WRI (World Resources Institute, in collaboration with UNEP and UNDP) (1992) World reources, a guide to the global environment, 1992–1993. New York, NY
Xiaowei J, Qi S, Yanfeng G (2011) New approaches to the green economy of China in the multiple crises. Energy Procedia 5:1365–1370
York J, Venkataraman S (2010) The entrepreneur–environment nexus: uncertainty, innovation, and allocation. J Bus Ventur 25(5):449–463
Yu C, Jong M, Dijkema GPJ (2014) Process analysis of eco-industrial park development—the case of Tianjin, China. J Clean Prod 64:464–477
Yusuf S, Nabeshima K (eds) (2007) How universities promote economic growth. The World Bank, Washington, DC
Zaccai E (2012) Over two decades in pursuit of sustainable development: influence, transformations, limits. Environ Dev 1:79–90
Zhu QH, Lowe EA, Wei Y, Barnes D (2007) Industrial symbiosis in China: a case study of the Guitang Group. J Ind Ecol 11:31–42
Zhu Q, Sarkis J (2004) Relationships between operational practices and performance among early adopters of green supply chain management practices in Chinese manufacturing enterprises. J Oper Manage 22:265–289
Zilahy G, Huisingh D (2009) The roles of academia in regional sustainability initiatives. J Clean Prod 17:1057–1066

Author Biographies

Begum Sertyesilisik is Assoc. Prof. and works in the Department of Architecture at Istanbul Technical University, Turkey. She has a Ph.D in the field of Construction Contracts from the Middle East Technical University, Turkey. She specializes in the areas of sustainability, project management, and contract and dispute management. She is a member of the Chamber of Architects in Istanbul and of the CIB. She has written books, articles, chapters and proceedings in the project management field. She is in the Editorial Board of various journals, and acts as a scientific board member in various international scientific conferences.

Egemen Sertyesilisik received a BA from the Political Science and Public Administration Department at the Bilkent University in 2007. He then received a MA from the program in Politics and the Mass Media at the University of Liverpool. In 2010 he obtained an MBA at the Yildiz Technical University. Sertyesilisik has been awarded his Ph.D from Political Economy in the Middle East program at the Marmara University.

The Relevance of Cultural Diversity in Ethical and Green Finance

Emese Borbély

Abstract

In the current paper the relevance of culture in the dissemination of ethical, especially green finance throughout the world is analysed. Different forms of ethical finance, such as ethical investment, shareholder advocacy and community finance, have become a novel instrument for those who were seeking alternatives for traditional investment tools. It is also clear that a great variety of ethical and environmental finance forms have evolved in different cultures. The question naturally arises whether typical ethical finance development patterns can be identified according to culture. In order to cluster different cultures, the principles of the comprehensive Global Leadership and Organizational Behaviour Effectiveness Research Program (GLOBE) was applied. It was found that the regional enhancement of different forms of sustainable economics, typical for certain regions, can be derived from GLOBE metrics for most cases. The preferential form and local evolution of ethical finance is also culturally determined to a great extent and this indicates that GLOBE clustering can be generalized for a broad set of sociocultural phenomena.

Keywords

Ethical finance · Green finance · Cultural clusters

E. Borbély (✉)
Keleti Faculty of Business and Management, Óbuda University, Budapest, Hungary
e-mail: borbely.emese@kgk.uni-obuda.hu

W. Leal Filho et al. (eds.), *Sustainable Economic Development*, World Sustainability Series, DOI 10.1007/978-3-319-45081-0_4

1 Introduction

Ethical finance refers to economic activities that do not focus primarily on financial gain, but instead consider social, welfare and ecological aspects. Green or environmental finance means the mobilisation of financial instruments to preserve natural resources and the environment.

The idea of sustainable economics in not a new phenomenon, Quakers and Methodists drew up the directive of not investing money in ethically questionable businesses already in the eighteenth century. Thus they distanced themselves from weapon manufacturing, slave-traffic and alcohol and tobacco factories (Kite and Kite 1818). Since this very first appearance of ethical finance, methods of ethical investments have progressed greatly and nowadays highly sophisticated techniques are available for responsible investors.

Among social finance, the most well-known and prevalent form is microfinance. Even though many similar institutions and cooperative banks have been operated throughout the world for decades, microfinance became widely popular after Muhammad Yunus received the Nobel Peace Prize for his comprehensive poverty eradication activities in this field (Yunus 2008).

The relevance and actuality of the research on ethical finance (sometimes also termed as sustainable finance) is greatly stressed by the recent economic crisis. The global crisis in 2008–2009 has pointed out the weaknesses of the current capital and money markets and many investors now question the sustainability of the system, thus ethical finance forms are significantly more appreciated. This is due to the fact that businesses which realised sustainable aspects in their business policy were more resistant to the economic turbulences caused by the crisis. Furthermore, political actors in low-GDP countries that were highly affected recognised the crisis management opportunities using the different tools of ethical finance. Additionally, as environmental pollution and resource scarcity are highly discussed topics, societies are becoming more interested in financial methods that help to solve or at least mitigate these problems.

It is also known that ethical finance tools that proved to be efficient, greatly differ from country to country and it is often not obvious, which additional or new sustainable financial method would be most appropriate for a given culture. Therefore, the question naturally arises whether typical ethical finance development patterns can be identified according to culture and whether this classification can help the development of the corresponding instruments.

In this paper, the relationship between culture and ethical finance is closely investigated. The aim of this paper is to identify cultural factors which influence the adaption and permeation of the diverse tools of ethical finance in different cultural clusters.

For clustering countries, findings of the GLOBE (Global Leadership and Organizational Behaviour Effectiveness Research Program) project were used as this was the most extensive study regarding culture in the beginning of this century. An international team collected by qualitative methods data at almost 1000 organisations in 61 countries. The outcome of the questionnaire, which analysed the culture of the societies in addition to the organisational cultures, was supported by archival data, interviews and media analysis as well. In the second part of the project, the results were empirically tested. In the end, GLOBE researchers identified nine cultural dimension that are able to describe national cultures and allows to define cultural clusters among the countries of the world (House et al. 2004).

The recent study aims to use the above mentioned cultural classifications to find out whether given cultural attitudes are of advantage or disadvantage for the prevalence of the different ethical financial tools in different cultures. Conclusions with concrete findings could help international organisations, financial institutions and governmental actors to use these financial tools in their decision process more efficiently.

After a short introduction of the standard forms of ethical (or sustainable) finance, the short history of ethical and ecological finance is described in order to set the context of the current study. The third part outlines the GLOBE project which is then used to classify the cultural attitudes that are of main importance for the prevalence of a given ethical financial tool in a given culture. The next section describes the GLOBE cultural clusters according to their ethical and ecological finance status which is followed by a short summary of the new findings.

2 Forms of Ethical Finance and Relationship to Green Finance

Typically three main forms of ethical finance are identified: ethical investment, engagement (or shareholder advocacy) and community finance. Here, as a general introduction, these categories are shortly reviewed.

Ethical investment focuses on financing of long-term innovative projects wherein economic return and moral profit go hand in hand. The very first sustainable investments were created for the different churches and operated mostly with knock-out criteria, namely, they deselected from their portfolio companies that did not conform with their values. Later on, KO-screening, or negative screening, was complemented by the positive screening method, where capital flows to companies that create a long-term benefit either for society or for the environment (Hellsten and Mallin 2006). Many investors realized that they might be excluding some companies from their portfolio based on their non-sustainable energy practices, even though these businesses are of high importance for the society. The 'best-in-class' approach tries to solve this problem by comparing corporations of a given sector to find those that perform best when considering ecology, social values and corporate governance. Nowadays, the above mentioned methods are usually used in combination to reach the best possible solution for an ethical investment portfolio.

Active shareholders make use of their voting rights connected to their shares to apply pressure on the management and to try to improve the ecological, social and corporate business policy of a corporation. They draft proposals and feature alternative solutions that are mainly presented at the annual shareholder meetings. The institutionalised form is called engagement, where a fund manager is trusted to collect the votes associated with the shares of many shareholders to be able to apply added pressure on the board.

Community finance incorporates many different subsets such as microfinance, cooperative banking or community investments. These financial forms aim to serve social classes in society that do not have access to conventional financial services or that are underrepresented. Informal financial institutions have a wide range of forms. Typically in low-GDP countries, societies spontaneously work out various monetary institutions or cooperations regarding their needs and demands.

The spread and penetration of the above described tools in different parts of the world depend not only on the economic frameworks but rely strongly on cultural attitudes. The main goals and targets and the evolution of ecological aspects in ethical finance differ from country to country. Typically, some cultures are very sensitive to ecological issues while others put an accent on social problems. To understand the background and frameworks for the dissemination of ethical finance, a short presentment of the history of ethical finance is presented.

3 Green Finance as a Form of Ethical Finance in History

The very first ethical financial practices were rooted in religiously motivated and/or human rights movements in the eighteenth and nineteenth centuries and were only later affiliated with other ethical, ecological or green aspects. Even though the idea of ecological responsibility has been present for several decades, the first half of the twentieth century did not favour green financial tools because countries were preoccupied with the restructuring of their societies and economies after the world wars and the 1929 financial crisis.

The only important milestone of ethical finance in the period of the world wars was the establishment of the world's first ethical fund on the demand of the Protestant Evangelists who wanted to avoid investing in "sin stocks"; alcohol, tobacco and weapon manufacturing companies. However, this fund was not available for the general public (Travinski 2008). Then, for a few decades, these activities remained hidden until the 1960s brought some new movements in human rights, corporate social responsibility and consumer awareness. The long-term effects of this line of thought in the United States can even be tracked in Ralph Nader's public criticism on the social and ecological irresponsible behaviour of multinational companies and his actions to make consumers and shareholders aware of their responsibility regarding their consumer behaviour and shareholder voting rights (Malkiel and Quandt 1971). His presidential campaigns also stressed these aspects in the first years of the new millennium.

Concerning organisationally supported activism, new sustainable funds were already created in the 1960s and 1970s, such as the Aktie-Ansvar in Sweden, the first open-public fund with mostly conventional screening strategies in 1965, and the US Pax World Fund in 1971 (Kreander et al. 2004). Also in the 1970s Charles Jacob, the Methodist Church's investment manager, worked out a new methodology on how to study a corporation as an aggregate unit including its manufacturing processes, employer-employee relationships and the overall contact to its ecological and social environment. This proposal included the first steps to the modern form of ethical finance; underlined the social and ecological responsibility of the companies and furthermore of its employees and customers and introduced the positive screening strategy (Stewardship Fund Prospectus 1973 in Kreander 2001 and Leahy 2008).

After the different political issues that concerned investors in the 1960s and 1970s, such as the Vietnam war and the South African Apartheid, the next decades brought a shift of the attention toward environmentally focused financial activities. Countries of the developed world set the concept of "ecological crisis" and as a first step to fight it, reacted mostly with "pragmatic legal-administrative response" (Berger et al. 2001). Different institutions were created to work up appropriate environmental policies and "end-of-pipe" solutions were introduced which prescribe the use of scrubbers and filters (Hajer 1996).

As a result of increased ecological attention, the World Commission on Environment and Development was created in 1983 by the United Nations for the popularisation of sustainable development and they saw "the possibility for a new era of economic growth, one that must be based on policies that sustain and expand the environmental resource base." Sustainable development is defined as meeting "the [human] needs of the present without compromising the ability of future generations to meet their own needs…" in addition it will be realized that "growth will come through better managing technology and social organization" as well Ecology and economy are becoming ever more interwoven locally, regionally, nationally, and globally into a seamless net of causes and effects." (WCED working paper 1981[1]) During the 1990s the European Union set an accent on the harmonisation of the member states' environmental policies on a global, national and regional level. Resource efficiency, internalisation of the ecological costs and industrial self-regulation were defined as the most important topics (Berger et al. 2001). In parallel, the Bush and Clinton administrations of the United States became again more aware of ecological problems, especially the ozone depletion, acid rain and hazardous waste. They imposed rules regarding emission levels and different taxes to put pressure on the green policies of corporations.

Ecological issues enforced also international collaboration, since the 1990s, many international summits were held in the topic of climate change, global warming, sustainable development and other environmental problems, where protocols and principles were drawn up as proposals. At the very last United Nations Climate Conference that was held in December 2015, the Paris Agreement was issued, which, according to the UNFCCC Executive Secretary Christiana Figueres,

[1]World Commission on Environment and Development (1987), p. 8.

"confirms the irreversible transition to a low carbon, safer and healthier world" (UNFCCC homepage).[2]

It is also worth mentioning that more recently, the World Business Council for Sustainable Development evolved new generations of green ethical funds. The Council is now an association of some 200 international companies and is a "CEO-led organization of forward-thinking companies that galvanizes the global business community to create a sustainable future for business, society and the environment. Through its members, the Council applies its respected thought leadership and effective advocacy to generate constructive solutions and take shared action to drive business action on sustainability…" (homepage WBDSD). Today, as a result of all of these positive changes in the way of thinking of global actors and societies "have locked the global economy and global ecology together in new ways" (Redclift 2005). Political and corporate decision makers, investors and consumers are all involved in a process to minimise ecological stress, to cooperate on the international level in order to curtail environmental costs on economic growth. In the last years, ethical forms of finance continually grow at a faster rate than the conventional asset management market in Europe (Eurosif 2014). These developments can be viewed as a process of globalisation of ethical and green finance. However, it has to be noted that this process is culturally differentiated, and exactly this feature will be analysed in the main part of this paper.

The increased public attention in ecological issues in recent decades set the stage for the development of different forms of ethical finance in different cultural circles. This evolution, however, can not be understood without the cultural specificity of the given social environment, therefore, cultural clustering has to be applied in order to properly highlight the mechanisms of the development of ethical and green finance in the world.

4 Cultural Clusters, Ethical Finance and Green Economy

Here, it will be shown how culture influences the prevalence of ethical finance, especially why given forms of ethical finance (e.g. ethical investment, microfinance or engagement) are well accepted and used in some parts of the world while others are neglected. Additionally, the objectives of ethical or ecological financial movements differ from one country to the other which is also culturally determined. This statement will be supported by the correlation of cultural clustering and typically observable forms of ethical finance. More specifically, for cultural clustering, the principles of the comprehensive Global Leadership and Organizational Behaviour Effectiveness Research Program (GLOBE) can be applied.[3]

[2]Homepage UNFCCC.http://www.cop21.gouv.fr/en/the-world-talks-about-parisagreement/ accessed on 2 Jan 2016.

[3]All Globe data in this paper are taken from the book Culture, Leadership, and Organizations: The GLOBE Study of 62 Societies.

The GLOBE project was a global research initiative of more than 170 researchers lasting for over a decade in the 1990s and beyond. The team collected vast amounts of data from 17,300 middle managers at 951 different organizations. Qualitative methods were used to assist the development of quantitative instruments. As a final outcome, nine cultural competencies had been clearly identified and more than 60 countries were grouped into ten clusters, representing common leadership attitudes. Here, a novel approach is presented where GLOBE findings are applied to characteristic features of ethical finance, therefore, as an initial step it is important to introduce the original principles and conclusions of the GLOBE study to set the framework for the correlation with ethical finance.

4.1 Cultural Clustering Principles of the GLOBE Project

The cultural clusters identified by the GLOBE project can be examined one by one by analysing their so-called dimensions that define the characteristics of cultural clusters. The dimensions used by the GLOBE project are the following:

Performance orientation is the degree to which a collective encourages and rewards group members for performance improvement.

Assertiveness defines to what extent individuals are assertive, confrontational and aggressive in their relationships.

Future orientation describes the importance of long-term thinking and systematic planning.

Humane orientation express the degree to which a collective encourages and rewards individuals for being fair, altruistic, generous, caring, and kind to others.

Institutional collectivism stands for collective distribution of resources and collective actions, while *in-group collectivism* describes the degree to which individuals express pride, loyalty, and cohesiveness in their organizations or families.

Gender egalitarianism characterises the rate of gender equality in a society.

Power distance is the degree to which members of a collective expect power to be distributed equally.

Uncertainty avoidance describes how a society or group relies on social norms, rules and procedures to minimise unpredictability of future events.

According to these cultural dimensions, ten clusters were identified: Latin Europe, Germanic Europe, Anglo cluster, Nordic Europe, Eastern Europe, Latin America, Confucian Asia, Southern Asia, Middle East and Sub-Saharan Africa.

Analysing these GLOBE dimensions and using its clustering strategies, it can be observed, which clusters are more sensitive to the traditional forms of ethical finance and are concerned more about social problems and which are those that are interested more about green issues. Some dimensions show the sensitivity towards ecological problems in the society to a great extent. One of these is assertiveness, where non-assertive societies tend to prize nature, the integrity of the environment and society and are concerned about ecological issues.

Selected GLOBE results on the self-evaluation of cultural dimensions in different clusters is shown in Table 1 (House et al. 2004) based on a scoring system with a potential range of values between 1 and 10. The world average is also shown for reference so that significant deviations from the world average are clearly visible. Section 4.2 contains a number of relevant findings with respect to ethical finance forms and practices that can be derived from these cultural features.

4.2 Forms of Ethical, Especially Green Finance Specific to Each Cultural Cluster

The **Nordic Europe**[4] cluster is the one with the lowest assertiveness ranking and this cluster indeed includes countries with outstanding ethical financial practices regarding ecological investment funds and other financial activities. In Sweden, for example, people are more attracted by ecological issues than social ones and a leading criterion in creating an investment portfolio is environmental awareness (Eurosif 2010, 2014). In this trend, however, the high economic standard of the Nordic economies plays also an important role. Interestingly, the engagement approach is also prevailing in the region; private and institutional investors are aware of their influence on the business and social/environmental policy of the corporations they hold shares of. The prevalence of ethical finance is also supported by high uncertainty avoidance scores which represent well-regulated and formalized countries that are often open for novel challenges. New ideas are welcome and people are ready to take a bit higher financial risk in order to support social and ecological solutions that might be of great positive impact in the future (Bengtsson 2008).

High future orientation and, interestingly, gender egalitarianism are two more dimensions that favour the popularity of green finance. People in these countries live in stable economies and therefore can afford to enter in eventually riskier, newly formed financial activities that hold promise of being of use for the wider society and global issues. They have expectations concerning the future and are convinced about the possibility to be able to influence future happenings with present actions and strategic planning. The high gender egalitarianism scores further shore up the attractiveness of ethical finance with its cooperative and intuitive attitudes.

The **Germanic cluster**[5] is also one that is well known to put a strong accent on environmental matters. This cluster has many similarities with the above mentioned Nordic cluster and shares similar societal values and interest in green finance. The high uncertainty avoidance score again makes the Germanic cluster societies driven by innovation and ready for higher risk-taking. When this comes together with the overall highest future orientation and very strong performance orientation scores and is additionally supported by strong economic background, ethical finance had and still has very good chances to become more widespread. In this cluster, a wide

---

[4]Denmark, Finland, Sweden.

[5]Austria, Germany, the Netherlands and the German speaking part of Switzerland.

Table 1 GLOBE research scores on a scale between 1 and 10 corresponding to the 9 cultural dimensions identified by the research for different country clusters

AS IS scores	Anglo	Nordic	Germanic	Latin Europe	Eastern Europe	Arabic	Southern Asia	Confucian Asia	Latin America	Sub-Saharan Africa	Average
Uncertainty avoidance	4.42	5.19	5.12	4.18	3.56	3.91	4.10	4.42	3.62	4.27	4.28
Future orientation	4.08	4.36	4.40	3.68	3.38	3.58	3.98	4.18	3.54	3.92	3.91
Power distance	4.97	4.54	4.95	5.21	5.25	5.23	5.39	5.15	5.33	5.22	5.12
Institutional collectivism	4.46	4.88	4.03	4.01	4.10	4.28	4.35	4.80	3.86	4.28	4.31
Humane orientation	4.20	4.17	3.55	3.71	3.85	4.36	4.71	3.99	4.03	4.42	4.10
Performance orientation	4.37	3.92	4.41	3.94	3.73	3.90	4.33	4.58	3.85	4.13	4.12
Group and family collectivism	4.30	3.75	4.21	4.80	5.53	5.58	5.87	5.42	5.52	5.31	5.03
Gender egalitariansm	3.40	3.71	3.14	3.36	3.84	2.95	3.28	3.18	3.41	3.29	3.36
Assertiveness	4.14	3.66	4.55	3.99	4.33	4.14	3.86	4.09	4.15	4.24	4.12

The last column contains the world average. Relevant conclusions that can be derived from these cultural leadership scores with respect to ethical finance are presented in the next Chapter range of different mutual funds are available (Tober and Vögele 2011), and green topics are very popular. There are different green or alternative banks in these countries that consider social and ecological aspects not only in their investment portfolio but also in their loan decision. These financial institutions often finance ecologically very useful projects that have a return on investment only on the long term and would not have access to capital on the conventional financial market (Eurosif 2010, 2014). Additionally, favourable legal frameworks support the prevalence of ethical investments, e.g. disclosure guidelines for pension funds, certificates and quality labels for sustainable funds and other products.

After the financial crisis, people of **Anglo**[6] cluster societies became disenchanted in conventional capital market that turned out to be unsustainable. They turned more to alternative financial and investment strategies, the demand for these new financial products rose within a short period of time (EIRIS 2010). Environmental issues came to the forefront and the establishment of the Green Investment Bank in the United Kingdom aimed to support the transition of the country into a low carbon economy. Additionally, many research houses and advisory centres were set up to help sustainable investors find the suitable financial method. One might think that the high performance orientation and individualism scores in the Anglo cluster may not favourable for ethical finance (House et al. 2004). High performance orientation countries, however, are the economically developed ones that put an accent on transparency in business that is mostly more ensured with sustainable financial practices. The high individualism is encouraging the prevalence of shareholder activism where investors try to influence business policies according to their moral values. Competitiveness is another attitude that goes hand in hand with the above-mentioned dimensions and leads to the development of new and attractive investment forms for interested investors.

Additionally, these scores combined with high future orientation made the cluster to become one of the pioneers in ethical finance that rarely represent short-time projects. Strategic planning belongs to future orientation that makes future happenings more predictable. People in these societies are aware of their responsibility regarding the shaping of the future for upcoming generations. In most cases this leads to the pronounced concerns on the subject of ecological issues (Chhokar et al. 2007). These facts all underline that beyond economic reasons, also cultural attitudes have been favourable for the prevalence of ethical finance in the cluster.

The **Latin Europe**[7] cluster is rather different from its other European counterparts regarding history, prevalence and forms of ethical finance. This comes as no surprise if one inspects the countries in the cluster. They show a very heterogeneous historical background, are quite different from the religious point-of-view and regarding their economies, some are highly developed where others have been badly affected by the economic crisis (Setem 2006). The cluster's countries have high in-group and family collectivism scores compared to other European countries

---

[6]Australia, Canada, New Zealand, the white sample of South Africa, United Kingdom, United States.

[7]France, Israel, Italy, Portugal, Spain, Switzerland.

in spite of the fact that worldwide ranking is around the average. The relatively early establishment of cooperative banks in the cluster reflects the cohesion in the society. Italy especially had very sophisticated agricultural, green and social issues oriented small financial institutions from the eighteenth century on that rely on solidarity, collaboration and democracy. This is further supported by low assertiveness figures (see Table 1), which again stands for cooperative societies; people try to live in harmony with their neighbours, the environment and values such as cohesion and equality are important. Other dimensions are mostly around the average in the cluster, without noticeable scores.

Regulatory frameworks of ethical finance are mostly well developed in the cluster and the social investment market is growing. Institutional investors are the key players in the field, however more and more private financiers are interested in socially and ecologically related financial products, especially in countries that have been facing more difficulties since the economic crisis (Eurosif 2014). People are looking for alternatives to conventional financial products and they have learned to concentrate more strongly on long-term planning and transparency, which highly favour the spread of ethical and also green investment tools. Switzerland, the cluster's (and one of the world's) leading financial centre plays also an important role on the sustainable market. Most financial institutions offer a large variety of sustainable financial products and additionally, Switzerland is one of the world's microfinance centres.

The **Eastern European**[8] cluster is lagging behind its Western European counterparts regarding ethical finance which can be explained by the economic and political transition they have been facing for decades (Eurosif 2010, 2014). These countries show low future and performance orientation scores—due to the planned economy in socialism—that do not favour ethical finance. The high power distance and collectivism scores can explain the lack of shareholder activities. At the same time, however, the high in-group collectivism and gender egalitarianism scores reflect the importance of empathy, commonality and other soft values in the cluster that might support other forms of ethical finance. In fact, there are some microcredit institutions all over the cluster and also local, pioneering financial institutions dealing with social and ecological problems (Kraemer-Eis and Conforti 2009). Moreover, international banks offer different investment alternatives that mostly focus on green issues.

The development of ethical finance in the **Southern Asia**[9] cluster is probably more influenced by economic than by cultural factors. There are many people living in poverty in the cluster that might be the reason for the success of community financial forms. Moreover, the cluster stands for the overall highest humane orientation scores and high collectivism scores supplemented with low assertiveness, definitely helping the dissemination of financial practices that are based on the moral values such as cooperation, mutual assistance and loyalty. As mentioned earlier, the first successful ethical financial forms were microcredit institutions that have plenty of different forms all around Asia developed according to the needs and

[8]Albania, Georgia, Greece, Hungary, Kazakhstan, Poland, Russia, Slovenia.

[9]India, Indonesia, Iran, Malaysia, the Philippines, Thailand.

demands of the poor (Yunus 2008). As the performance orientation dimension is high in the cluster, these institutions are mostly very cost-effective and represent a really efficient tool in the reduction of poverty. The high power distance scores, that mostly correlate with rigid hierarchical systems and the again overall highest in-group and family collectivism scores probably repress individual actions that shareholder activity requires, so this form is almost non-existent in the cluster.

Regarding green issues, climate change became an important issue for responsible investors "and targeted investments to tackle environmental challenges are growing in importance. 62 % of respondents state that climate risk will be more important in the next 2 years" (ASrIA 2003).[10] The three most important targets are climate change, energy and water security. Recently, the Association for Sustainable and Responsible Investment in Asia, leading political decision makers of the region, international experts of climate change and financial organisations arranged a meeting to find options for alternative financial strategies and for "climate finance solutions."[11] On the whole, there is a conscious process of these countries to turn their economies into a low carbon one and they try to use various means to reach this goal, including the tools of green finance.

The other Asian cluster is comprised of very strong economies: the **Confucian Asian**[12] cluster's countries are becoming the leaders of ethical finance in Asia, even though the ethical money market is still in its early stages. However, demographic expansion caused severe ecological and social problems that must be treated both on the political and societal level. Most activities are supply-driven, on the one hand because of the international pressure, on the other hand because governments realised that these financial tools can assist them in mitigating the above-mentioned problems. According to the latest research of ASrIA,[13] their "findings also confirmed that investors in Asia are increasingly concerned about climate risk. Importantly, climate change is considered an issue for both dedicated sustainable investment managers but also for more mainstream asset managers."[14] Cultural dimensions are strongly influenced by religion. These countries are highly performance-oriented and masculine. This means that depositors are usually focused on investments that bring higher profit in the short run, soft values and long term social or ecological profit represent negligible factors. As mentioned above, this has been changing in recent years as people and investment managers are directly facing the social and ecological problems and realise the fact that social investments can make the same profit as conventional ones (SIFJ 2009). This, also

[10]http://asria.org/asias-sustainable-investment-market-is-robust-and-growing/ accessed on 23. April 2015.

[11]http://asria.org/transforming-economies-how-to-fund-green-growth-in-southeast-asia/ accessed on 23. April 2015.

[12]China, Hong Kong, Japan, South Korea, Singapore, Taiwan.

[13]Association for Sustainable and Responsible Investment in Asia.

[14]http://asria.org/asias-sustainable-investment-market-is-robust-and-growing/ accessed on 23. April 2015.

complemented by high collectivism and future orientation scores indicate the strong prevalence of these financial methods in the future.

The **Middle East**[15] cluster's ethical financial market is very different from the above mentioned ones due to the fact that financial decisions are strongly influenced by Islam. However, Islamic finance operates on a similar principle to sustainable investments; helping the poor and solidarity play an important role. The Dow Jones Islamic Market Index launched in Bahrain in 1999 for example screens for Islamic compliance (Hassan and Mahlknecht 2011) and the Egyptian Corporate Responsible Index is the first index in the region that concentrates on social and ecological aspects (EIoD[16] homepage). Climate change and food security are those global problems that concerns these societies the most (MENA 2009). Important steps were made by the leading private equity fund, the Abraaj Capital that created the Ethical Framework for Investors and introduced a number of sustainable investment practices. The high net worth individuals (who usually invest in equity funds) are increasingly interested in renewable energy and clean technology projects. GLOBE dimensions describe the cluster as highly group and family collectivist which is probably the reason for the different forms of community finance developed in the region, mostly according to Islamic principles. These institutions often support families (rather than women as in the conventional microfinance world) and allow only transfer of goods (instead of money) to exclude interest rates. Additionally these institutions focus more on charity enabled by donations prescribed by religion (Costa et al. 2010).

In the **Sub-Saharan Africa**[17] cluster, ethical finance initiatives come from above (from international development organisations and the governments) and as many people live in poverty, diverse forms of community finance are very popular (Cleene and Wood 2004). High in group collectivism, humane orientation and assertiveness scores are conducive to the spread of ethical financial practices. People believe in the positive correlation of input and result and humane aspects play an important role in the decision making process. These societies are not formalised, it is easy to enter in small financial interactions, take a small loan and start a new business. The regulation of this by the governments is however of great importance, since many usury-based programs run under the flag of microfinance.

Regarding ethical investment, South Africa is the centre for sustainable investment in the region and the very first responsible index in emerging markets was also launched here. The most important exclusion criteria of the fund interestingly focus on environmental issues which shows the awareness of citizens (having a higher standard of living) on the long-term negative effect of climate change and water/air pollution (International Finance Corporation 2011). The index tries on the one hand to encourage corporations to integrate environmental and social aspects in their business policy and additionally convince private but mostly institutional investors to consider sustainable principles (International Finance Corporation 2011).

[15]Egypt, Kuwait, Morocco, Qatar, Turkey.

[16]Egyptian Insitute of Directors, accessed on 2 Jan 2016.

[17]Namibia, Nigeria, South Africa, Zambia, Zimbabwe.

The **Latin American**[18] cluster drew up early interest in responsible finance. Compared to other emerging markets, ethical investments and community finance are both developed in the region (Kumar and Siddy 2009). Informal relations, familiar cohesion and tradition all play an important role in these clusters and this is demonstrated by the low uncertainty avoidance and performance orientation scores along with high family collectivism and gender egalitarianism. Overall cultural frameworks (except for the low future orientation) with its humane ideals create a really favourable environment for ethical finance. There are several microfinance institutions and cooperative banks in the region (Meagher et al. 2006) that are not only important self-organisation groups but also strongly supported by the government and to some extent they are also regulated. Regarding ethical investments, pension funds play the most decisive role as investors. Issues targeted represent mainly human rights, poverty and green issues. Own development banks (like in Brazil), social investment mutual funds and corporate sustainability indices all support the further evolution and expansion of responsible financial markets in Latin America.

5 Conclusions

Here, it was argued that even though green finance has spread globally, the observable, significant country-by-country variations of ethical instruments are culturally determined. Beyond pure economic background, cultural frameworks are the most important factors that influence the acceptance, prevalence and evolution of ethical finance. The cultural dimensions introduced and quantified by the GLOBE study have been found to correlate strongly with the ethical-ecological finance characteristics of cultural clusters. It was shown that most of the GLOBE dimensions introduced are helpful tools for understanding the evolution and current state of green finance in different countries that can be clearly classified into a GLOBE cluster. The most important cultural dimensions were identified as humane and future orientation, gender egalitarianism, performance orientation and assertiveness. These dimensions fundamentally determine attitudes toward ethical finance target groups (e.g. the poor) and target objectives (environment protection, air pollution, lack of water etc.). The resulting financial activities and prevailing ethical finance instruments were then found to be specific to cultural clusters.

Obviously, local grassroots initiatives regarding ethical finance are automatically compatible with the cultural framework and their natural evolution is mostly secured by the prompt response of a society for internally identified needs and demands. However, when development and green projects are initiated by external actors like international development organisations or banks, the proper adaptation to the given culture is of great importance and specific cultural aspects have to be

[18]Argentina, Bolivia, Brazil, Columbia, Costa Rica, Ecuador, El Salvador, Guatemala, Mexico, Venezuela.

considered for efficient use of resources, and eventually for the success of the project itself. Conclusions drawn from the cultural clustering of the GLOBE Research Program and adapting its results to ethical finance structures can greatly assist in defining the proper ethical finance instruments for a given cultural circle. The methodology that is based on and developed according to this finding will certainly lead to more efficient utilization of ethical finance practices all over the world.

References

Association for Sustainable and Responsible Investment in Asia (ASrIA) (2003) SRI in Asian emerging markets. Centre for Social Markets online, India

Bengtsson E (2008) A history of scandinavian socially responsible investing. J Bus Ethics 82 (4):973

Berger G, Flynn A, Hines F, Johns R (2001) Ecological modernization as a basis for environmental policy: current environmental discourse and policy and the implications on environmental supply chain management, innovation. Eur J Soc Sci Res 14(1):56

Chhokar JS, Brodbeck FC, House RJ (2007) Culture and leadership across the world, the GLOBE book of in-depth studies of 25 societies, Lawrence Erlbaum Associates, 297 passim

Cleene S, Wood C (2004) Sustainability banking in Africa. African Institute of Corporate Citizenship, Johannesburg, pp 18–20

Costa V, Makhlouf H, Mazaud P (2010) Women's empowerment through Islamic microfinance in Egypt. MESCI 2009–2010, Rome, p 9

Eurosif (2010) European SRI study, 2010 online, 8 passim

Eurosif (2014) European SRI study, 2014 online, 36 passim

Experts in Responsible Investment Solutions (2010) Climate change tracker: Europe. EIRIS, London

Hajer MA (1996) Ecological modernisation as cultural politics. Sage, London, p 248

Hassan K, Mahlknecht M (2011) Islamic capital markets: products and strategies. Wiley, Chichester, pp 129–130

Hellsten S, Mallin C (2006) Are ethical or socially responsible investments socially responsible? J Bus Ethics 66(4):397

House RJ, Hanges PJ, Javidan M, Dorfman P, Gupta V (2004) Culture, leadership, and organizations: the GLOBE study of 62 societies. Sage Publications, Thousand Oaks. All Globe data are taken from this book

International Finance Corporation (2009) Sustainable investment in the middle east and North Africa—region report. IFC, Washington, DC, pp 22–45

International Finance Corporation (2011) Sustainable investment in Sub-Saharan Africa, investment practitioners views of sustainable investment of private equity and asset management in South Africa, Nigeria and Kenya. IFC, Nariobi, pp 60–78

Kite B, Kite T (1818) The works of John Woolman in two parts, 5th edn. Philadelphia Publisher, Philadelphia, p 75

Kraemer-Eis H, Conforti A (2009) Microfinance in Europe—a market overview. European Investment Fund, Luxembourg, p 13

Kreander N (2001) An analysis of European ethical funds. Certified Accountants Educational Trust, London, p 16

Kreander N, MecPhail K, Molyneaux D (2004) God's fund managers—a critical study of stock market investment practices of the Church of England and UK Methodists. Acc Audit Acc J 17 (3):409–410

Kumar R, Siddy D (2009) IFC sustainable investment country reports: sustainable investment in Brazil 2009. International Finance Corporation, Washington, DC pp 25–34

Leahy J (2008) Socially responsible investing. Acc Irel 40(6):47
Malkiel BG, Quandt ER (1971) Moral issues in investment policy. Harv Bus Rev 48(2):39
Meagher P, Campos P, Christen RP, Druschel K, Gallardo J, Martowijoyo S (2006) Microfinance regulation in seven countries: a comparative study, IRIS Center, New Delhi, pp 16–19, 41–46
Redclift M (2005) Sustainability—critical concepts in the social sciences. Routledge Abingdon, p 5
Setem Federation (2006) http://inanzaseticas.org/pdf/EthicalFinanceinItaly.pdf. Accessed on 2 Jan 2016, p 1–6
Social Investment Forum Japan (2009) Review of socially responsible investment in Japan, SIF-J, Tokyo, pp 2–6, 22–29
Tober DS, Vögele G (2011) Marktbericht Nachhaltige Geldanlagen 2011—Deutschland, Österreich und die Schweiz, Forum Nachhaltige Geldanlagen e. V., Berlin, pp 6–7
Travinski M (2008) Private sector employee pension plans: advocating in socially responsible investing in 401(k) plans. J Pension Benefits Issues Admin 15(3):63
World Business Council for Sustainable Development accessed on 20. April 2015. www.wbcsd.org
World Commission on Environment and Development (1987) From one Earth to one World: An overview. Oxford University Press, Oxford, p 8
Yunus M (2008) Banker to the poor: Micro-lending and the Battle Against World Poverty. PublicAffairs, New York

Author Biography

Emese Borbély holds a master's degree in business administration and a PhD in social and economic science from the Vienna University of Economics and Business, Austria. She has been teaching economics at the Óbuda University of Budapest, Hungary for several years. Her main research interests are ethical finance and sustainable financial methods in different cultures. She is founder of the cross-cultural initiative Taita Foundation (Hungary) which has been developing and funding sustainable projects for orphanages in Kenya since 2006 mainly based on self-developed Hungarian community funding schemes. She is currently on maternity leave.

Green Agriculture in Hungary: The Factors of Competitiveness in Organic Farming

Csilla Mile

Abstract

In Hungary there are only 1455, mostly small sized, organic farms, which cover 2.5 % of the total agricultural land. More than 90 % of the farmers are involved in the agri-environmental program, which provides financial support for organic farming. Only a small proportion of the farmers are competitive without a compensatory payment, and mostly those who sell more value-added products directly to consumers. The Hungarian strategy involves an extension of agricultural areas under organic farming, but under the present circumstances this is only conceivable with the restructuring of the support system. Organic farming, however, requires vocation. According to the unanimous opinion of concerned producers the organic market is based on confidence, where efficiency cannot be achieved without the development and maintaining appropriate market relations. Producers ignoring these characteristics of the market will be spilled out of the competition after a while, but more importantly will undermine the confidence of the entire organic market. The Hungarian regulatory and support system should consider these intangible factors in the same way as the proper implementation of the production.

Keywords

Organic farming · Competitiveness · Agri-environmental policy · Financial subsidies · Market of confidence

C. Mile (✉)
Department of Management Studies, Kodolányi János University of Applied Sciences, Szekesfehervar, Hungary
e-mail: mile@kodolanyi.hu

W. Leal Filho et al. (eds.), *Sustainable Economic Development*,
World Sustainability Series, DOI 10.1007/978-3-319-45081-0_5

1 Introduction

Organic farming is the fulfilment of sustainable agriculture (Baranski et al. 2014). It is not a new concept either in professional or consumer circles. The term "organic" was brought to life as a consequence of the intensive agricultural production practices based on external industrial inputs at the beginning of the sixties (Buday-Sántha 2009). At that time environmental aspects did not appear in either the Hungarian or the European agricultural policy. The time of nutrition and crop production based on cheap petroleum have resulted a wasteful and highly polluting agricultural practice.

Parallel to the first occurrence of the problems consumer demands started to undergo a slow and gradual transformation. By today this process has significantly been accelerated. The demand of additive-free, safe and environmentally friendly products is growing all over the word (Baji 2015). The phenomenon is naturally stronger in developed countries, which is due to the fact that environmental and health issues can receive greater attention after basic needs are covered (FAO 2015).

Out of the world's nearly two million organic producers 334,870 are European farmers and only 1500 are Hungarian (Fibl and Iofam 2015). In Hungary only 2.5 % of the agricultural land is covered by organic farmland. 80–85 % of the organic products are exported, while around 90 % of the Hungarian organic consumption goods are imported. The question is how the Hungarian demand and supply of organic products could be aligned, and what the key factors of competitiveness of the Hungarian organic producers are. It is also useful to know how the current regulatory system can strengthen these qualities, and what pitfalls should be paid attention to. The study seeks to answer these research questions.

2 Research Methodology

The research is based mainly on primary data. The secondary data are collected by the two Hungarian institutions responsible for organic inspection and certification, namely Biokontroll Hungária Nonprofit Ltd, and Hungária Ökogarancia Ltd. Furthermore, the databases of Research Institute of Organic Agriculture (Fibl) were used. Within the framework of primary qualitative research somein-depth interviews as well as a case study were made. Among the interviewees there were organic producers and a leading co-worker of Biokontroll Hungária Nonprofit Ltd. The content of the questionnaire was compiled on the basis of information from the interviews, focusing mainly on crop production. In order to provide simplicity for responders the survey has been edited online and it includes both unstructured and structured questions. The latter includes selectively closed, alternative and scaled questions as well. The scale questions contain comparative constant sum scales, non-comparative Likert scales and semantic differential scales.

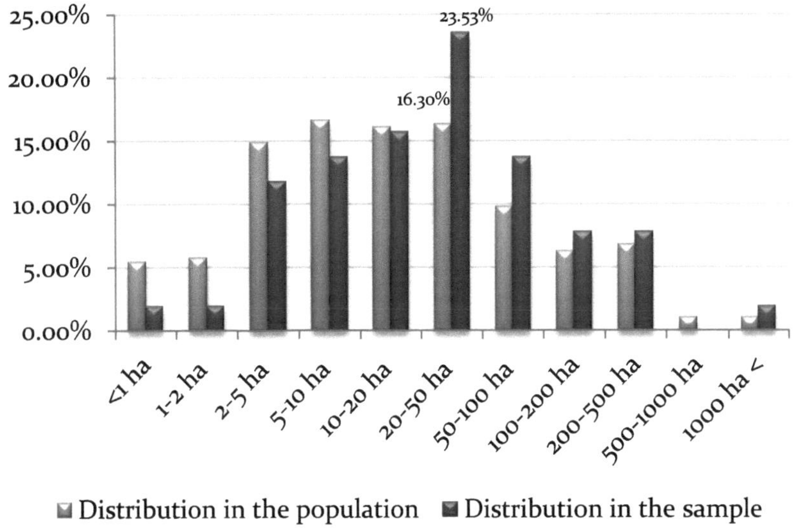

Fig. 1 Distribution of organic farms by size in the population and in the sample. *Source* Based on the database of Biokontroll Hungaria Ltd. and primary survey

The questionnaire has been sent to 830 farmers active engaged in organic farming, with 70 persons responding, which represent a 8.4 % response rate. The spatial distribution of the sample is close to the distribution of the whole population: farmers filled the form from almost every county (except for two of the nineteen). As to the distribution ratio of cultivated plants, the sample is different from the population: while the share of meadows and pastures for example is more than 50 % of the whole Hungarian area under organic farming, it is under-represented in the sample, containing only 8 %. Those cultivations, however, where market relations and competitiveness affect efficiency more than the average, appear with a higher proportion in the sample: fruity and berry plantations represent 13 % (3 % in the population), field vegetables represent 8 % in the sample (1 % of the total population).

According to the distribution by farm size the sample represents the population almost properly (Fig. 1).

We can find a significant difference only in the category of size between 20 and 50 ha where the number of farms is over-represented in the sample by 7.23 %.

Due to the nature of the survey basic statistics and cross tabulation analysis have been carried out.

3 Results and Discussion

3.1 Organic Farming in the World

At the end of 2013 a total 43.1 million hectares of agricultural land were used for organic farming, which was a nearly fourfold increase compared to the 1999 base

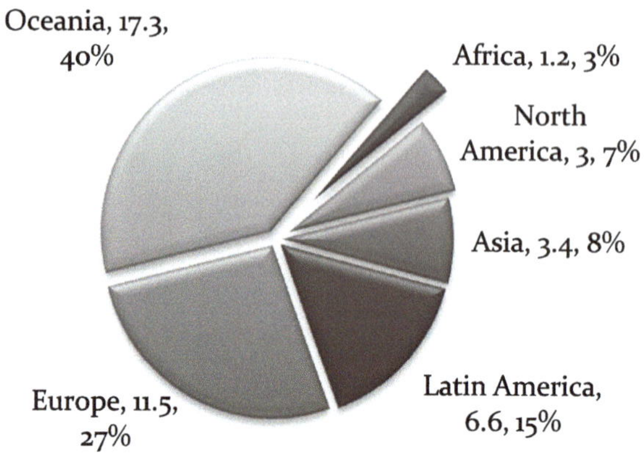

Fig. 2 Size (hectare) and distribution (%) of organic agricultural land by regions, 2013 (Fibl and Iofam 2015)

year (Organic World 2013). The growth appears very uneven when disaggregated by country. Oceania has the largest areas with 17.3 million hectares, followed by Europe (11.5 million hectares) and Latin America (6.6 million hectares), finally Asia and North America with under four million hectares (Fig. 2).

This 43.1 million hectares is still only 0.98 % of the world's agricultural land, although some countries have significant areas cultivated organically. It is noteworthy that the majority of the countries located in the forefront of the list are European. According to the size of the areas of course a different rating would result as only in Australia the size of ecologically cultivated land is nearly twice of the total area of Hungary (Fig. 3).

In the last decade there has been a significant increase in the size of organic agricultural areas in India, China, Columbia and Romania. The whole world is basically characterized by a greater or lesser rate of increase of organic areas; there are only a few places where downward trend was registered (Fig. 4).

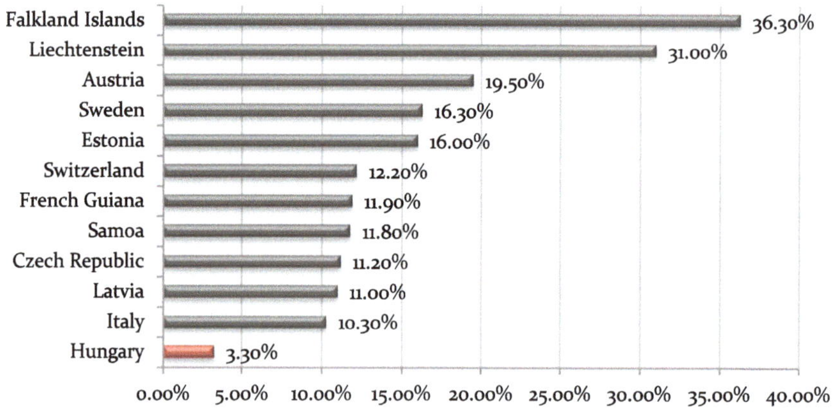

Fig. 3 Shares of organic agricultural land—leading countries over the world and Hungary (Fibl and Iofam 2015; Hivatal 2015)

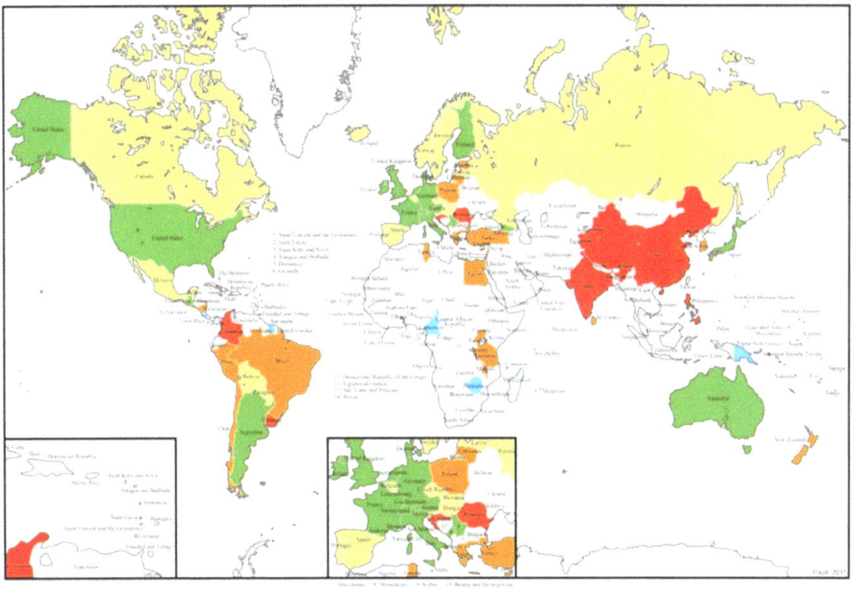

Fig. 4 World map of organic agriculture growth rates for 71 countries over a decade (2001–2011). *Red* triple digit growth; *orange* double digit growth; *yellow* single digit growth (above average); *green* single digit growth (below average); *blue* negative growth (Paull 2011)

3.2 Organic Farming in Hungary

3.2.1 Regulation and Support

Legal background of organic farming in Hungary is ensured by the Council Regulation No 834/2007, Commission Regulation No 889/2008, Commission Regulation No 1235/2008 and the 34/2013. (V.14.) VM Regulation about the certification, producing, labelling and inspection of organic products.

Certification and inspection of organic farming are managed by two organizations in Hungary, namely Biokontroll Hungária Ltd, and Hungária Ökogarancia Ltd. The latter has been operating only since 2006, while Biokontroll Hungária Ltd. (together with its predecessor organization) has been active since 1986. The older organization covers most of the market (Fig. 5).

The number of Hungarian organic farms shows sudden increases in 2004 and in 2009, and after these years there is a slow but steady decline observable. This is clearly attributable to the effects of the beginning of agri-environment programs.

The Hungarian agri-environmental program was established in 1999 based on the European model. Its introduction was due in 2000, but finally the program started only in 2002. Some authors explained that the reason of the delay was the lack of financial support, while others pointed out the lack of political will (Szabó 2003). The first 2 years of the program was supported from the national budget. It

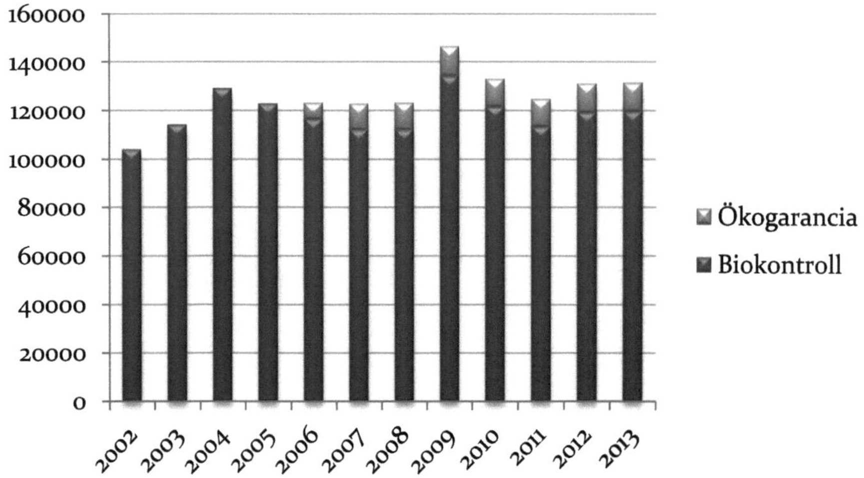

Fig. 5 Verification and certification of land under organic farming in Hungary (hectares) (Biokontroll Hungária Kft 2013; Hungária Ökogarancia Kft 2013)

was followed with great interest; Numerous applications had to be rejected due to the lack of resources.

The year 2004 was a milestone in the history of the program, since Hungary's EU accession made it possible for Hungarian producers to have access for the EU community's support. By the EU support, approximately HUF 44 billion was allocated in 2004, which was a multiple increase in comparison with the previous years' budget. This allowed a significant increase in payment amounts as well as the number of participating producers (Mile 2006). The program covered a 5-year period; the next opportunity to apply for support was published in 2009.

In the program launched in 2009 the amount of support increased again compared to the previous period. The size of payment was diversified according to the nature of plantation (Nemes 2010). The financial aid that was available in the 5-year programs motivated producers who previously were not engaged in organic farming. The only reason for a small number of dropouts after 2004 and 2009 was that some of the farmers were hardly or not able to meet the strict criteria (Solti 2012). The last program ended up in 2014, and currently there is no financial support available. Since more than 90 % of Hungarian organic farmers are affected by the lack of provisions, this can result in serious difficulties in the sector.

3.2.2 Competitiveness of Hungarian Organic Producers

Basic Features of the Sample

The competitiveness of Hungarian organic farmers was researched with the help of a professional questionnaire which had been completed by 70 producers. The survey was conducted for a month; the sample is considered representative according to farm size and geographic localization. Respondents have been

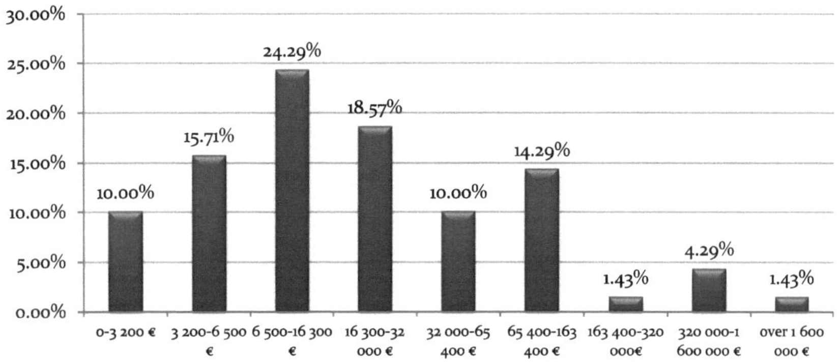

Fig. 6 Annual revenue of the farms in the sample

engaged in organic farming for 15 years on average, and nearly 65 % continue management of family tradition. The average farm size is 136 ha, out of which an average of 84 ha are rented. 26 % of the respondents carry out traditional farming in parallel with the organic method.

The annual revenue is less than HUF 10 million (approximately EUR 32 thousand) in the case of nearly 70 % of the sample. 63 % of the revenue of those producers who also continue traditional farming comes from organic farming (Fig. 6).

Key Factors of Motivation

In contrast to prior assumptions, organic farmers are not primarily motivated by the financial supports available in agri-environmental programs. Rather, they are committed to protecting natural resources, biodiversity and human health. The question exploring this topic was a multiple choice type question where respondents were allowed to select more than one answer. In their answers, nearly 80 % of the farmers take into account environmental and human health issues. Naturally, this does not mean that farmers do not consider financial aspects. 31 % of the respondents are motivated by financial support and by a so-called organic premium.[1] Among the "other" factors mostly family tradition and the rural way of life was noted (Fig. 7).

Crosstab analysis reveals that those producers who uphold family traditions are more strongly motivated by the organic premium than those who have just started their land management. This is probably attributable to a more thorough knowledge of farming and market opportunities.

A great majority (80 %) of respondents believe that organic farming is not suitable for being the prevailing method of agricultural production. The main reason of this seems to be the difficulties of producing certain plants without chemical plant protection. In the case of these plants the yield loss may be so great that

[1]The price difference as much the organic products can be sold more expensive than conventional products.

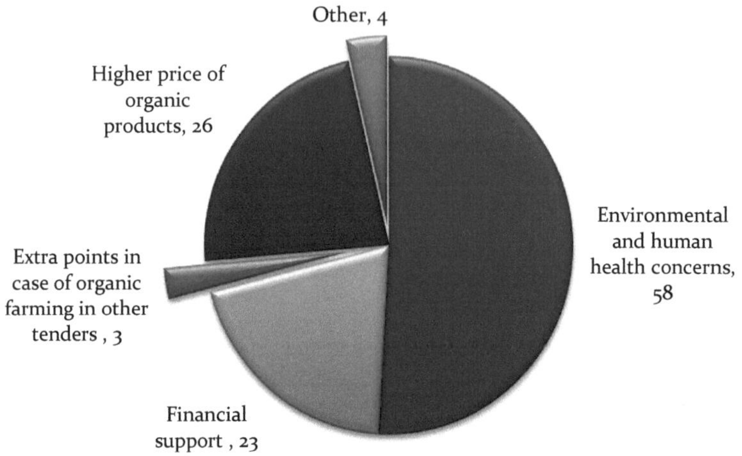

Fig. 7 Factors of motivation of organic farming—responding frequency (n = 70)

production cannot be profitable. At the same time world population growth strengthens the quantitative aspects of food production to the detriment of quality. Many producers mention the strength of supplier industry lobby, which prevents the environmentally friendly, organic production.

Respondents have been asked about the amount of importance they attach to certain factors in achieving competitiveness and about the amount of importance they attach to the same factors in case of their own management. In accordance with the diagram presenting motivational factors, farmers have assigned conviction and vocation as the two most important factors of success. Figure 8 shows that there are only two factors valued by the respondents higher in case of their own management than the importance attached to it concerning competitiveness. Both factors are considered intangible assets, namely conviction and professional knowledge. In the case of all the other factors, their importance is scored higher than when considering the farmer's own capabilities.

Expenditures and Revenues

The survey confirmed that organic farming is very labour-intensive and mechanization demanding. These two items make 60 % of the total cost, while costs of pest management and fertilization are typically lower than in the case of usual conventional agriculture. The results are most likely influenced by the fact that a significant proportion of organic plant protection is mechanical, and respondents classified these costs in the categories of labour and machine work (Fig. 9).

In the questionnaire farmers have been asked to compare the items of total cost in organic farming with the traditional one. Most respondents believe that there is no cost difference between organic and conventional products in case of storage, transportation and packaging. The difference is greater concerning the costs of labour and machine work—according to 70–80 % of the producers these costs are

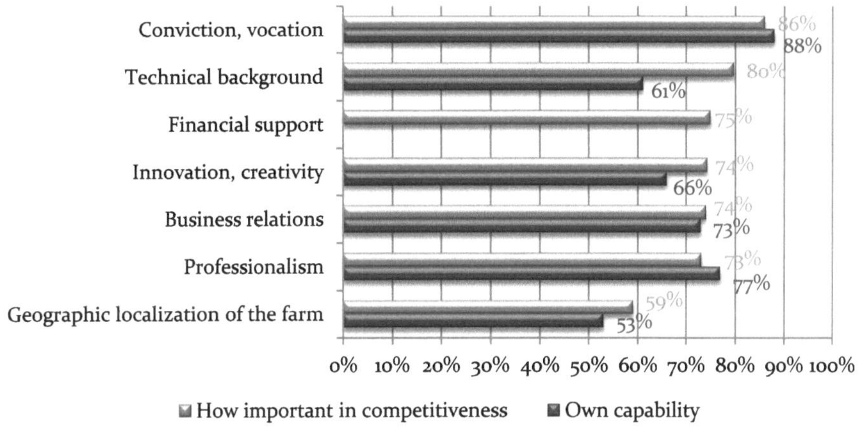

Fig. 8 Key factors in competitiveness—importance and capability

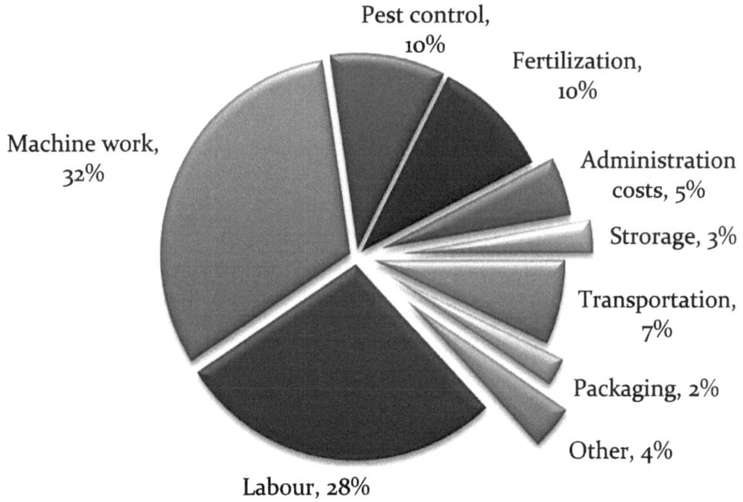

Fig. 9 The ratio of components of total cost in organic farming

20–30 % higher in organic farming. Administrative costs are also considered higher, which is probably due to the higher administrative background of organic inspection and certification, and the participation in agri-environmental programs.

Organic farmers have to finance the costs of control and certification as well. The minimum fee at Biokontroll Ltd is 83,500+vat (27 %), in the case of Hungária Ökogarancia Ltd. is 50,000+vat (27 %). Above a certain size of property per hectare pricing comes into effect. According to a quarter of farmers the costs are unreasonably high, only 22 % of the respondents consider the minimal fee appropriate (Fig. 10).

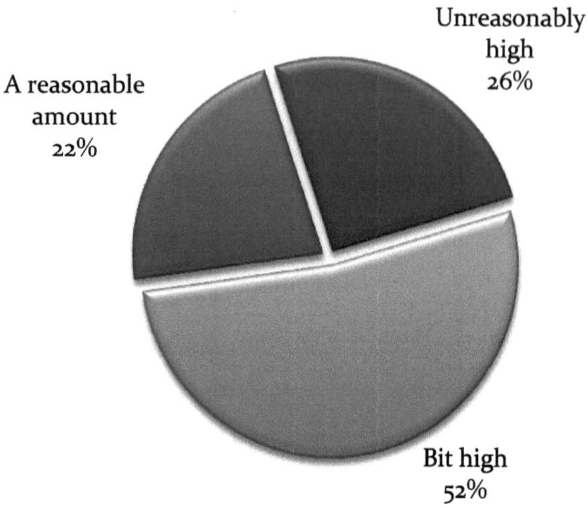

Fig. 10 Costs of inspection and certification according to the respondents

In addition to the high costs farmers have to cope with greater yield loss compared to conventional farming. Organic production is obviously more sensitive to adverse weather conditions or pest invasion; organic farmers occasionally face even 50 % greater yield loss than producers in conventional production. This is generally typical in fruit production, but losses are also great in case of vegetables, and corns, sunflowers or even cereals. There were only 26 % of the respondents who did not experience difference in yield losses between organic and conventional farming, mostly tending meadows, pastures or alfalfa.

In a certain part of organic products producers are not able to realize organic premium. Most of the respondents can sell their products at 20–30 % higher prices; there are only a few farmers who can enforce 50–100 % higher prices. The latter category typically includes those who sell products directly to consumers (Fig. 11).

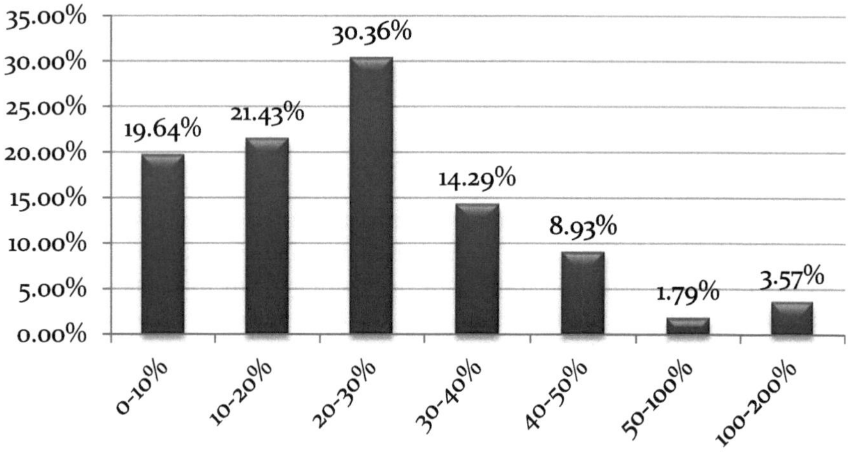

Fig. 11 Size of available organic premium on the market

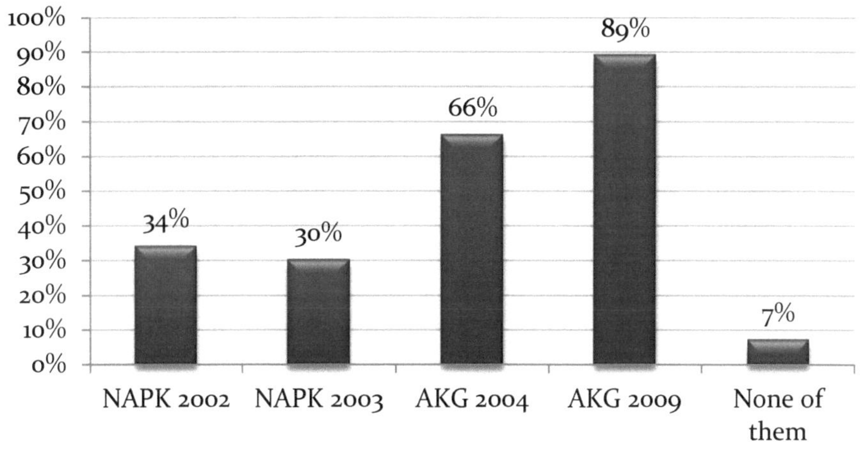

Fig. 12 Participation in agri-environmental programs

Participation in the Agri-Environmental Programs

Organic farming can be simplistically characterized by higher costs and lower yields (Urfi and Kormosné Koch 2011), therefore it is hardly surprising that most of the farmers are involved in agri-environmental programs and take advantages of the financial benefits. Based on the developments of costs and yields we can define these grants as compensatory payments. One third of the farmers in the sample was involved in the program in 2002 already, and the rate of participation increased to 90 % by 2009. This rate is the same as the turnout nationally registered (Fig. 12).

While only 22 farmers (31 % of the respondents) have declared to be motivated by the financial aid available in agri-environment programs, the participation was much higher: 89 % of the respondents took part in the last program between 2009 and 2014. Cross tabulation analysis shows that those farmers who were mostly motivated by financial aid would stop their organic producing in the lack of support. There have been only 16 persons (23 %) from the sample who believe his/her activity would be profitable even without financial aid (Fig. 13).

Even more interesting results have come from the question if the temporary cessation causes any financial difficulties for the farmers. Figure 14 shows that the lack of financial support is a problem for 65 % of the farmers. Three percent of them have already finished organic farming and an additional six % is thinking about exiting from the sector and continue his/her activity by producing conventional products.

Market Relations

To be able to sell organic products on the market not only the production process but also the product itself has to be qualified. The inspection and certification of the product mean an additional cost. As a result, some organic farmers do not or only partially sell organic products. The following figure illustrates this ratio (Fig. 15).

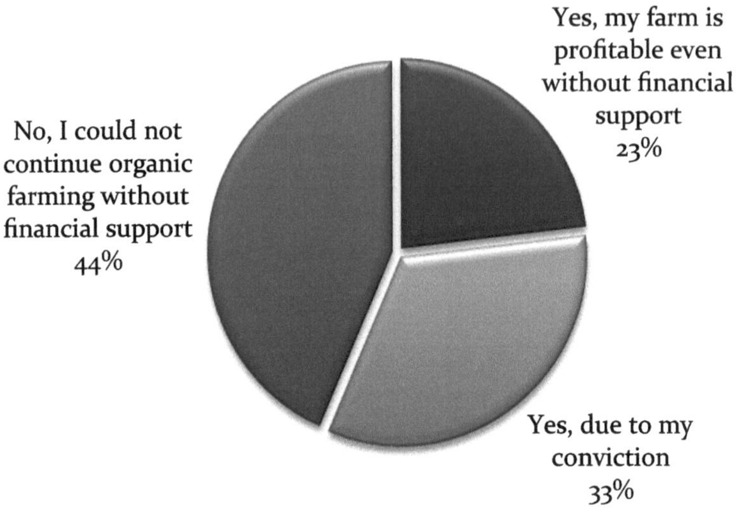

Fig. 13 Would you go on organic producing in the lack of financial support?

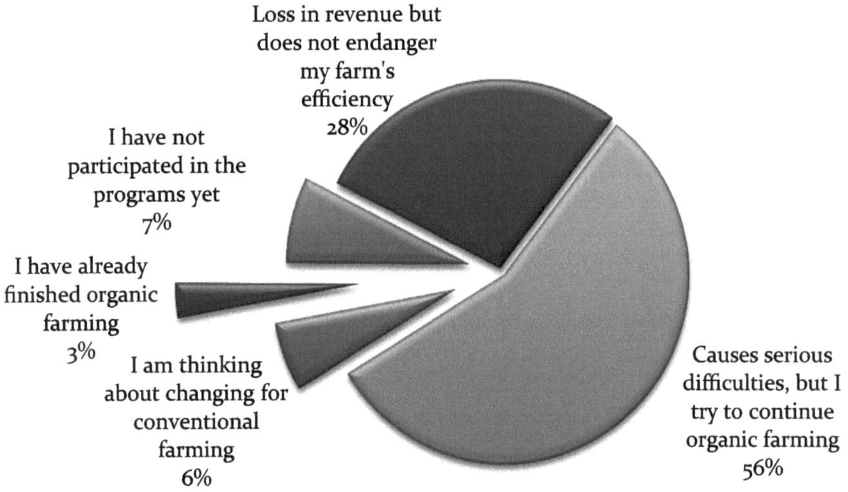

Fig. 14 Consequences of the temporary suspension of the agri-environmental program

The main reason of selling products without the organic label is the lack of business relations, but respondents also mentioned the underdeveloped distribution system and the high costs of organic certification.

77 % of the farmers sell their products unprocessed. Opinions are divided on whether the foreign buyers required processed products. According to some points of view, raw products are preferred since the added value occurs at the purchaser. However, the increase of transport costs makes the transportation of large amount

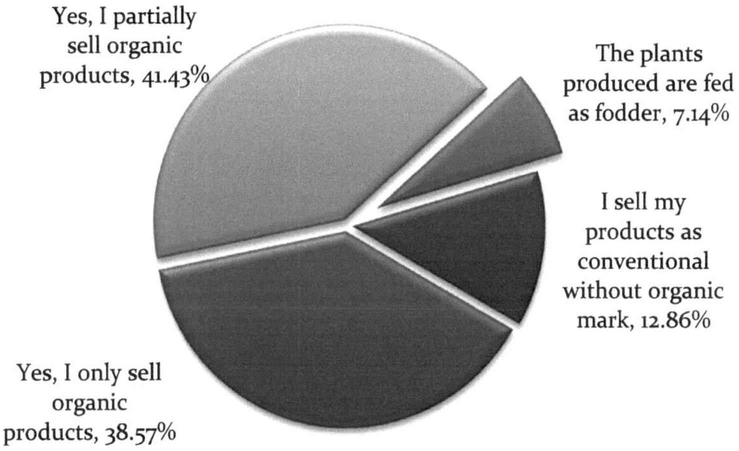

Fig. 15 Do you sell your products as organic?

of raw material more expensive, while the processed form, lighter and more easily transportable, could be more profitable for the purchaser (Fig. 16).

Those producers who sell their products unprocessed mostly refer to the lack of capital necessary to install a processing plant, and to the uncompetitive quantity of production. According to 14 % of the respondents the market does not require high level of procession; raw materials are easy to sell. This smoothness is predominant in the sector; only 16 % of the respondents reported frequent difficulties in sales, for the rest of them sales are usually or always trouble free (Fig. 17).

Producers think that there is hardly any competition among Hungarian farmers. On a seven-graded Likert scale, competition strength was marked as 2.76 on

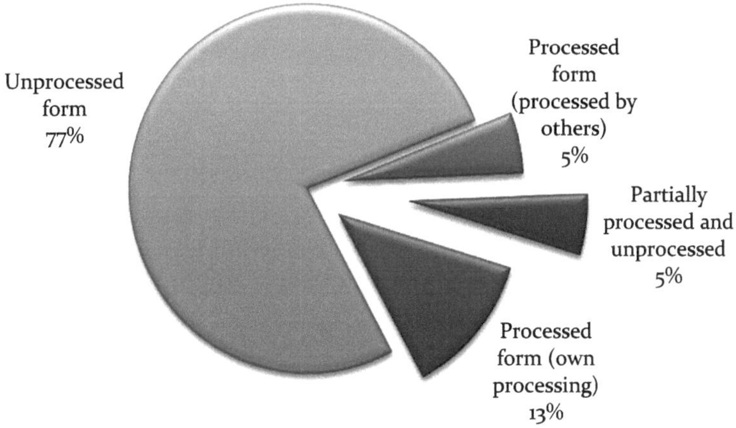

Fig. 16 Level of processing of sold organic products

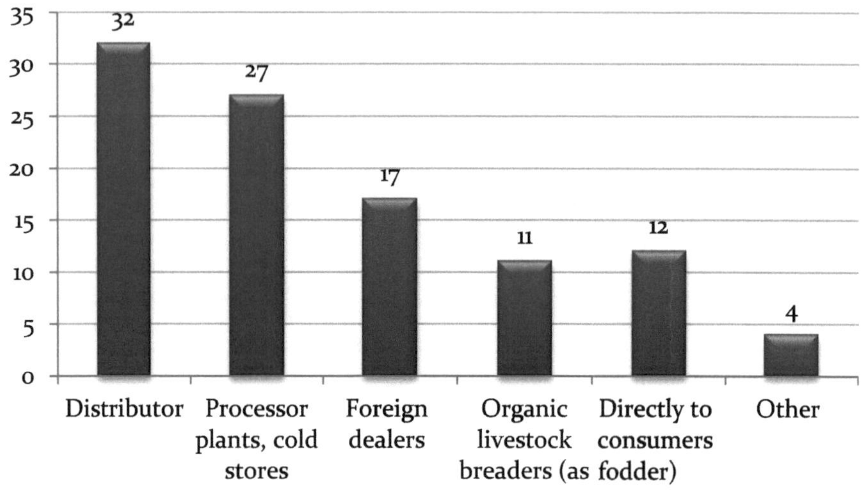

Fig. 17 Purchaser (contracting partner) of organic products—frequency of mention

average. Respondents perceived competition stronger (3, 89) if imported products are also taken into consideration.

As Farmers See the Hungarian Organic Market

Most of the Hungarian organic producers partially or completely consume organic products. This supports the vocation and an environmentally sensitive way of thinking, which is considered the most important motivation in organic farming. Only 5.71 % of the respondents say that they do not consume organic products, which might be due to the nature of their production (Fig. 18).

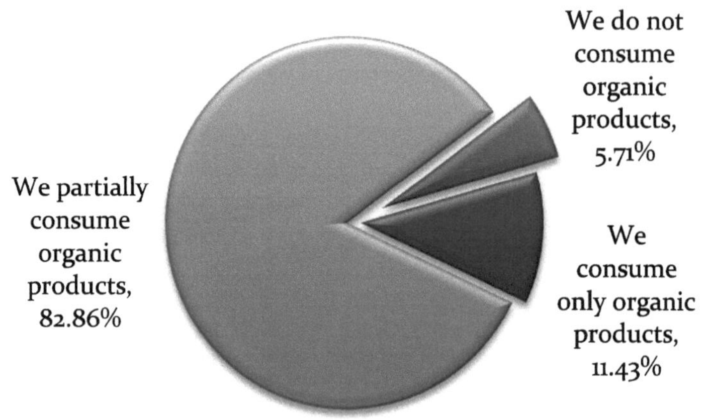

Fig. 18 Do you and your family consume organic products?

Three quarters of the farmers partly or wholly agree that the presence of cheap import makes it difficult to get on the domestic market. 70 %, however, also agree in part or whole that Hungarian consumers have a preference for domestic products. More than half of the farmers are convinced that substantial growth cannot be expected in the consumption of organic products in Hungary, and there is only a small proportion of consumers who always prefer organic products to conventional ones.

In order to reach a substantial expansion of organic production and market, farmers suggest increasing financial supports, reducing bureaucracy and inflexibility of the system of inspection and certification, and most importantly, a change in attitude, the formation of thinking green, which requires active participation of the family, education and the media. Finally, the necessity of growth of purchasing power can be also mentioned, since the market for organic products is not the market of cheap prices, rather the market of quality.

3.3 Case Study: Association of Organic Producers in Szabolcs-Szatmár-Bereg County

The case study analyses an association of organic farmers, which is a bottom-up initiative of 21 independent organic producers in Szabolcs-Szatmár-Bereg County, Hungary. The main profile of the organization is fruit production, mainly apple but there is a significant proportion of cherry, walnut and pear as well. Crop production can also be found among the activities.

Similarly to the producers' cooperatives, the association is supposed to strengthen the members' bargaining power on the market. However, members have not committed themselves in any legal form, thus in effect creating a non-governmental organization, where membership fee is only HUF 3000.[2] Due to the single market activity members can take greater advantages of the opportunities unavailable without cooperation. Members are not obliged to sell their products through the association, but if they use this opportunity they are supposed to pay one per thousand of the net value. There is no mandatory contribution to assets or to operational costs; neither any other service fees, nor compulsory purchases nor sales.

The association is not liable to value added tax and income tax and it is not required to maintain self-management. The research for market opportunities, negotiations and signing contracts are managed by the voluntary leader of the association, who is also an organic farmer. Nevertheless, his legal independence has disadvantages as well. In the case of investments, for instance, only a smaller amount of financial aid is available (maximum 35 % compared with 45–60 % in the case of producer cooperatives), as members can only apply individually, and there is no legal opportunity for the association for single action. Members, however, will do everything in order to enforce the rights of the association and have the same advantages, which are currently available only for cooperatives. They have

[2] 10 euros.

achieved a number of remarkable results: thanks to the minimal cost, asset accumulates where it is most needed promoting local employment and increasing local tax revenue. Traceability of products is fully guaranteed like this, so farmers sense and accept greater responsibility for their products.

The common cohesion of the association stems from the deep conviction and commitment for organic farming. Only producers may become members; neither processors nor traders are allowed to enter. The reason for this is the incidental conflict of interest that would prevent a sufficient advocacy. Although the association has been operating only for a year, it has already achieved significant results. Right in the first year they managed to contract with a German processor and sell 500 tons of organic apple. In 2015 the same purchaser indicated his demand for another 500 tons of apple and 150 tons of sour cherry. The success of the negotiations is clearly reflected in the difference between the market price and the contracted price of the fruits.

The association would be pleased to have the opportunity to sell the organic product in the domestic market, but currently export seems to be the only realistic solution. The primary reason is the geographically unbalanced Hungarian demand of organic products. The biggest market is in the capital, yet producers find it really difficult to enter the group of sellers there. The other problem is the relatively small market: organic food represents only 0.5 % of all food consumption in Hungary (Szente et al. 2011). Due to the marginal demand the conditions of economies of scale cannot prevail, which leads the Hungarian organic producers straight to foreign markets. The mandatory use of organic products in mass catering could be a solution, but it can be expected not earlier than 2020, and only in 30 %—according to the national action plan for organic farming (VM 2014). A change of attitude in the Hungarian health care and medical recommendation could also help a lot to promote the consumption of organic products. Under current circumstances, however, there is no chance to compete with the Hungarian pharmaceutical lobby—according to the leader of the association.

Neither the legal framework nor the support system helps effectively enough to "green" Hungarian agriculture. For instance, in the agri-environmental program between 2009 and 2014 the grants for organic and those for integrated schemes only slightly differed from each other, and this encouraged farmers to apply for the support of integrated farming (MTI 2012), in which the requirements are easier to meet, there is no yield loss, while its environmental benefit is highly debatable.

Many farmers face with difficulties due to the temporary suspension of the agri-environment program in the year 2015. At the time of the completion of the present manuscript, there is minimal information available about the new system of support. According to the forecasts, a significant increase of the aid can be expected, even a three time difference could occur in the payments compared with the last program. Organic production will get out of agri-environmental programs and get under individual legislation.

The leaders of the association fear a dilution of the organic market, where new entrants are not necessarily committed to the environment, but occasionally guided by financial considerations (Scheftsik 2013). Confidence is the key issue in the organic market. Obtaining consumers' confidence is a long and arduous process, and this fragile balance can be easily disrupted in the lack of conviction and commitment. According to the proposal of the association's members, a mentoring system could solve this problem. In this system new entrants would be helped by a senior organic farmer, making it easier to enter the market: mentors could also work as a filter among the new producers. Members claim that the most efficient professional information exchange takes place within the organization. The widely available agricultural advisory system provides for no or only minimal information on organic farming methods, problems and solutions, since consultants are mainly proficient in chemical crop protection and fertilization. Thus the other significant advantage of the association beside tracing and treatment of market relations is the professional support among the members.

The activity of the association is exemplary, surrounded by more and more attention by non-member organic farmers. Environmental, economic and even social benefits are far beyond its costs; therefore it could serve as a model for organic farmers' cooperation in other regions.

4 Conclusions

Currently, Hungarian organic production can only be competitive with financial support. Competitiveness is primarily understood in relation to the production of traditional agricultural products, which does not mean that competition in the organic sector cannot be detected, but that it is less noticeable in Hungary. Independence is not impossible; there are organic producers who can be profitable even in the lack of subsidies, but this is only a small segment of the producers. The benefit depends on the size of the organic premium; higher premium is only available on the consumer market. For higher prices, however, mainly processed products are required, but processing in many cases is prevented by the lack of capital. In addition, farmers are poorly motivated as long as foreign traders create continuous demand for organic raw materials. In this case, the price of the products cannot compensate the higher costs and loss of yield. This implies that more than 90 % of Hungarian organic farmers are involved in agri-environmental programs.

The recipe for success at organic farming is complex. For long-term success, farmers should own the suitable professional knowledge and efficient market relations, but they also need to be committed to protecting the environment. The really successful producer identifies himself with the production method he applies and the products he produces. This belief is therefore the main motivating factor in this sector burdened with difficulties.

According to the unanimous opinion of the farmers, organic market is a market of confidence built on a long-term cooperation and trust between seller and buyer. The development of this relationship is an arduous and time-consuming process, and the delicate equilibrium is very easy to disrupt by market dilution and the incidental negative behaviour of new entrants. Tightening the conditions, however, is not necessarily a good solution. The growing administrative burden increases the costs, thereby displacing some of the farmers out of production. The real solution would be the change of values and strengthening of confidence, and to achieve this a long time, and perhaps a generation change might be needed.

References

Baji B (2015) Permakultúra és önellátó biogazdálkodás. Első lánchíd Bt, Kiadó

Baranski M et al (2014) Higher antioxidant concentrations and less cadmium and pesticide residues in organically-grown crops: a systematic literature review and meta-analyses. Br J Nutr 112(5):794–811

Biokontroll Hungária Kft (2013) Annual report. http://www.biokontroll.hu/cms/images/downloads/eves_beszamolok/eves_jelentes_.pdf

Buday-Sántha A (2009) Környezetgazdaságtan. Dialog-Campus Kiadó, Budapest, Pécs. ISBN 9789637296833

Drexler D (2013) Roppant kevés a biogazdálkodás Magyarországon. http://www.uzletresz.hu/penzugy/20131009-roppant-keves-a-biogazdalkodas-magyarorszagon.html

Fibl and Iofam (2015) The world of organic agriculture, statistics and emerging trends 2015. https://www.fibl.org/fileadmin/documents/shop/1663-organic-world-2015.pdf

Food and Agriculture Organization of the United Nations (FAO) (2015) Environmental and social standards. http://www.fao.org/environmental-social-standards/en/

Horváth Z (2013) Termelői Értékesítő Szervezetek szerepe a magyar kertészeti ágazatban. Őstermelő Gazdálkodók lapja, XVII/3, pp 16–20

Hungária Ökogarancia Kft (2013) Annual report. http://www.okogarancia.hu/pdf/eves_jelentes/2013.pdf

Hivatal KS (2015) Tájékoztatási Adatbázis. http://statinfo.ksh.hu/Statinfo/haViewer.jsp. Központi Statisztikai Hivatal

Mile C (2006) Az agrár-környezetgazdálkodás EU-konform lehetőségei Magyarországon. Doctoral dissertation, Kaposvár

MTI (2012) Csökkent a bioterületek nagysága. ElemzésKözpont. http://www.elemzeskozpont.hu/content/cs%C3%B6kkent-bioter%C3%BCletek-nagys%C3%A1ga

Nemes G (2010) Az Agrár-környezetgazdálkodás Implementációja az Új Magyarország Vidékfejlesztési Programban. Budapest, Magyar Tudományos Akadémia, Közgazdaság- és Regionális Tudományi Kutatóközpont

Organic World (2013) The world of organic agriculture. In: Global organic farming statistics and news. http://www.organic-world.net/fileadmin/documents/yearbook/2013/web-fibl-ifoam-2013-25-34.pdf

Paull J (2011) The uptake of organic agriculture: a decade of worldwide development. J Soc Dev Sci 2(3):111–120 (ISSN 2221-1152)

Scheftsik T (2013) Biogazdálkodás és vegyszerek. In: National Geographic. Sanoma Media Budapest Zrt, Budapest

Solti G (2012) Az ökológiai gazdálkodás helyzete Magyarországon. Sárközi Péter Tudományos Emlékülés. http://www.sarkozybio.hu/sarkozy/emlekulesek/images/2012/az_okologiai_gazdalkodas_helyzete_magyarorszagon.pdf

Szabó G (2003) A nemzeti agrár-környezetvédelmi program pályázatainak elemzése. Gazdálkodás XLVII. évfolyam 1. sz. Nemzeti Agrárszaktanácsadási, Képzési és Vidékfejlesztési Intézet, Budapest, pp 26–39

Szente V, Szakály Z, Széles G (2011) Ökoélelmiszerek megítélése Magyarországon. Gazdálkodás 55/5, pp 512–517

Urfi P, Kormosné Koch K (2011) Az ökológiai gazdálkodás költség-jövedelem viszonyai. Gazdálkodás, 55/1. Budapest Agrárszaktanácsadási, Képzési és Vidékfejlesztési Intézet, pp 28–38

Vidékfejlesztési Minisztérium (VM) (2014) Nemzeti Akcióterv az Ökológiai Gazdálkodás Fejlesztéséért (2014–2020). http://videkstrategia.kormany.hu/download/3/c8/90000/Nemzeti%20Akci%C3%B3terv%20az%20%C3%96kol%C3%B3giai%20Gazd%C3%A1lkod%C3%A1s%20Fejleszt%C3%A9s%C3%A9%C3%A9rt_vegleges.pdf

Author Biography

Csilla Mile Ph.D. works as a professor at the Department of Management Studies of Kodolányi János University of Applied Sciences, Hungary, where she teachers environmental economics, micro- and macroeconomics, economic policy and marketing. Her main research areas include sustainability in the agricultural sector. She coordinates the research team focusing the competitiveness of Hungarian organic farmers and the consumer habits of consumption of organic products.

Why Sustainable Consumption Is Not in Practice? A Developing Country Perspective

Farzana Quoquab and Nurain Nisa Sukari

Abstract

In recent years, sustainable consumption became a global concern in order to attain sustainable development. The natural resources are diminishing in alarming rate which initiated the discussion on sustainable consumption among the academicians and practitioners. However, it seems that the developing countries are comparatively far behind of adopting and practicing sustainable consumption. The present paper attempts to shed some light on this issue from developing country perspective like Malaysia. More particularly, this study seeks to answer the research question: Why sustainable consumption is not in practice? What are the reasons behind this slow movement? In order to gain deeper understanding regarding the issue, 15 in-depth interviews were conducted. It is hoped that, the findings from this study will illuminate the understanding of the sustainable consumption practice.

Keywords

Sustainable consumption · Problems and prospects · Malaysian citizens · Developing country perspective

F. Quoquab (✉) · N.N. Sukari
International Business School, Universiti Teknologi Malaysia,
Kuala Lumpur, Malaysia
e-mail: fqbhabib@ibs.utm.my

W. Leal Filho et al. (eds.), *Sustainable Economic Development*,
World Sustainability Series, DOI 10.1007/978-3-319-45081-0_6

1 Introduction

The notion of sustainable consumption is interconnected with the sustainable development movement. The scholars, policymakers and practitioners started to highlight about sustainable consumption since early 1970s (Holt 2012). However, it received significant attention since the Rio "Earth Summit" in 1992. It encompasses a broad spectrum including maintaining a balance in consumption in order to protect the right of future generation, to consider the environmental welfare and to attain the quality life. Although this concept has gained considerable research attention, still it is comparatively a new topic in the developing countries. The importance to study 'sustainable consumption' issue is due to the reason that, this is one of the major components to overcome the challenges (e.g., optimizing the natural resource, climate changes and environmental pollution) related to attain nation's sustainability as a whole. Indeed, the necessity to adopt sustainable consumption in daily life is due to the responsibility to assure future generation's wellbeing (OECD 2006).

Sustainable consumption issues increasingly getting attention in policy making mainly in developed countries such as United Kingdom, United States, Denmark, and Germany. The researchers from these countries are also advanced in studying this phenomenon from various perspectives. For example, Tanner and Kast (2003) studied the determinants of green purchases among Swiss consumers, while Seyfang (2007) focused on the shopping attitude of UK consumers. Again, Hornibrook et al. (2015) discussed about exploring the role of carbon labeling in retail (supply chains) in United Kingdom. Since this issue is widely highlighted in most of the European and American countries, it is also important to understand sustainable consumption practice based on developing country perspective like Malaysia. Nonetheless, the reasons behind such reluctance of adopting this best practice among developing countries are yet to be explored. The present study is such an effort to fill this gap in the existing literature. Considering this, the present study aims to answer the question: Why sustainable consumption is not in practice? What are the reasons behind this slow movement?

Since Malaysia is moving towards to its vision 2020, it is utmost necessary to consider sustainable consumption to attain the sustainable development. This view is reflected in the following statement which was stated in National Policy of the Environment Malaysia (2002), "The National Policy on the Environment aims at continuous economic, social, and cultural progress and enhancement of the quality of life of Malaysians, through environmentally sound and the sustainable development" (Ministry of the Science, Technology and the Environment, Malaysia, 2002, p. 2). To make it happen, the policy making process of Malaysia has to go through several stages such as identifying the arising problems, planning, recommending alternatives, and legitimizing policies, implementing the suitable action policy, coordinating various events to suit the established policy and finally evaluating the effectiveness of such policy. As such, it is expected that the findings from this study will be helpful for the Malaysian government as well as for the policy makers in order to understand the issue deeper.

The rest of the paper is structured in the following way. First, relevant literature is reviewed pertaining to sustainable consumption. Next, the adopted methodology is briefly mentioned followed by qualitative data analysis. In the following section the findings are discussed. The paper ends with conclusion, managerial implications and future research directions.

2 Literature Review

In this section, the notion of sustainable consumption is discussed in brief. Moreover, its problems and prospects are highlighted.

2.1 Sustainable Consumption

The ideology of the 'sustainability' encompasses both the environmental (pollution, waste, resource use), economic (proper utilization of resources), and social (health, welfare) aspects. It focuses on consumption by households as well as governments (OECD 2006). However according to Peattie and Collins (2009), sustainable consumption viewed by some researcher as an oxymoron, because the dictionaries stated that to 'consume' something is to use it up or to destroy it, which is the complete opposite to 'sustain' something. It also requires a different perspective on 'consumption' which advocates caring for the greater welfare of the society and mankind on the whole.

The word 'sustainable' refers to maintain, upheld or defended (The Oxford Dictionary 2013), while the word 'consumption' is referred as the action of using the resource. On the whole, it advocates for maintaining a balanced consumption which is economically, socially as well as environmentally desirable for a sustainable development for any nation. A cursory review of relevant studies reveals that, different authors have defined this phenomenon in different ways. For example, Lee (2014) stated that, it is an ecologically and socially responsible consumption habit whereas, Hornibrook et al. (2015) defined it as the consumption practice that respond to basic needs and bring a better quality of life. The definitions from past studies are summarized in the Table 1. The definitions given are always meant to convey the message about the perspective or theme of the research and the objectives of the research. Nevertheless, it is obvious that, there is a lack of consensus about defining this variable. This is due to the fact that different authors have focused different aspects of the sustainable consumption.

The definitions provided in Table 1 generate the insight that, some researchers perceived sustainable consumption from environmental perspective (Hornibrook et al. 2015; Lorek and Fuchs 2013; Wolff and Schonherr 2011). This stream of studies advocates for maintaining such consumption patter which will reduce the

Table 1 Summary of definitions of sustainable consumption phenomenon in selective past studies

Sources	Definitions
Hornibrook et al. (2015)	Sustainable consumption can be defined as "the use of goods and services that respond to basic needs and bring a better quality of life, while minimizing the use of natural resources, toxic materials and emissions of waste and pollutants over the life cycle, so as not to jeopardize the needs of future generations" (p. 268)
Lee (2014)	Sustainable consumption entails concerns for the well-being of other human beings and animals (such as being for fair trade and against animal testing), future studies should investigate other dimensions of environmental concern (especially the altruistic orientation) to attain a more comprehensive understanding of the relationship between environmental concern and sustainable consumption behavior (p. 235)
Lorek and Fuchs (2013)	Sustainable consumption is used to refer to sustainable resource consumption, taking into account the complete product life cycle. Sustainable resource consumption involves the consumption patterns of industries, governments, households, and individuals (p. 36)
Alvarez-Suarez et al. (2011)	Sustainable consumption within the framework of education can be defined as the way the consumer behave that redefining the model of society as one that contributes to sustainability by reducing consumption, and education has a major role to play in this regard (p. 4)
Wolff and Schonherr (2011)	Sustainable consumption is defined as a more ecologically but also socially premised way of buying, using, and disposing of goods and services (p. 45)
Stevens (2010)	Sustainable consumption is an inefficient route to sustainable production and, ultimately sustainable development. The need for integrated government approaches which simultaneously address consumption and production is needed to correct the market and system failures that impede progress on sustainability (p. 17)
Zhao and Scroeder (2010)	Sustainable consumption "is increasingly recognized as an analytical perspective to understand the complex social, economic and political drivers of global environmental change, including the current challenge of global climate change. It offers a multitude of practical approaches and policy tools to achieve a resource efficient and low carbon economy and to support societies in global and local environments" (p. 5)
Peattie and Collins (2009)	Sustainable consumption is such consumption practice "whereby each person consumes only their "earthshare", equivalent to 2.1 gha', as this situation would be both socially equitable and ecologically sustainable" (p. 108)
Bennett and Collins (2009)	… Minimising the use of capital while reducing waste and pollution is good, that 'environmentally friendly' or 'green' product consumption should be encouraged, and that the current generation should reduce its demands in the interests of future generations (p. 48)

(continued)

Table 1 (continued)

Sources	Definitions
Mont and Bleischwitz (2007)	Sustainable consumption is generally defined as the consumption of goods and services that meet basic needs and quality of life without jeopardizing the needs of future generations… This broad definition includes not only consumption by private consumers, but also by institutions (businesses and organisations); not only consumption of resources (appropriation of resources from nature), but also final consumption (p. 62)

Source Compiled by the researchers

waste as well as the over-use of the natural resources. Conversely, some others focused on considering not jeopardizing the need of the future generations (Bennett and Collins 2009; Lee 2014; Mont and Bleischwitz 2007).

Based on the past studies definition compilation, the authors came out with the holistic view of sustainable consumption. Sustainable consumption as refers as the continued act of controlling desire by avoiding from making extravagant purchases and rationalizing use of goods and services that satisfy the basic needs. Sustainable consumption goes beyond the environmental concern by ensuring and managing the existing resources that, not only able to meet the current demand, but also without jeopardizing the future generation.

2.2 Problems and Prospects Pertaining to Sustainable Consumption

The sustainable consumption phenomenon is espoused with some benefits as well as challenges. There are several aspects that are needed to be taken care of to develop sustainably for the long run. It includes: (a) sustainable building and constructions, (b) sustainable public procurement, (c) sustainable tourism, (d) sustainable products and (e) sustainable consumption. It is believed that, consumers play key role in sustainable production and sustainable development. It is needed to pint that, some of the sustainable consumption activities are limited to purchasing organic foods. However, the sustainable consumption goes beyond of its true sense.

The debate on SC is not new. It started during 1970s and the idea has grown older through the passage of time. There are lots of benefits highlighted in favor of adopting sustainable consumption practices. For instances, it is considered as one of the key indicators for sustainable development. Additionally, it can be the mechanism for reducing worldwide poverty and so on. Unfortunately, most of these studies focus on the advantages of adopting sustainable consumption practices. Nevertheless, yet to the authors' knowledge, no studies have highlighted the barriers that are associated with adopting sustainable consumption. Moreover, despite of accepting the positive aspect of sustainable consumption, not many consumers are willing to embrace such practice. More importantly, there is not much discussion how to inculcate such behavior among the individuals, particularly in the developing countries.

3 Methodology

Sustainable Consumption is a socially constructed phenomenon. As such, the best way to answer the research question for this study (which is "why sustainable consumption is not in practice?") would be conducting a qualitative study. Furthermore, the research question consists of a 'why' component which certainly requires to utilize qualitative approach (Yin 2013). 15 in-depth interviews have been carried out to answer the research question. Eight respondents were male whereas, the rest were female. All participants were between 29 and 48 years old. In terms of ethnicity, 7 participants were Malay Muslim, 5 participants were Chinese followed by 3 Indians.

Contacts have been made through personal calls since all of the participants were known to the researchers. Few interviews were conducted at researchers' house, while some other interviews were conducted in coffee shops as well as at respondents' house based on their convenience. At first, participants were asked whether they were aware about the sustainable consumption phenomenon. The researchers then briefed the participants about the meaning of sustainable consumption. Semi structured interviews were considered and the interview protocol has been prepared before the interviews were conducted. In order to assure the reliability of the answers, two basic questions were asked to all respondents:

- Have you ever heard about sustainable consumption? If yes, from where you have heard about it?
- What is your opinion about such consumption practice?
- Do you practice sustainable consumption? If not, why you do not practice such consumption habit?
- In your opinion, why sustainable consumption is not practiced well among Malaysian consumers?

All interviews were recorded upon the permission of participants. Next, transcriptions have been made for each interview in order to facilitate the content analysis. The transcriptions were then reviewed by the participants to assure validity. Content analysis enabled the researchers to generate relevant themes in order to answer the research question.

4 Key Findings and Discussion

The present study sought to answer the research question: Why sustainable consumption is not in practice? In this regard, four major themes were generated as the outcome of content analysis: ignorance, no clear policy, lack of awareness development programs and compliance. The findings are discussed in the following paragraphs.

4.1 Ignorance

Most of the participants stated that the concept of sustainable consumption is not widely spread yet in Malaysia. There is still lack of understanding about this phenomenon among the greater Malaysian citizens. Some of the statements in support of this view are as follows:

> …Could you please make me more clear about this sustainable consumption thing? I am hearing it for the first time. Actually, not only me I think many Malaysians will not be able to understand about what you said about it…

> …It seems interesting. But not so sure how you are going to make people aware about it…

> …Don't you think that it is too much to expect that all Malaysians will be caring about this issue? Some people do not have food to eat for two times a day…

> … To me, not only Malaysia, all developing countries are lagging behind in this regard… It is not because we don't care, this is because our socio-economic structure and cultural pattern is quite different than the developed countries…

> … When we are suffering from inflation, job insecurity, high unemployment rate, what do you expect? You are fighting in the rate race to survive, then how you will think for the nature and future generation's wellbeing?… Do you see my point or not?

> … I think only highly educated people are aware about sustainable consumption…

4.2 Lack of Policy and Regulations

The second issue is pointed as lack of regulatory instruction as well as governmental policy implications. Following are some of the statements that the interview participants highlighted.

> … If it is so crucial, there should be some policy and regulations from the government side. So that, though some individuals are not aware about it they will be forced to maintain it automatically…

> … In fact, it can be included in the school text book… Come on… we are already old, but our children are the future of the nation. You need to educate them first if you really want a great impact in this regard…

> … The triple bottom effect has to be there… Not only government, but also marketers have to take some initiatives… They will show all those alluring advertising and then finally you will expect us not to be extravagant in our shopping!

4.3 Lack of Awareness Development Programs

The third reason was mentioned as absence of awareness building programs is also mentioned as the reason for not adopting sustainable consumption practices.

… You see, there is not much discussion about it in the public media… If you are only writing the research papers, how it will reach to the general consumers or all citizens of the country…

… I think, television and radio should take some initiatives to disseminate this information in more friendly way or in more interesting way… For example, they can make a cartoon to

Table 2 Themes generated from the content analysis

Ignorance	No clear policy	Lack of awareness development programs	Compliance
It is not a widespread concept among Malaysian citizens	Government in Malaysia yet to imposed any clear policy about sustainable consumption	Due to no exposure from early stage about the needs of knowledge about sustainable consumption, the grown up adults in Malaysia do not aware about sustainable consumption	Most of the citizens are too comfortable with existing culture and too lavish in consumption without thinking about the effect on others
Not only Malaysia, but also most of the developed countries are not aware about it	It can be included in the school and college text books to make the new generation about it	Public media should discuss about it more to create awareness among mass citizens	Youths are more extravagant in expenditure than the old generation
The socio-economic structure of the developing countries are different than the developed countries	Need cooperation from marketers as well. The attractive advertisements allure consumers to buy more which goes against sustainable consumption concept	Use of television and radio can help	Certain individuals are more concern about the consumption about house and office utility, rather than fashion oriented products
Highly educated people are more aware about it		To make the young citizens aware about it in their early stage of life, cartoon series can be created to highlight this issue	
Due to the fighting for survival, people do not think about environment or future generation's well-being		At the school level, some competition or exhibition can be arranged in this regard	

highlight this issue… I am sure that in this way, our children will get the message without any effort… I am telling you, this kind of learning is more effective…

… Even the awareness programs can be conducted for the children. At the school level, some awareness building competition or exhibition can be arranged too…

4.4 Compliance with Conventional Purchase Habit

Last but not the least, compliance to the present consumption practice was stated as the last reason for not adopting sustainable consumption practices.

…Honestly speaking, I am aware about it, but do not practice actively… For water and electricity use, yes, I am conscious about not to waste. But when I go for shopping it is hard to maintain… More particularly, for buying dress and tudung (scarf)…(laughter)…

… I am also very conscious about not to misuse the water, electricity and other utility services no matter it is for my house or for my office… But yes, I am not so particular about buying products… You know what I mean…

… My opinion is different. It is easy to practice when you are old, I guess. For example, I usually careful about my purchase and use of products, not too great extent though… But I find it difficult to make my children maintain it…

In a nutshell, the situation portrayed in the above mentioned discussion highlights the necessity for more practical initiatives from both government and marketers' sides in order to inculcate the sustainable consumption practice among the greater number of Malaysian citizens. The summary of the findings is provided in the Table 2.

5 Conclusion and Implications

The aim of this study was to gain in-depth understanding about the issue pertaining to the barriers that are associated with sustainable consumption practice among the Malaysian consumers. However, the progress of adopting such behavior is still not much promising. This is mainly because, not many people are aware about the sustainable consumption phenomenon be it Malaysia or other developing countries. The findings from this study reveal that, there are four big reasons that are creating barriers to embrace such noble behavior. As mentioned above the mostly cited reasons are found as ignorance, lack of clear policy and regulations, lack of awareness building programs and comfort in maintaining the traditional consumption practices.

In order to align with the vision 2020, the Malaysian citizens need to adopt sustainable consumption practices. As such, the policy makers should take social marketing campaign in order to make people aware about the consequences of excessive consumption. Opinion leader can be also used in this regard.

For example, famous movie stars or singers can be used to create a short term buzz in the national television channels to make people aware about this issue.

Also, as suggested by the interviewees, the government of Malaysia also needs to take some actions to include the discussion on sustainable consumption in the school and college text books in order to make the new generations aware about the necessity of practicing such consumption behavior. Since it is found that the youngsters are more prone to carefree consumption style, incorporation of such discussion in the early childhood text book may provide proper understanding about the necessity for being careful in the consumption pattern.

Although, this study provides some useful insight about the reason behind not adopting the sustainable consumption among Malaysian consumers, it is not beyond of its limitations. For example, in-depth interviews were carried out among 15 individuals which limits the generalizability of the study findings. It is suggested that, the future researchers may conduct more exploratory studies with a bigger sample size in order to get more concrete findings.

It is hoped that the present study enlightens the understanding of sustainable consumption practice by exploring the phenomenon from Malaysian country perspective. By utilizing the findings from this study will benefit the Government as well as practitioners. Not only this, it is expected that this study will open up the avenue for future studies in this field.

References

Alvarez-Suarez P, Vega-Marcote P, Mira RG (2011) Sustainable consumption: a teaching intervention in higher education. Int J Sustain High Educ 15(1):3–15

Bennett J, Collins D (2009) The policy implications of sustainable consumption. Aust J Environ Manag 16(1):47–55

Holt DB (2012) Constructing sustainable consumption from ethical values to the cultural transformation of unsustainable markets. Ann Am Acad Pol Soc Sci 644(1):236–255

Hornibrook S, May C, Fearne A (2015) Sustainable development and the consumer: exploring the role of carbon labelling in retail supply chains. Bus Strategy Environ 24:266–276

Lee K (2014) Predictors of sustainable consumption among young educated consumers in Hong Kong. J Int Consum Mark 26(3):217–238

Lorek S, Fuchs D (2013) Strong sustainable consumption governance—precondition for a degrowth path? J Clean Prod 38:36–43

Mont O, Bleischwitz R (2007) Sustainable consumption and resource management in the light of life cycle thinking. Eur Environ 17:59–76

Ministry of the Science, Technology and the Environment, Malaysia (2002) National policy on the environment. http://www.doe.gov.my/portalv1/wp-content/uploads/2013/01/dasar_alam_sekitar_negara.pdf. Accessed 3 Nov 2015

Organisation for Economic Co-Operation and Development (OECD) (2006) Education at a glance 2006—highlights. https://www.oecd.org/education/skills-beyond-school/37376068.pdf. Accessed 1 Nov 2015

Peattie K, Collins A (2009) Guest editorial: perspectives on sustainable consumption. Int J Consum Stud 33(2):107–112

Seyfang G (2007) Growing sustainable consumption communities: the case of local organic food networks. Int J Sociol Soc Policy 27(3/4):120–134

Stevens C (2010) Linking sustainable consumption and production: the government role. Nat Resour Forum 34:16–23

Tanner C, Kast SW (2003) Promoting sustain able consumption: determinants of green purchases by swiss consumer. Psychol Mark 20(10):883–902

The Oxford Dictionary (2013) Definition of sustainable. http://www.oxforddictionaries.com/definition/english/sustainable. Accessed 1 Nov 2015

Wolff F, Schonherr N (2011) The impact evaluation of sustainable consumption policy instruments. J Consum Policy 34(1):43–66

Yin RK (2013) Case study research: design and methods. SAGE, Thousand Oaks

Zhao W, Scroeder P (2010) Sustainable consumption and production: trends, challenges and options for the Asia-Pacific region. Nat Resour Forum 34(1):4–15

Authors Biography

Dr. Farzana Quoquab is a senior lecturer at International Business School, UTM. She has received her Doctorate degree from Universiti Kebangsaan Malaysia. She has presented papers at various international and national conferences and published articles in peer-reviewed international journals such as Asian Case Research Journal, Asia Pacific Journal of Marketing and Logistics, International Journal of Economics and Management, International Journal of Business Governance and Ethics, Emerald Emerging Markets Case Studies, Asian Academy of Management Journal and Journal of Islamic Marketing. Since 2008, she has produced 53 international conference proceedings and eight book chapters. She is one of the editorial board members of 'Case Studies in Business and Management' and 'Journal of Economic and Administrative Science'.

Ms. Nurain Nisa Sukari is a post-graduate student as well as research assistant at International Business School, Universiti Teknologi Malaysia. Her current research interest is sustainability marketing.

Managing Sustainable Consumption: Is It a Problem or Panacea?

Farzana Quoquab and Jihad Mohammad

Abstract

With the growing concern pertaining to the problems related to environmental issues, sustainable consumption became the central research priority in recent years. Sustainable consumption is being consistently associated with green and ethical consumerism which advocates for wise and careful consumption as well as efficient use of goods and services. It goes beyond the environmental concern by ensuring and managing the existing resources in order to meet the current need without jeopardizing the need of the future generation. Clearly, it calls for a paradigm shift from the existing extravagant consumption pattern to a more future oriented eco-friendly and eco-concerned consumption. By considering such importance, the present study attempts to shed some light on problems and prospects related to adopting sustainable consumption behaviour among consumers. Furthermore, the sustainable consumption construct is defined holistically in order to highlight the depth and breadth of the construct. This study utilized scientific way of reviewing literature as the means to gather required information. Literature pertaining to sustainable consumption is analysed, compared and synthesized to fulfil the research objective. In a nutshell, the paper presents an overview of sustainable consumption concepts, its problems and prospects in order to advance the knowledge among academicians and practitioners. It is expected that, this study will motivate future researchers to conduct more research in this field.

Keywords

Sustainable consumption · Ethical consumerism · Problems and prospects

F. Quoquab (✉) · J. Mohammad
International Business School, Universiti Teknologi Malaysia,
Kuala Lumpur, Malaysia
e-mail: fqbhabib@ibs.utm.my

W. Leal Filho et al. (eds.), *Sustainable Economic Development*,
World Sustainability Series, DOI 10.1007/978-3-319-45081-0_7

1 Introduction

Sustainable consumption is an integral part of sustainable development (UNEP 2014). For any nation, unless and until sustainable consumption is assured, the sustainable development will be hindered to reach its desired state. Considering its importance and contribution to the socio-economic growth and welfare, this topic has received significant research attention in recent years.

Sustainability can be considered as "…a global approach towards securing lasting welfare for entire human race" (Nkamnebe 2011: 222). On the other hand, sustainable consumption can be considered as adopting such consumption behaviour that enables sustainable development. It advocates for change of consumption pattern in order to mobilize the sustainable development. It is argued that, although the sustainable consumption concept is very appealing, the core content of sustainable consumption remains obscure (Nwankwo et al. 2009). According to Nkamnebe (2011), the discussion of sustainable development and consumption is rhetorical and sceptical. Till to date, this discussion is unable to provide a clear guide to action.

Many studies in the field of consumer behaviour have discussed issues pertaining to sustainability marketing and green marketing (Kirchgeorg and Winn 2006; Jones et al. 2008; Polonsky 1994, 2011). Nevertheless, Most of the articles equate the notion of sustainability marketing with green marketing concept (Iles 2008; Rettie et al. 2012; Peattie 2001). Moreover, *green branding* (Hartmann et al. 2005), *green brand equity* (Chen 2010), *green supply chain* (Srivastava 2007), *ecolebeling* (Rex and Baumann 2007), *green purchase behavior* (Chan 2001; D'Souza et al. 2007; Kim and Choi 2005), *environmental citizenship behaviour* (Dobson 2007), are also discussed which considers ethical aspect of consumption and marketing activities. However, little has been discussed about sustainable consumption in its true sense.

The notion of sustainability consumption is in its infancy. Although this issue has been raised and discussed since last two decades, the concept of sustainable consumption still did not reach its maturity. As such, the present study aims to understand the sustainability consumption phenomenon by reading all available articles in hand. Based on this review, an attempt has been made to provide a comprehensive definition of the construct. Additionally, the major three pillars of sustainable consumption are also discussed. Furthermore, this study also highlights the challenges that are associated with sustainable consumption practices. It is important to note that, most of the discussion related to sustainable consumption is carried forward by developed countries' researchers and practitioners. It clearly emphasizes the need to consider this phenomenon from developing country perspective. This study is such an effort to address this gap by discussing the sustainable consumption issue from developing country perspective.

In the following sections data gathering process has been discussed followed by a brief discussion on sustainable consumption phenomenon. Next, three facets of sustainable consumption are briefly highlighted and its problems and prospects are discussed. Lastly, implications and future research directions are also provided.

2 Gathering Information Through Critically Reviewing Literature

In regard to answer the research questions, this research critically reviewed existing literature. Articles were downloaded from various databases such as Science Direct, Emerald, ProQuest and the like. In addition to it, Google Scholar search engine also utilized in order to gather conference proceedings, journal articles and e-Books pertaining to the topic. The search result yielded 61 scholarly articles related to sustainable consumption that are published from different corners of the world. Among these, 41 are journal articles and the rest consists of conference proceedings and book chapters. A cursory review was made to review the articles in regard to sustainable consumption from socio-economic perspective. It is found that most of the articles on sustainable consumption topic are published during early 20s.

3 The Notion of Sustainable Consumption: Understanding and Definition

The notion of sustainable consumption was incepted in Oslo Symposium in 1994. In this Symposium, the working definition of sustainable consumption was proposed as "the use of goods and services that respond to basic needs and bring a better quality of life, while minimising the use of natural resources, toxic materials and emissions of waste and pollutants over the life cycle, so as not to jeopardise the needs of future generations" (Oslo Symposium 1994). This definition provides a holistic picture of sustainable consumption which provides the idea that, sustainable consumption is not limited to direct consumption; rather it encompasses the whole consumption pattern of the individuals. It emphasizes on improving quality of life rather than materialistic outlook.

Different researchers have defined sustainable consumption in different ways. Most of the definitions provided by past studies discuss sustainable consumption from macro aspect, in which the whole economic and societal aspects receive the main focus. However, it is also crucial to understand the concept from consumer's perspective. Table 1 depicts some of the definitions (arranged by year) of sustainable consumption suggested by past researchers.

The definitions compiled in Table 1 provides the insights that sustainable consumption suggests for practicing wise consumption habit in which individuals need to consider the post-consumption consequences for the present as well as for the future. It is a socially and environmentally concerned way of buying, using and disposing goods and services. It advocates for wise and careful consumption pattern as well as efficient use of goods and services. It refers to the act of avoiding over indulging in purchase and careful use of goods and services that satisfy the basic

Table 1 Definitions of sustainable consumption

Author and the year	Definitions
Lee (2014)	Sustainable consumption focuses on environmental concern in which individuals make choices in their private consumption decision. It is needed to be an ecological and socially responsible citizen in order to care for the society as well as environment.
Jones et al. (2013)	Sustainable consumption requires integration from all aspects; from individuals for everyday decision making; from marketers for formulating appropriate business strategy; and from all levels of the organization. It also requires proper monitoring and public reporting process.
Hornibrook et al. (2013)	Sustainable consumption refers to the proper use of goods and/or services to meet basic needs and to avail better quality of life. Also, it minimizes the use of natural resources, toxic materials and emissions of waste and pollutants over the life cycle. It also considers the needs of future generations.
Wolff and Schönherr (2011)	Sustainable consumption is a socially and ecologically concerned way of buying, using, and disposing of goods and services.
Zhao and Scroeder (2010)	Sustainable consumption is an analytical perspective to comprehend the complex social, economic and political drivers of global environmental change, including the global climate change. It offers practical approaches to achieve a resource efficient and low carbon economy.
Bennett and Collins (2009)	Sustainable consumption supports the continued ability of capital to generate well-being for consumers.
Peattie and Collins (2009)	"Sustainable consumption would meet our current wants and needs at a level and in a form that could be continued indefinitely, without impoverishing future generations and the planet's ability to meet their wants and needs (p. 107)".
Stevens (2009)	Sustainable consumption is the way to attain sustainable production as well as sustainable development. It requires proper integration of government approaches in order to identify the system failures the hinder the progress of sustainability and to find solution to solve those problems.
Haron et al. (2005)	Sustainable consumption dealt with the economic activity of choosing, using and disposing goods/services. It concerns about bringing social change to attain environmental benefits.
Fuchs and Lorek (2005)	Sustainable consumption requires technological improvements to acquire efficiency in consumption and the eco-efficiency of consumption. Improving the efficiency of consumption can be seen as the necessary prerequisite for achieving sustainable consumption.
Kates et al. (2005)	Sustainable consumption provides the ground for greater efficiency in energy and resource use and to minimize waste generation. It also assists individuals and households to make environmentally sound purchase decisions and reinforcing values that support sustainable consumption.
Seyfang (2005)	Sustainable consumption can be referred as the efficient way of consuming goods/services. The individuals who adopt such consumption behavior are the driving force of market transformation,

(continued)

Table 1 (continued)

Author and the year	Definitions
	incorporating both social and environmental concerns when making purchasing decisions.
Hobson (2004)	Sustainable consumption takes into consideration of the impact of consumption on environment. It requires environmentally friendly consumer choices that are both widely available and affordable. It also requires supports from government as well as marketers to support such consumption pattern.
Seyfang (2004)	Sustainable consumption meets the basic needs and brings a better quality of life. It helps minimizing the use of natural resources and toxic materials as well as the emissions of water and pollutants over the life cycle. In such way, it assures protection of the needs of the future generation.
Southerton et al. (2004)	Sustainable consumption is the act that focuses on proper utilization of resources in order to meet the need of individuals while taking care of the natural resources in order to avoiding jeopardizing the need of the future generations.
Dolan (2002)	"The prospects of sustainable consumption must be connected to the cultural frameworks of consumption, and it should be acknowledged that modern consumption is rational within those cultural frameworks (p. 5)".
Hobson (2002)	Sustainable consumption can be referred as the 'rationalization' of lifestyle practices, which helps the consumption more efficient shapes it based on the logic of instrumental rationality.
Sanne (2002)	Sustainable consumption is the way to achieve sustainable development. Individuals are the catalyst of such change in behavior. From economics point of view, sustainable consumption is assumed to be based on consumers' decisions on rational and deliberate considerations of how to satisfy their preferences.
Heiskanen and Pantzar (1997)	Sustainable consumption meets the needs of present generations without compromising the needs of future ones.

needs. While it meets the basic needs of the present consumers, it does not jeopardize the need of the future generation.

Another important aspect of sustainable consumption is that, it enhances the quality of life by offering practical approaches to achieve a resource efficient and low carbon economy. It minimizes the use of natural resources, toxic materials and emissions of waste and pollutants over the life cycle.

Indeed it requires change of customer value, social value as well as ecological values (Elkinton 1999). Moreover, such holistic approach of consumption requires support from all stakeholders of the society including the government, organizations (both public and private) as well as the individuals.

4 Three Pillars of Sustainable Consumption

Based on Table 1, it is understood that sustainable consumption has three main aspects: (i) environmental concern, (ii) considering need of the future generation and (iii) meeting basic needs wisely. For example, using biodegradable package may be considered as such consumption behaviour that reflects environmental concern. A consumer who practices sustainable consumption, he/she would be willing to bear extra expense in order to purchase and use environmentally product compared to traditional value oriented consumers. Again, reducing excess consumption in order to save the natural resources may be considered as care for future generation. Last but not the least, making a habit of recycling and re-using the used products or packages is an example of meeting the needs and wants in a better way.

The notion of sustainable consumption not only considers social aspects but also it incorporates behavioural aspect as well as attitudinal and cognitive aspects. Behavioural aspect exhibits individual's action whereas attitudinal aspect shows positive attitude towards this practice. On the other hand, cognitive aspect represents firm commitment to continue this action in future.

In short, it can be concluded that, sustainable consumption is a multidimensional construct which incorporates attitudinal, cognitive and behavioural aspects. It implies that, the sustainable consumption practice cannot be assured only by its behavioural aspect, it also requires individuals' positive intention and deep commitment. It is argued that customer's preference structure of a particular brand or product is manifested in his/her behaviour (Jacoby 1971; Jacoby and Chestnut 1978). This is because behaviour reflects the ultimate action of individuals. Again, observing the behavioural outcome is insufficient to explain the sustainable consumption phenomenon in full and thus, it is necessary to investigate and to explain the attitudinal aspects of customer's mind (Aydin and Ozer 2005). The support for such argument is that, attitudinal aspect may predict the future intention in a better way (Rundle-Thiele and Bennett 2001). Also, attitudinal measures are less sensitive to short-run fluctuations and able to predict true brand-loyal consumers (Mellens et al. 1996). On the other hand, inclusion of cognitive aspect is also crucial since the individuals should have firm commitment towards sustainable consumption practices.

Generally it refers as the continued act of controlling desire by avoiding extravagant purchases and rationalising use of goods and services that satisfy the basic needs. Sustainable consumption goes beyond of the environmental concern by ensuring and managing the existing resources that, not only able to meet the current demand, but also without jeopardizing the need of future generation. It ensures at least three aspects: quality of life, protecting and preserving the environment, and keeping the natural resources useful for the future generation. A brief discussion on these three aspects is given below.

4.1 Quality of Life

In most cases, marketers offer goods and services that satisfy the immediate need of the consumer. Consumers also reciprocate by purchasing and using such offering without considering the consequences. Not only this, they appreciate those companies and brands that satisfy their extravagant wants without any question. This myopic exchange is fuelled by marketing slogan "consumer is the king". As a consequence, the quality of life deteriorates in the long run. Practicing sustainable consumption may help in assuring the quality of life by practicing good consumption habit.

4.2 Environmental Protection

Sustainable consumption also may help in protecting the environment to some extent. It is argued that, the present world is increasingly devoting to reckless consumption habit which profoundly ignores considering the negative effects on the ecosystem. For example, to meet up the extra need of the consumers, manufacturers and marketing managers are over utilizing the natural resources. In turn, the factories are contributing to water, land and air pollution. Certainly there is a price to pay for the growing uncontrolled consumption. Whatever an individual consumes is produced by utilizing the natural resources either by extracting or by cutting, fishing, farming or mining (Magdoff and Foster 2010). However, the resources on this planet are limited. The excess consumption leads to over-extraction of resources—forests, fish, soil, minerals, water which results in degraded and collapsing ecosystems, habitats and species.

If it continues in this manner, certainly the entire world would diminish to a point that the ecosystem may barely sustain lives. In this instance, it is utmost necessity to adopt sustainable consumption for the entire world. This is due to the reason that sustainable consumption suggests considering environmental aspect while meeting the current needs.

4.3 Meeting Needs of Future Generation

Indeed there are several individuals who indulge in extravagant expenditure, whereas some other individuals from other part of the society suffer from having less than what they need. Excess consumption should be condemned since it ignores the right of other parties (directly and indirectly) involved. It stimulates over use of marketing efforts which in turn over uses the natural resources. For example, fashion shows to promote jewellery, ladies accessories and dresses are inducing individuals for excess consumption. World Centric (2014) reported that, 80 % of the world's resources are used by a minority of the world's population (17 % approximately) for luxuries product consumption such as perfume ($15 billion), ocean cruises ($14 billion), ice-cream ($15 billion) and makeup ($18 billion). These resources could be utilized to provide the basic necessities of food, water, health, sanitation etc. for the rest of the world's population.

In this way, lack of sustainable consumption practice is posing threat on present and future generation. It is reflected in the worldwide concerns for natural resource preservation and protection since the source of natural resource is depleting in its alarming rate. It is expected that, if the general population make sustainable consumption as a habit, it may reduce the economic burden as well as may help future.

5 Problems Lie in Adopting Sustainable Consumption

Although sustainable consumption can be considered as the solution of saving the natural resources, the adoption rate of such consumption habit is very slow. However, adopting the sustainable consumption practice is very slow in developing countries. The barriers that are hindering the adoption process are discussed in brief below.

First, there is a lack of awareness among general population in regard to sustainable consumption. Most of the consumers are not aware of the negative consequences of not practicing the sustainable consumption. Generally educated people are more conscious about this phenomenon.

Second, there is a lack of media aggression to educate the general population about the need of such consumption pattern. Furthermore, there is no policy implementation at the moment.

Third, changing consumption pattern requires motivation to embrace the new habit which seems to be difficult for the inert group of people. Therefore, unless it becomes the social norm, not many people will try to change themselves.

Fourth, certain level of costs is associated with sustainable consumption. This is basically due to the fact that, environmentally friendly products are sometimes costly than the traditional goods. As such, low income group of people may not find it attractive.

Lastly, sometimes individuals are not morally motivated to sacrifice or change for others. Weak ethical and moral filter creates barriers to sustainable consumption.

6 Conclusion, Implications and Future Research Directions

Sustainable consumption is widely regarded as the way of sustainable production and ultimately sustainable development (Kates et al. 2005; Sanne 2002; Seyfang 2004). As stated by Tukker et al. (2010: 1), "the future course of the planet depends on humanity's ability to provide a dignified quality of life for a prospective 9 billion people without exhausting the earth's resources or irreparably damaging its biochemical systems". In this regard, sustainable consumption can be the panacea of such future challenge. Not only this, it can be the answer of poverty reduction worldwide (Connoly and Prothero 2003). Therefore, it is needless to say that researchers, policy makers and practitioners all need to devote to understand the theoretical aspects as well as its practical aspects to be able to materialize it.

Indeed, sustainable consumption demands a system perspective in which production, consumption and disposal of wastes will work together as a system (Nkamnebe 2011). It requires a great deal of cultural change (Jones et al. 2011). All sectors should put their hand together to make it happens (Srinivas 2015). It targets everyone, across all sectors and all nations, from the individual to marketers and governments. Marketers need to formulate proper business strategy in order to accelerate the sustainable consumption practice while consumers need to control and monitor their everyday consumption decision making. On the other hand, government needs to identify the barriers that impede the progress of sustainable consumption and to provide solution for that.

This study provides a comprehensive definition of sustainable consumption as well as discusses three facets (environmental concern, considering need of the future generation and meeting the basic needs and wants) of sustainable consumption. These facets work as integral part of sustainable consumption: environmental protection, quality of life and Meeting needs of future generation. In addition, the positive effect of sustainable consumption is also highlighted. Furthermore, this study emphasizes that in order to become successful in implementing (from the perspective of government and policy makers) and adopting (from the perspective of consumers) it is crucial to understand the challenges associated with such consumption practice.

Last decades, most of the research priorities went for understanding and developing proper measures to control environmental pollution and hazards (Mont and Plepys 2008). However, little attention has been paid in regard to developing measures for making sustainable consumption practice as a success. It is suggested that, sustainable consumption can be achieved if the consumers shift their consumption pattern to wise consumption behaviour and reduce their consumption level (Fuchs and Lorek 2005). While the concept of sustainability is in the centre of discussion in developed nations, it is neither widely understood nor well explained in developing countries. Put together, sustainable consumption is such an issue that requires more efforts to understand the phenomenon more detail to its depth and breadth. It certainly calls for more research initiatives in this field.

References

Aydin S, Ozer G (2005) The analysis of antecedents of customer loyalty in the Turkish mobile telecommunication market. Eur J Mark 39(7–8):910–925

Bennett J, Collins D (2009) The policy implications of sustainable consumption. Aust J Environ Manage 16:47–55

Chan RYK (2001) Determinants of Chinese consumers' green purchase behavior. Psychol Mark 18(4):389–413

Chen YS (2010) The drivers of green brand equity: green brand image, green satisfaction, and green trust. J Bus Ethics 93(2):307–319

Connoly J, Prothero A (2003) Sustainable consumption: cpmsumption, consumers and the commodity discourse. Consumption Markets Cult 6(4):275–291
Dobson A (2007) Environmental citizenship: towards sustainable development. Sustain Dev 15 (5):276–285
Dolan P (2002) The sustainability of "sustainable consumption". J Macromark 22(2):170–181
D'Souza C, Taghian M, Lamb P, Peretiatko R (2007) Green decisions: demographics and consumer understanding of environmental labels. Int J Consum Stud 31:371–376
Elkinton J (1999) Cannibals with forks: The triple bottom line of 21st century business, Oxford, London
Fuchs D, Lorek S (2005) Sustainable consumption governance: a history of promise and failures. J Consum Policy 28(3):261–288
Haron SA, Paim L, Yahya N (2005) Towards sustainable consumption: an examination of environmental knowledge among Malaysians. Int J Consum Stud 29(5):426–436
Hartmann P, Ibáñez VA, Sainz JF (2005) Green branding effects on attitude: functional versus emotional positioning strategies. Mark Intell Plan 23(1):9–29
Heiskanen E, Pantzar M (1997) Toward sustainable consumption: two new perspectives. J Consum Policy 20(4):409–442
Hobson K (2002) Competing discourses of sustainable consumption: does the 'rationalisation of lifestyles' make sense? Environ Politics 11(2):95–120
Hobson K (2004) Environmental justice: an anthropocentric social justice critique of how, where and why environmental goods and bads are distributed. Environ Politics 13(2):474–481
Hornibrook S, May C Fearne A (2013) Sustainable development and the consumer: exploring the role of carbon labelling in retail supply chains. Bus Strategy Environ 24:266–276
Iles A (2008) Shifting to green chemistry: the need for innovations in sustainability marketing. Bus Strategy Environ 17(8):524–535
Jacoby J (1971) A model of multi brand loyalty. J Advert Res 11:25–30
Jacoby J, Chestnut RW (1978) Brand loyalty measurement and management. Wiley, New York
Jones P, Clarke-Hill C, Comfort D, Hillier D (2008) Marketing and sustainability. Mark Intell Plan 26(2):123–130
Jones P, Hillier D, Comfort D (2013) In the public eye: sustainability and the UK's leading retailers. J Public Aff 13(1):33–40
Kates R, Parris TM, Leiserowitz AA (2005) What is sustainable development? Environt 47(3):9–21
Kim Y, Choi SM (2005) Antecedents of green purchase behavior: an examination of collectivism, environmental concern. Adv Consum Res 32:592–599
Kirchgeorg M, Winn MI (2006) Sustainability marketing for the poorest of the poor. Bus Strategy Environ 15(3):171–184
Lee KM (2014) Predictors of sustainable consumption among young educated consumers in Hong Kong. J Int Consum Mark 26(3):217–238
Magdoff F, Foster JB (2010) What every environmentalist needs to know about capitalism. Home, 61(10). http://monthlyreview.org/2010/03/01/what-every-environmentalist-needs-to-know-about-capitalism/
Mellens M, Dekimpe MG, Steenkamp JBEM (1996) A review of brand loyalty measures in marketing. Tijdschrift voor Econoniie en Management 41(4):507–533
Mont O, Plepys A (2008) Sustainable consumption progress: should we be proud or Alarmed? J Cleaner Prod 16:531–537
Nkamnebe AD (2011) Sustainability marketing in the emerging markets: imperatives, challenges, and agenda setting. Int J Emerg Mark 6(3):217–232
Nwankwo S, Chaharbaghi K, Boyd D (2009) http://homepages.uel.ac.uk/D.A.C.Boyd/IJTMSD%20Multidisciplinary%20paper.pdf. Accessed 31 May 2015
Oslo Symposium (1994) http://www.iisd.ca/consume/oslo004.html. Accessed 31 May 2015
Peattie K (2001) Towards sustainability: the third age of green marketing. Mark Rev 2(2):129–146
Peattie K, Collins A (2009) Guest editorial: perspectives on sustainable consumption. Int J Consum Stud 33(2):107–112

Polonsky MJ (1994) An introduction to green marketing. Electron Green J 1(2):1–10
Polonsky MJ (2011) Transformative green marketing: impediments and opportunities. J Bus Res 64:1311–1319
Rettie R, Burchell K, Riley D (2012) Normalising green behaviours: a new approach to sustainability marketing. J Mark Manag 28(3–4):420–444
Rex E, Baumann H (2007) Beyond ecolabels: what green marketing can learn from conventional marketing. J Clean Prod 15:567–576
Rundle-Thiele S, Bennett R (2001) A brand for all seasons? A discussion of brand loyalty approaches and their applicability for different markets. J Prod Brand Manag 10(1):25–37
Sanne C (2002) Willing consumers—or locked-in? Policies for a sustainable consumption. Ecol Econ 42(1–2):273–287
Seyfang G (2004) Time banks: rewarding community self-help in the inner city? Community Dev J 39(1):62–71
Seyfang G (2005) Shopping for sustainability: can sustainable consumption promote ecological citizenship? Environ Politics 14(2):290–306
Southerton D, Chappells H, vanViet B (2004) Sustainable consumption: the implications of changing infrastructures of provision. Edward Elgar, Aldershot
Srivastava SK (2007) Green supply-chain management: a state-of-the-art literature review. Int J Manag Rev 9(1):53–80
Srinivas H (2015) http://www.gdrc.org/sustdev/concepts/22-s-consume.html
Stevens C (2009) Guest editor: special issue on gender and sustainable development. Int J Innov Sustain Dev 4:2–3
Tukker A, Cohen MJ, Hobacek K, Mont O (2010) Editorial—sustainable consumption and production. J Ind Ecol 14(1):1–2
UNEP (2014) Education for Sustainable Consumption (ESC). http://www.unep.org/resourceefficiency/Consumption/EducationLifestylesandYouth/EducationforSustainableConsumptionESC/tabid/101302/Default.aspx. Accessed 31 May 2015
Wolff F, Schönherr N (2011) The impact evaluation of sustainable consumption policy instruments. J Consum Policy 34:43–66
World Centric (2014) Vision for a sustainable world. http://worldcentric.org/conscious-living/increasing-consumption
Zhao W, Scroeder P (2010) Sustainable consumption and production: trends, challenges and options for the Asia-Pacific region. Nat Resour Forum 34:4–15

Authors Biography

Dr. Farzana Quoquab is a senior lecturer at International Business School, UTM. She has received her Doctorate degree from Universiti Kebangsaan Malaysia. She has presented papers at various international and national conferences and published articles in peer-reviewed international journals such as Asian Case Research Journal, Asia Pacific Journal of Marketing and Logistics, International Journal of Economics and Management, International Journal of Business Governance and Ethics, Emerald Emerging Markets Case Studies, Asian Academy of Management Journal and Journal of Islamic Marketing. Since 2008, she has produced 53 international conference proceedings and eight book chapters. She is one of the editorial board members of 'Case Studies in Business and Management' and 'Journal of Economic and Administrative Science'.

Dr. Jihad Mohammad is a senior lecturer at International Business School, UTM. He has received his DBA degree from University Kebangsaan Malaysia. He has presented papers at various international and national conferences and published articles in peer-reviewed international journals. His area of reseach interest includes organizational citizenship behavior, innovation.

Brazilian Public Policies and Sustainable Development that Influence the National Bioindustry

Elias Silva Gallina, Lianne de Oliveira Cruz and Fernanda Matias

Abstract

Brazil is an emerging country member of the BRICS, the seventh world economy and considered one of the countries with mega-biodiversity. However, it has a duality between economic development and sustainable use of natural capital also showing social and environmental fragility. Public policies try to solve these problems improving science, technology, and innovation by-laws (tax incentives; innovation; science, technology, and Innovation framework) and papers (White; Green). One of the strategic areas considered in these policies is biotechnology, particularly when associated with the development of a green economy, like biofuels. Biotechnology is an area of intense scientific production, and policies that stimulate research and cooperation between public and private sectors to promote innovation, reducing institutional barriers and facilitating access to tax incentives for private companies are substantial achievements for bioindustry development. On the other hand, national policy related to genetic resources access and traditional knowledge associated prevented progress in the commercialization of local biodiversity during 15 years. Policies created by

E.S. Gallina (✉) · L. de Oliveira Cruz · F. Matias
Animal Sciences Department, Laboratory of Bioreactors, Nanobiotechnology and Innovation, Universidade Federal Rural Do Semi-Árido, Mossoró, RN, Brazil
e-mail: eliassgallina@hotmail.com

L. de Oliveira Cruz
e-mail: lianne.oliveira@hotmail.com

F. Matias
e-mail: fernandamatias@ufersa.edu.br

W. Leal Filho et al. (eds.), *Sustainable Economic Development*,
World Sustainability Series, DOI 10.1007/978-3-319-45081-0_8

Brazilian governments aligning economic development and natural resources, through genetic resources access and traditional knowledge associated law; are necessary for the national bioindustry development and as a model for other countries.

Keywords
Biotechnology · Green economy · Public policies · Bioindustry

1 Introduction

In 1972, the United Nations adopted the term "sustainable development" being defined in the Brundtland Report in 1987 as the model of development that meets the needs of present generations without compromising the needs of future generations. Sustainable development was the central focus of the United Nations Conference on Environment and Development (Rio-92), held in Rio de Janeiro in 1992. The Rio-92 resulted in the formulation and signing of important documents, such as the Charter of the Earth, Agenda 21 and the Convention on Biological Diversity. While the first two dealt with the sustainable development, the Convention on Biological Diversity, opened for signature at the event, addressed the conservation and sustainable use of biodiversity.

In 2012, the city of Rio de Janeiro hosted the United Nations Conference again having as theme Sustainable Development (Rio + 20). This conference had as objective the renovation of global political commitments to sustainable development assumed in the Rio-92. One of the central themes of the conference was the Green Economy, which results in "improvement of human wellbeing and social equality while significantly reducing environmental risks and ecological scarcities" (UNEP 2011). To implement green economy are needed some "facilitators" provided by governments consisting mainly of incentives and subsidies policies, and international cooperation, like Brazil is doing. In the early 2000s, were established in the country programs that selected priority areas for investment to accelerate the concomitant social and economic development with environmental preservation. Among the priority sectors are biotechnology and, consequently, the national bioindustry that makes direct use of natural capital, especially biodiversity, developing through intense scientific research, numerous products, processes, and services. Some Brazilian laws have recently being changed to improve private innovation and bioindustry in the country. Some of policies and incentives provided by Brazilian government are papers (White and Green) and laws (Tax incentives; Innovation; Science, Technology, and Innovation framework; genetic resources access and traditional knowledge associated). These policies try to create a more sustainable and competitive development that could be used in other countries as public policies model.

2 Methodology

Given the Evidence-Based Practice, this work is characterized by an integrative systematic literature review (Whittemore and Knafl 2005) and a narrative type (Whittemore 2005) drawing a parallel between the development of the country and laws of the development of science, technology, and innovation, especially about biotechnology. For that were analyzed not only legislation, decrees, and subsidies, as well as human development. This parallel was drawn mainly to demonstrate the development of inequalities in the country and the generation of new ventures. A systematic review was chosen because it plans and critically evaluates the study (Broome 2000).

Brazil is positioned as the seventh world economy (World Bank 2015) and stands out as the wealth of natural resources and biological diversity due to its vast territory, linked to its geographical position. Brazil also hosts an amazing socio-biodiversity, with more than 200 indigenous people and local communities, e.g. quilombo and riverine people, rubber tappers and picturesque coastline, which hold a priceless collection of traditional knowledge on biodiversity (Brazil 2014a). In addition to its ecological importance, biodiversity has an enormous representation in the Brazilian economy (31 % of exports) (Brazil 2014a). Although the country has high natural and economic resources, it still faces a reality of social inequality and is in a duality between accelerated growth and sustainable use of its natural capital.

3 Brazil Geopolitics

Brazil is one of the BRICS members given its economic ascension. The BRICS consists of developing countries with outstanding economic power being: Brazil, Russia, India, China and South Africa. The historical context of the emergence of BRICS was the global financial crisis of 2007/2008, which allowed the de-globalization and the decreasing dependence of emerging countries to developed countries (Sally 2011 cited Asuelime and Jethro 2013).

Unlike the European Union that is an economic bloc, BRICS is the group of countries representing a model of economic growth which have global influence due to its broad natural resources, population, and territory (Radulescu et al. 2014). The BRICS' aim is to reduce the economic barriers between member countries, increase participation in international negotiations, besides favoring the constant economic and social growth. Therefore, new public policies and programs are created to develop capital and society mutually in these countries. Some of the challenges to be overcome are poverty reduction, curbing corruption, which creates a negative image of the member countries, and the quality of economic growth. However, the economic growth is focused without investing in education, technology and innovation (Radulescu et al. 2014). Recently, these regional and global socio-economic challenges have been recognized by member countries that, to change this reality, signed a memorandum establishing a strategic model for cooperation in Science, Technology & Innovation (ST&I) (Brazil 2015a).

4 Biotechnology and Sustainability

The Human Development Index (HDI) in Brazil grew significantly, going from 0.665 in 2000 to 0.730 in 2012 (Fome 2013). Although the HDI of the country is considered high, there are huge differences between regions. Table 1 shows the values for several indexes by region and it is noticeable that the southern regions are more populous, prosperous and have lower illiteracy rate than the northern regions (IBGE 2013, 2014). When comparing the postgraduate population, the difference between regions becomes more pronounced. According to an analysis made in 2010 by the Management and Strategic Studies Center, there were 10,705 doctors in Brazil in 2008, where the majority of doctors were concentrated in the southeast, especially in the state of São Paulo (45 % of Ph.D.) (CGEE 2010). Although the total number of postgraduate is relatively small compared to the total population, it should increase due to policies that prioritize the education and professionalization of Brazilians. A recent search on the web for biotech companies in Brazil showed 87 new firms in the last 5 years (Table 1), most of them in the Southeast region which is the fourth region in illiteracy, the second in GDP *per capita*, the first in population and GDP participation per cent.

Because they are more developed, South and Southeast are the regions exhibiting the lowest environmental degradation in the country (Pinto et al. 2014), which is in agreement with the assumption that the more developed is the region the more concern and environmental sensitivity it has. In the case of the green economy, these regions concentrate most of the projects. The State of São Paulo is the one that stands out with the largest and most diverse green market economy of the country (Investe SP 2015).

To overcome Brazilian inequalities, especially about the concentration of capital and intellectual resources in the South and Southeast, it is necessary new forms of development, such as those that allow green growth. Among the emerging technologies that aim and create ways to achieve this desired green perspective, there is

Table 1 Social and economical indicators in Brazil by region

Region	GDP per capita (R$)	Illiteracy	Population (%)	GDP participation (%)	Biotech firms
Brazil	22,645.86	8.7	100	100	87
North	14,179.48	10	8.5	5.3	0
Northeast	11,044.59	17.4	27.8	9.8	7
Midwest	29,843.65	6.7	7.5	13.5	2
Southeast	29,718.34	4.8	42	55.2	59
South	25,633.53	4.4	14.2	16.2	19

Biotechnology, which provides solutions and alternatives not only locally, but worldwide, reducing environmental stress caused by humans. Some of these solutions are biofuels, like alcohol, the primary and oldest example of Brazilian Clean Development Mechanisms (CDM). CDM practices reduce the emission of pollutants, carbon dioxide to the atmosphere and consequently greenhouse gasses.

4.1 Biofuels

The key biofuels produced in Brazil are biodiesel and ethanol, both first-generation productions. According to Statistical Review of World Energy, in 2013 the country was the second largest producer of biofuels, accounting for 24 % of world output, of which 10 % was biodiesel (BP 2014). This participation reflects the national obligatory addition of 7 % biodiesel to diesel (Law No. 13,033/14), and a previous government program launched in 2002, ProBiodiesel (Brazil 2004a) which stimulated the research, development, and production and ensured market for producers. The country has 50 producing biodiesel companies, using as primary raw material soybean oil, corresponding to 74.72 % of the manufacture of biodiesel (ANP 2015a).

Due to the oil crises in the 1970s, the Brazilian government launched the Pro-Alcohol program that aimed to reduce national dependence on fossil fuels. The program stimulated the production of ethanol and made the country one of the leaders in this market. The first-generation ethanol is critical for Brazilian energy matrix, used directly as a fuel as well as it is added to gasoline in a percentage that ranges from 18 to 27.5 % (Law 13.033/2014). Currently, there are 292 companies in Brazil producing this fuel (ANP 2015b).

Besides the first-generation biofuels, the country shows a high degree of technological expertise in the development of second-generation biofuels (Köhler and Walz 2014) and shows industrial production (Table 2) (Bacovsky et al. 2013). The

Table 2 Installed and planned companies that are producing second-generation biofuels in Brazil

Company	Start-up	Feedstock	Biofuel
Amyris Biomin	2010	Sugarcane	Diesel-type hydrocarbons
Amyris Paraiso	2012		
Amyris São Martinho	2013		
Amyris pilot and demonstration plant	2009		
JBS	2013	Bovine fat	Biodiesel
Fertibom; Bertim	2007	Bovine fat, vegetable oils	
Minerva biodiesel	2011		
Petrobras	2007	Sugarcane bagasse	Ethanol
Raizen	2014		
(Graal Bio) Granbio; CTC/São Manoel	2014	Sugarcane bagasse and straw	

Modified from Bacovsky et al. (2013)

possibility of use of sugarcane bagasse and straw (by-products of first-generation ethanol production) allows an increase in production capacity without increasing planted area, making the sugar-energy industry more profitable and sustainable.

Recently biofuels have reached the aviation sector, and Brazil is positioned at the forefront of the development of jet biofuel with partnerships and international cooperation, especially with the United States (Köhler and Walz 2014). Ethanol, biodiesel, and Kerosene are an alternative to fossil fuels (Cremonez et al. 2015; Moraes et al. 2014). Although still in early stages of development, jet biofuel is an achievement for the reduction of emissions of greenhouse gasses.

The biodiesel industry looks for new forms of production as the third-generation biofuels production to reduce the competition with the food industry, which are derived from microalgae and cyanobacteria biomass. These photosynthetic microorganisms can be used to extract lipid composition for biodiesel production, and the remaining biomass may be utilized in second-generation ethanol (Parmar et al. 2011).

To reduce CO_2 and toxic gasses emissions, the use of biofuels can bring economic stability to the national energy sector reducing dependence on fossil fuels. Also, it can provide a better income distribution and encourage the development of poor regions of the country. In the case of biodiesel production, the National Program for Production and Use of Biodiesel promotes tax breaks for companies that obtain raw material from family-based agriculture, with greater tax incentives in the poorest regions (Brazil 2015b).

5 Public Policies Affecting the Development of National Bioindustry

Due to the position of Brazil in the global scenario, there is a greater focus on its social and economic realities which contains disparities in the country. Brazil found in bioindustry a form of sustainable use of its natural capital while seeking economic growth. Public policies have been established for bioindustry in the form of papers, laws, decrees and regulations that together aim the technological development and innovation.

5.1 Green and White Papers

The Green Paper on Science, Technology, and Innovation presents a broad discussion about knowledge and innovation and its importance in accelerating the social and economic development process in the country. Moreover, it gives the general guidelines consistent with national needs established in the critical areas of social and economic impact to strengthen research and to promote technological development. Biotechnology is considered a strategic area in Brazil for its multidisciplinary aspect and its participation in several key sectors for the development of a nation, e.g. health, agriculture, environment, and various industrial processes. The Green Paper in all its content addresses biotechnology, pointing its importance and says that "the field of biotechnology will have a significant role in determining the competitiveness, economic development and quality of life" (Silva and Melo 2001). With these assumptions, the Paper presents a set of guidelines for the development of biotechnology and consequently the bio-industry. Among the guidelines are the training of specialized human resources; expansion of biotechnological knowledge; support tools for research development and the biotechnology industry; to encourage sustainable use of biodiversity by ensuring the preservation of the environment and preventing biopiracy; international cooperation; and prospective studies in the area for analysis of market trends.

Furthermore, the White Paper proposes the implementation of a national system for Science, Technology, and Innovation addressing, again, biotechnology as a priority area. Its discussion is focused on the development and consolidation of a stimulating and inducing environment for innovation to accelerate the development of the country. The primary thrust of this Paper is to increase private sector participation in research and investment in innovation. Such involvements also aim to promote regional integration of the country through the distribution of investment considering the potential of each region, providing them with a scientific and technological structure capable of supporting productive and socioeconomic development (Brazil 2002).

These Papers were the benchmark for the development of new and encouraging regulatory policies to research and development in both public and private sectors.

5.2 Innovation Law and Tax Incentives Law

To have a more efficient development of technology and innovation, Brazil adopted laws that manage and enable a favorable environment to innovation. These two laws, 10,973/04 and 11,196/05, also known as the Innovation Law and Tax Incentives Law or Law of Good, respectively, were the pioneers in the Brazilian legislation when dealing with innovation. In the Innovation Law fomentation is intended to cover the cost of innovation activity, including third-party services, personnel, patents, conservation of the immovable property for the area and so on. In exchange, the Tax Incentives Law supports the payment of researchers with titles of Masters and Doctors hired by the companies (FINEP 2010). These laws are a way to create incentives for cooperation between companies and universities, as well as tax incentives to firms that conduct Research and Development (R&D) and technological innovation in the country.

The Innovation Law is divided into seven chapters, in these are addressed collaboration between companies and Scientific and Technological Institution (STI); provision of financial resources in the form of subsidy for businesses and; the encouragement to the independent inventor as well as the percentage of participation in profits secured to the creator. The signing of contracts and agreements between STI's and companies allows conducting research together, sharing technology, empowering and enabling the autonomy of the country in several areas, respecting the exploitation and the right to use. The law also provides that the basic and technological research will be priority areas and encompasses the responsibility to encourage R&D for the Union, the Federal District and the municipalities, which may sign cooperation agreements with public or private entities for research projects execution. It provides incentive grants to the server, military or STI public employee during the implementation of activities related to innovation. The law mentions the need to sign a contract to define the ownership of intellectual property and participation in the search results, ensuring the parties involved in the contract the right to license. The percentage of participation of the creator in profits is at least 5 % and a maximum of 1/3 of earnings by STI. These economic profits can be earned through technology transfer agreements and licensing, as royalties, compensation or any financial benefits. Moreover, in some cases, the STI may assign the rights of the creation to the creator. Additionally, it is expected the assistance to private national nonprofit entities focused on research activities, as well as charges for operational and administrative expenses that may appear during the contract. The government researcher may be temporarily removed to provide collaboration with another STI. If the government researcher wants to start a company based on innovation, it is given a license up to 3 years, renewable for the same period (Brazil 2004b). On January 2016 was sectioned the new legal framework for science, technology, and innovation (Brazil 2016). The Law 13,243/2016 will facilitate the interaction between government, companies and STI's reducing bureaucracy. It will also allow professors under an exclusive dedication work system to get paid in the private sector and more time to spend in this activity (from 120 to 416 h/year or 8 h/week). The Union would finance and have social participation in companies, and businesses could have the intellectual property of the process or product resulting from research.

One of the sources of an economic subsidy for innovation in the country is the Financier of Studies and Projects (FINEP), a public company under the Ministry of Science, Technology and Innovation (MCTI). FINEP seeks to promote economic and social development by financing ST&I to increase the competitiveness of the business sector. Thereby, it provides a subsidy through the Economic Subsidy Program to companies that meet the prerequisites defined in notices. The grant is a policy adopted in developed countries and introduced in Brazil in 2006. This type of monetary support applies not refundable resources directly in companies, sharing the costs and risks of the economic activity. In the 2011 report about the profiles of companies benefited from the grant, FINEP conducted an assessment of the Brazilian market. Regarding the geographical distribution of businesses that received the subsidy (Table 3), there was a concentration of enterprises in the Southeast. One of the reasons that justify and proves the importance of Innovation Law is provided in its Article 27 "the prioritization of the less developed regions of the country and the Amazon in research areas that look for increasing human resources and technological capacity".

Also, the report shows that the companies benefited have focused projects in the areas of information technology and public security, which together count for 48 %

Table 3 Number of companies by region

Region	Number of companies
Northeast	61
North	10
Midwest	17
Southeast	288
South	124

Modified from FINEP (2011)

of the projects. Although it is small the involvement of biotechnology projects in the program (12 % of 695 projects), this participation will increase since the biotechnology sector remains a priority area in Brazil.

Moreover, the Tax Incentives Law creates the unique system of taxation and addresses the incentives for technological innovation. Tax benefits can be summarized as a deduction of up to 34 % in the Corporate Income Tax and Social Contribution on Net Profits; 50 % reduction in taxes on the purchase of machinery and equipment and; depreciation and accelerated amortization of such assets, which reduce taxation. For companies would use these incentives, the investment in R&D is required. In this Law, there is also the differentiated fomentation for the various regions of Brazil, where the subsidy amounts are up to 60 % for companies operating in the Northeast and Amazon, and up to 40 % for other regions. This differentiation of incentives is consistent with the guidelines of the Innovation Law discussed above (Brazil 2005).

5.3 Genetic Resources Access and Associated Traditional Knowledge Legislation

In 1992, a legislation to biodiversity access was validated through the Convention on Biological Diversity—CBD, one of the greatest advances achieved during the Rio-92 that Brazil became a signatory in 1994. Since the CBD genetic resources are State property, the exploitation and conservation of these resources must be governed by domestic environmental policies (UNCED 1992). From this date, many Bills have been elaborated, but only in 2001 the country honored legislation through Provisional Measure (PM). In 2000, it was made a controversial deal between the Social Organization Bioamazônia and the multinational Novartis Pharma giving to the company exclusive access to the Amazonian biodiversity and bringing only a few benefits to the country. Because of social pressure, the agreement was canceled, and a PM that regulates genetic resources access was prepared in a hurry. PM 2,052 from 2000, currently in effect under number 2,186-16/2001, gave to indigenous people and local communities ownership of their traditional knowledge associated (TKA), preventing unauthorized people to explore. Also, the PM ensures that the benefits from the use of TKA are fairly shared with the holder.

According to the PM 2,186-16/2001, benefit-sharing derived from the economic exploitation of products or processes developed from genetic resources or TKA should be conducted with an Agreement of Genetic Heritage Use and Sharing of Benefits. This agreement should be established between the institution authorized to make access and representatives of the TKA or the owner of the area where the genetic resource is. The sharing of benefits could occur as profit sharing, royalty payments, access and transfer of technology, training resources, among others, it should be fair and equitable and the validation of the Agreement depended on the approval of a deliberative agent (Brazil 2001).

The competent authority for resolutions and norms of PM was the Council of Genetic Patrimony Management (CGEN). However, this Council only started operations in April 2002, which created a situation of uncertainty about the possibility of doing some researchers in the country and difficulties regarding the exchange of biological material for scientific purposes, from June 2001 to April 2002. Furthermore, CGEN made the process of access to genetic resources bureaucratic, intricated and even unviable, locking all the research and development of Brazilian biodiversity given the rejection of applications for research and the long time required to access grant analysis, which in some cases can reach years. From 2003 to 2013, only 37 % of cases were resolved among all the requests sent to the CGEN.

It is observed that the PM has created an obstacle to the development of national bioindustry, penalizing and fining companies in many millions of dollars. Also, there was a disagreement between the government and the scientific community of what would be "bioprospecting". There was a professional disqualification in the preparation of the PM and the CGEN management which compromised the development of Brazilian bioindustry. Many of Brazilian researchers did not know this PM and ended up being fined for breaking the law. In legal terms, the PM was unconstitutional since it would have a maximum period of 90 days for preparation and approval of a law by the Congress. This PM was valid until November 2015 when new legislation became effective.

The new law, 13,123/2015, brings back many concepts and problems solved in MP, mainly related to TKA (Brazil 2015c). The law keeps CGEN but makes it more representative with 40 % of civil society participation and 60 % of public sector involvement. It facilitates the use of the genetic resource, mainly microorganisms, when simplifies registration and removes the prior consent. But when dealing with TKA, it becomes a problem. There is the limitation of pecuniary benefit-sharing from 0.1 to 1 %, without knowing the value of the final product and just in the original product. All derivatives would be excluded from payment. The benefit-sharing can be non-monetary what is aligned with the innovation law and tax incentives law, making a breakthrough in science, technology, and innovation. It is given amnesty for companies fined previously by CGEN facilitating, in some cases, the human exploitation, especially in TKA. Another problem that can occur is the exemption from benefit-sharing for small and micro enterprises. If it encourages the development of new businesses and can make companies re-invest in the national natural capital providing the strengthening of local bioindustry, otherwise it encourages big business to use these small firms just not to pay the proper value, mainly when related to TKA.

6 Conclusion

Although Brazil is one of the countries with the greatest biodiversity, this biodiversity does not generate income as a high-tech product, still being an export product. Also, there is a significant social and economic difference in the country that implies a further development of bioindustry in the southeastern and southern regions, respectively. Policies relating to the least favored regions are necessary for the development and reduction of inequalities in the country. Papers Green and White were the guidings of public policies directing the priority regions and areas, among which are biotechnology, from 2000s years. These papers were the base for the creation of public policies that would encourage national development through innovation. In contrast, the system of genetic resources access, unconstitutional and rushed, inhibited the development of a bioindustry for 15 years. Recent changes in legislation with Science, Technology and Innovation framework and genetic resources access and Traditional Knowledge Associated law promise to reduce the remained problems to improve science, technology and innovation and partnerships between private and public sectors. Brazilian mistakes and successes in legislation should be taken by countries that are promoting local innovation and believe that biotechnology is one of the priorities areas, particularly developing countries.

References

ANP—Agência Nacional de Petróleo, Gás Natural e Biocombustíveis (2015a) Boletim Mensal do biodiesel (Fevereiro). http://www.anp.gov.br/?pg=74914&m=&t1=&t2=&t3=&t4=&ar=&ps=&1430403516911. Accessed 20 Mar 2015

ANP—Agência Nacional de Petróleo, Gás Natural e Biocombustíveis (2015b) Boletim do etanol n ° 03/2015. http://www.anp.gov.br/?pg=74078&m=&t1=&t2=&t3=&t4=&ar=&ps=&1430403443938. Accessed 20 Mar 2015

Asuelime L, Jethro L (2013) Building with BRICS: a burgeoning dynamic for South Africa. JRPGD 3(3):29–37

BP (2014) BP Statistical review of world energy. http://www.bp.com/content/dam/bp/pdf/Energy-economics/statistical-review-2014/BP-statistical-review-of-world-energy-2014-full-report.pdf. Accessed 20 Mar 2015

Brazil (2001) Provisional Measure No. 2,186-16, de agosto de 2001. http://www.planalto.gov.br/ccivil_03/mpv/2186-16.htm. Accessed 20 Nov 2014

Brazil—Ministério da Ciência e Tecnologia (2002) Livro Branco: Ciência, Tecnologia e Inovação, Ministério da Ciência e Tecnologia, Brasília. http://www.cgee.org.br/arquivos/livro_branco_cti.pdf. Accessed 25 Oct 2014

Brazil—Ministério da Ciência e Tecnologia (2004a) Comunicação Nacional Inicial do Brasil à Convenção das Nações Unidas Sobre Mudança do Clima, Ministério da Ciência e tecnologia. http://www.mct.gov.br/upd_blob/0005/5586.pdf. Accessed 28 Apr 2015

Brazil (2004b) Law No. 10,973, de 02 de dezembro de 2004. http://www.planalto.gov.br/ccivil_03/_ato2004-2006/2004/lei/l10.973.htm. Accessed 12 Nov 2014

Brazil (2005) Law No. 11,196, de 24 de novembro de 2005. http://www.planalto.gov.br/ccivil_03/_ato2004-2006/2005/lei/l11196.htm. Accessed 18 Nov 2014

Brazil—Ministério do Desenvolvimento Social e Combate à Fome (2013) Estudo Técnico No. 06/2013—Resenha crítica sobre a situação do Brasil no Relatório de Desenvolvimento Humano 2013, intitulado "A Ascensão do Sul: Progresso Humano em um mundo diversificado". Brasília, DF. http://acervodigital.mds.gov.br/xmlui/bitstream/handle/123456789/299/ETEC_06_%202013%20Resenha%20cr%C3%ADtica.pdf?sequence=1. Accessed 02 Apr 2015

Brazil—Ministério do Meio Ambiente (2014a) Biodiversidade Brasileira. http://www.mma.gov.br/biodiversidade/biodiversidade-brasileira. Accessed 02 Apr 2015

Brazil (2015a) II Reunião de Ministros de Ciência, Tecnologia e Inovação do BRICS. http://www.itamaraty.gov.br/index.php?option=com_content&view=article&id=8342:ii-reuniao-de-ministros-de-ciencia-tecnologia-e-inovacao-do-brics-documentos-aprovados-brasilia-18-de-marco-de-2015&catid=42&lang=pt-BR&Itemid=280/. Accessed 24 Mar 2015

Brazil—Ministério do Desenvolvimento Agrário (2015b) Programa Nacional de Produção e Uso do Biodiesel (PNPB). http://www.mda.gov.br/sitemda/secretaria/saf-biodiesel/sobre-o-programa. Accessed 27 Apr 2015

Brazil (2015c) Law No. 13,123, de 20 de maio de 2015. http://www.planalto.gov.br/ccivil_03/_Ato2015-2018/2015/Lei/L13123.htm. Accessed 14 Jan 2016

Brazil (2016) Law No. 13,243, de 11 de janeiro de 2016. http://legis.senado.leg.br/legislacao/ListaTextoIntegral.action?id=250114&norma=269577. Accessed 14 Jan 2016

Bacovsky D, Ludwiczek N, Ognissanto M, Wörgetter M (2013) Status of advanced biofuels demonstration facilities in 2012: a report to IEA bioenergy. http://task39.sites.olt.ubc.ca/files/2013/12/2013_Bacovsky_Status-of-Advanced-Biofuels-Demonstration-Facilities-in-2012.pdf. Accessed 18 Nov 2014

Broome ME (2000) Integrative literature reviews for the development of concepts. In: Rodgers BL, Knafl KA (eds) Concept development in nursing, 2nd edn. W.B. Saunders Company, Philadelphia, pp 231–250

CGEE—Centro de Gestão e Estudos Estratégicos (2010) Doutores 2010: Estudos da demografia da base técnico-científica brasileira. Brasília, DF. http://www.cgee.org.br/publicacoes/doutores.php. Accessed 30 Mar 2015

Cremonez PA, Feroldi M, Oliveira CJ, Teleken JG, Alves HJ, Sampaio SC (2015) Environmental, economic and social impact of aviation biofuel production in Brazil. New Biotechnol 32:263–271

FINEP—Financiadora de Estudos e Projetos (2010) Manual de Programa Subvenção Econômica à Inovação Nacional. http://download.finep.gov.br//manuais/manualSubvencao2010.pdf. Accessed 02 Apr 2015

FINEP—Financiadora de Estudos e Projetos (2011) Perfil das Empresas Apoiadas pelo Programa de Subvenção Econômica 2006 a 2009. Rio de Janeiro. http://www.finep.gov.br/programas/subvencao_perfil_2006_2009.pdf. Accessed 01 Apr 2015

IBGE—Instituto Brasileiro de Geografia e Estatística (2013) Síntese de Indicadores Sociais: Uma análise das condições de vida da população brasileira. http://biblioteca.ibge.gov.br/visualizacao/livros/liv66777.pdf. Accessed 25 Mar 2015

IBGE—Instituto Brasileiro de Geografia e Estatística (2014) Contas Regionais 2012. http://loja.ibge.gov.br/contas-regionais-do-brasil-2012.html. Accessed 27 Mar 2015

Investe SP (2015) Economia verde. http://www.investe.sp.gov.br/setores-de-negocios/economia-verde/. Accessed 23 Mar 2015

Köhler J, Walz R (2014) Lead markets in 2nd generation biofuels for aviation: a comparison of Germany, Brazil and the USA. EIST 10:59–76

Moraes MAFD, Nassar AM, Moura P, Leal RLV, Cortez LAB (2014) Jet biofuels in Brazil: sustainability challenges. Renew Sustain Energ Rev 40:716–726

Parmar A, Singh NK, Pandey A, Gnansounou E, Madamwar D (2011) Cyanobacteria and microalgae: a positive prospect for biofuels. Bioresour Technol 102:10163–10172
Pinto NGM, Coronel DA, Conte BP (2014) Mapping of environmental degradation in regions and states of Brazil. WSEAS Trans Bus Econ 11:442–453
Radulescu IG, Panait M, Voica C (2014) BRICS countries challenge to the world economy new trends. Proc Econ Financ 8:605–613
Silva CG, Melo LCP (2001) Ciência, tecnologia e inovação: desafio para a sociedade brasileira—livro verde, Ministério de Ciência e Tecnologia. http://livroaberto.ibict.br/handle/1/859. Accesed 16 Oct 2014
UNCED—United Nations Conference on Environment and Development (1992) Convention on biological diversity. https://www.cbd.int/doc/legal/cbd-en.pdf. Accessed 03 Apr 2015
UNEP—United Nations Environment Programme (2011) Towards a green economy: pathways to sustainable development and poverty eradication—a synthesis for policy makers. http://www.unep.org/greeneconomy/Portals/88/documents/ger/GER_synthesis_en.pdf. Accessed 20 Mar 2015
World Bank (2015) Brazil overview. http://www.worldbank.org/pt/country/brazil/overview. Accessed 02 Apr 2015
Whittemore R (2005) Combining evidence in nursing research: methods and implications. Nurs Res 54:56–62
Whittemore R, Knafl KA (2005) The integrative review: updated methodology. J Adv Nurs 52:546–553

Authors Biography

Elias Silva Gallina Undergraduate in Biotechnology at Universidade Federal Rural do Semi-Árido, RN, BR.

Lianne de Oliveira Cruz Undergraduate in Biotechnology at Universidade Federal Rural do Semi-Árido, RN, BR.

Fernanda Matias Ph.D. in Biotechnology from University of Sao Paulo, SP, BR, is Adjunct Professor at Universidade Federal Rural do Semi-Árido, RN, BR. She is a researcher at Animal Sciences Department, where she also teaches Nanobiotechnology, Biorreactors and Biofactories and Legal frameworks on biotechnology and patents. Her main research subjects include innovation, legal framework and analysis of companies involved in Biotechnology in Brazil.

Sustainable and Economical Production of Biocellulose from Agricultural Wastes in Reducing Global Warming and Preservation of the Forestry

Ida Idayu Muhamad, Norhayati Pa'e and Khairul Azly Zahan

Abstract

Preservation of the forestry, particularly trees, is essential in managing global warming. However, excessive use of trees for cellulose-based products has continuously depleted world's forest resources. Nowadays, around 14 % of deforestation is caused by logging for the production of wood cellulose-based products. Many scientists predicted that in 2030, only 10 % of the mature tropical forest will remain. In the era of declining forest resources and expansion of industrialization, it is a worthwhile effort to consider an alternative source of plant cellulose i.e. bacterial cellulose or biocellulose. Biocellulose produced by bacterial fermentation method using various substrates has been proven to be a remarkably versatile biomaterial for a variety of cellulose-based products. Interestingly, the highest cellulose worldwide demands are in the pharmaceutical sector with an annual demand of 30,000 tons. For Malaysia, as a country with

I.I. Muhamad (✉) · N. Pa'e · K.A. Zahan
Department of Bioprocess and Polymer Engineering, Faculty of Chemical and Energy Engineering, Universiti Teknologi Malaysia, 81310 Johor Bahru, Johor, Malaysia
e-mail: idayu@cheme.utm.my

N. Pa'e
e-mail: hayatipae@gmail.com

K.A. Zahan
e-mail: khairulazly@unikl.edu.my

I.I. Muhamad
IJN-UTMCardio Engineering Centre, V01 FBME, Universiti Teknologi Malaysia, 81310 Johor Bahru, Johor, Malaysia

K.A. Zahan
Malaysian Institute of Chemical and Bioengineering Technology, Universiti Kuala Lumpur, Vendor City Taboh Naning, 78000 Alor Gajah, Melaka, Malaysia

W. Leal Filho et al. (eds.), *Sustainable Economic Development*,
World Sustainability Series, DOI 10.1007/978-3-319-45081-0_9

vast agricultural areas, the use of generated agricultural wastes such as palm oil mill effluent, pineapple and sugarcane as a fermentation medium can not only avoid environmental pollution but also reduce the production cost. In fact, the use of agricultural wastes as a fermentation medium has proven to reduce 20 % of carbon sources supply for fermentation. In this respect, biocellulose production can play an important role in reducing global warming and preservation of nature.

Keywords
Preservation of forestry • Reducing global warming • Cellulose-based products • Biocellulose • Fermentation • Use of agricultural wastes

1 Introduction

Natural polymers such as cellulose have attracted much attention due to the effect of environmental pollution of the synthetic polymer (Pei et al. 2013). Cellulose and its derivatives such as cellulose acetate are known for their applicability of being utilized as a membrane. The development of biopolymer started to gain attention due to its high potential applications in many fields, including bioseparation, tissue engineering and food processing (Liang et al. 2007). However, the use of trees or its cellulose for the production of paper, biofuels and construction materials has continuously depleted forest resources (deforestation) which lead to global warming problems. In this case, it is important to consider a new alternative for plant cellulose that is bacterial cellulose.

Bacterial cellulose produced by bacteria is the best choice as it can be produced within a few days of fermentation while a tree needs an average of more than 30 years to achieve full growth. Thus, the use of bacterial cellulose as the main ingredient in cellulose-based products will be environmentally friendly, green and sustainable.

In recent times, bacterial cellulose started to be used to substitute plant derive cellulose because of its excellent and promising properties (Ashori et al. 2012). These were proven by several researchers that successfully produced bacterial cellulose composite membranes such as cellulose acetate membranes reinforced with bacterial cellulose sheet (Gindl and Keckes 2004). Sokolnicki et al. (2006) suggested several novel uses for bacterial cellulose, which may include membrane systems for tissue growth, cell-based therapies and drug delivery.

Bacterial cellulose can form in many media such as coconut water, fruit juice, and domestic waste product or a nutrient medium. Traditionally, bacterial cellulose has been produced in static trays by using coconut water, sugarcane, pineapple juice and others. In 2008, several mechanisms had been initiated to produce bacterial cellulose using agricultural waste (Kongruang 2008). These mechanisms can be seen as an effort to reduce production cost, at the same time support zero waste campaign for a sustainable environment.

2 Deforestation

Deforestation is the permanent destruction of forests to make the land available for other uses. It is considered to be one of the contributing factors to global climate changes. There are many causes of deforestation. The WWF reports that half of the trees illegally removed from forests are used as fuel. Other common reasons are the harvesting of timber in order to produce commercial items such as paper, furniture and homes as well as to create ingredients that are highly prized consumer items or cellulose derivatives products such as the microcrystalline cellulose (MCC) and carboxymethyl cellulose (CMC).

Starting from the year 1852, global deforestation has sharply accelerated (Roth 1983). It has been estimated that about half of the Earth's mature tropical forests that covered the planet have been destroyed compared to the total area found in 1947 (Wilson 2002). Some scientists have predicted that by 2030, there will only be 10 % of the forestry remaining, with another 10 % in a degraded condition and 80 % will have been lost, and with them, hundreds of thousands of irreplaceable species (Ron 2006).

As of 2007, less than 1 % of Haiti's forests remained. Mexico, India, the Philippines, Indonesia, Thailand, Burma, Malaysia, Bangladesh, China, Sri Lanka, Laos, Nigeria, the Democratic Republic of the Congo, Liberia, Guinea, Ghana and the Ivory Coast have all lost large areas of their rainforests. Several countries, notably Brazil, have declared their deforestation a national emergency (John 2005). The World Wildlife Fund's eco-region project catalogues habitat types throughout the world, including habitat loss such as deforestation, showing, for example, that even in the rich forests among the many parts of Canada (e.g.: the Mid-Continental Canadian forests of the prairie provinces) half of the forest cover has been lost or altered. In 2011, Conservation International listed the top 10 most endangered forests (Table 1), characterized by having lost 90 % or more of their original habitat, and each harbouring at least 1500 endemic plant species (Bergen 2011).

Table 1 Top 10 most endangered forests in 2011 (Bergen 2011)

Endangered forest	Region	Remaining habitat (%)
Indo-Burma	Asia-Pacific	5
New Caledonia	Asia-Pacific	5
Sundaland	Asia-Pacific	5
Philippines	Asia-Pacific	7
Atlantic Forest	South America	8
Mountains of Southwest China	Asia-Pacific	8
California Floristic Province	North America	10
Coastal Forests of Eastern Africa	Africa	10
Madagascar and Indian Ocean Islands	Africa	10
Eastern Afromontane	Africa	11

2.1 Effect of Deforestation

Deforestation is a threat to life worldwide. Deforestation may have profound effects on global warming, as well as the climate change, and is often cited as one of the major causes of an enhanced greenhouse effect. Approximately 15–20 % of global greenhouse gas emissions are caused by degradation and deforestation. Trees play a huge role in the carbon cycle. They convert the CO_2 in the air to oxygen through the process of photosynthesis, and consequently, they are viewed as a natural regulator of carbon dioxide. The existence of a larger amount of trees will no doubt lead to the reduced presence of carbon dioxide in the atmosphere and the increased existence of oxygen. Unfortunately, deforestation is preventing this process from being fully accomplished, and with half of the Earth's forests gone in addition to the four million trees being cut down each year just for the production of paper, the amount of carbon dioxide is progressively rising. With more carbon dioxide in the atmosphere, more of the sun's radiation is being reflected back to Earth instead of being released into space, and this is causing the planet's average temperature to rise. Accordingly, deforestation is a major issue when it comes to global warming. According to data from NASA, the global temperature in 2013 averaged to about 58.3 °F (14.6 °C), roughly a degree warmer than the twentieth-century average while the amount of heat-trapping carbon dioxide (CO_2) in the atmosphere due to emissions from burning fossil fuels and forests has increased, peaking at 400 parts per million in 2013, as shown in Fig. 1 (Marijnissen et al. 2004).

According to the United Nations Framework Convention on Climate Change (UNFCCC) secretariat, logging contributed 14 % of deforestation (UNFCCC 2007). In the European Union, 428 million m^3 of timbers were harvested, including the new member states, (EU27) in 2006. Furthermore, 163 million m^3 of raw woods

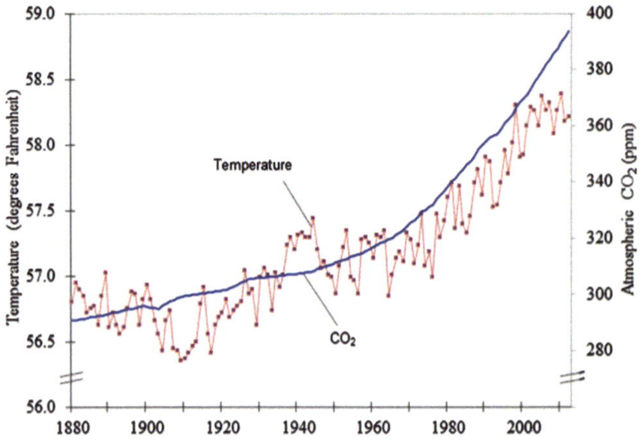

Fig. 1 Average global temperature and atmospheric carbon dioxide concentration from 1880 to 2013. *Source* NASA GISS; NOAA ESRL; Worldwatch (Marijnissen et al. 2004)

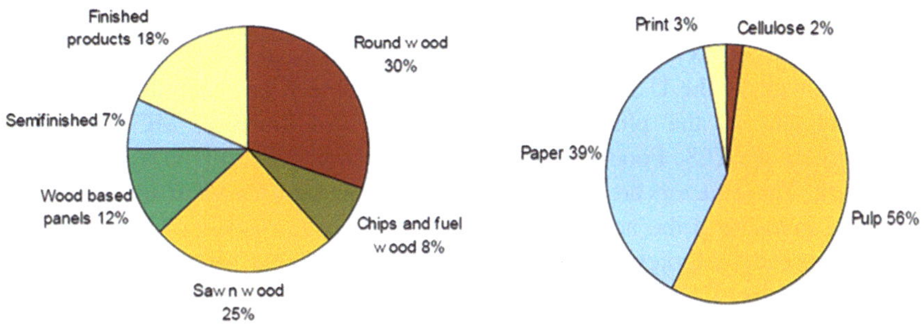

Fig. 2 Tree-based products in the European Union in 2006 (UNECE 2007)

are required to manufacture the wood-based products imported from non-EU countries as the wood products represent 57 % and pulp and paper 43 % of total wood used (UNECE and FAO 2007) (Fig. 2).

3 Cellulose-based Products

Forests and forest-related goods have become important raw materials and sources for all developing and developed countries. Although the production of wood-based goods such as houses, furniture, building material and bio-plastics usually requires far less energy or oil input than the processing of steel, metal, concrete and plastic products for similar purposes, it can cause adverse environmental implications. In particular, the creation of a large new global market for wood for bioenergy could increase overall pressures on forests as well as on other ecosystems (Ernsting 2012).

Cellulose is an organic compound with the formula $(C_6H_{10}O_5)_n$. Cellulose is mainly used to produce paperboard and paper. Smaller quantities are converted into a wide variety of derivative products such as cellophane and rayon. Cellulose fibers are fibers made with ether or esters of cellulose, which can be obtained from the bark, wood or leaves of plants, or from a plant-based material. Besides cellulose, these fibers are the compound of hemicellulose and lignin, and different percentages of these components handle different mechanical properties observed. The main applications of cellulose fibers are in the textile industry as a chemical filter, food industry as thickeners and binders, in the paper industry and fiber-reinforcement composite.

3.1 Microcrystalline Cellulose

Microcrystalline cellulose (MCC) is a term for refined wood pulp and is used as a texturizer, anti-caking agent, fat substitute, emulsifier, extender, and bulking agent in food production (Kirk 1993). The most common form is used in vitamin supplements or tablets. It is also used in plaque assays for counting viruses, as an alternative to carboxymethyl cellulose (Matrosovich et al. 2006). MCC has also been approved by the U.S. Pharmacopeia Convention. It has been deemed to be a safe excipient in the pharmaceutical industry. Additionally, MCC has been approved by the U.S. Food and Drug Administration (U.S. FDA) as a safe food ingredient. These factors are expected to drive the demand for MCC in the food and beverage industry in the next few years.

Transparency Market Research has released a new market report titled "*Microcrystalline Cellulose (MCC) Market—Global Industry Analysis, Size, Share, Growth, Trends and Forecast 2014–2020*". According to the report, the global MCC market was valued at US$ 632.9 million in 2013 and is projected to reach US $ 936.3 million by 2020, expanding at a CAGR of 5.8 % from 2014 to 2020. Demand growth is centred in developing the Asia markets, as markets in North America, Europe, and Japan are mature. The pharmaceutical field was the largest end-user segment of the global MCC market in 2013, owing to the utility of MCC as a binding agent in tablets and other medical formulations. In terms of volume, the pharmaceutical segment accounted for more than 35 % of the global MCC market in 2013.

Commercially available MCC is derived from both *gymnosperms* (generally conifers) and other softwoods, and from hardwood dicotyledons. These woods differ considerably in chemical composition (proportions of cellulose, hemicelluloses, and lignin) and structural organization which affect the composition of the α-cellulose extracted as well as the composition and crystallinity of MCC that is finally produced (Landin et al. 1993). Besides the wood pulp as a source of cellulose and its derivatives, the purified cotton linters obtained from *Gossypium* species are also a common source (Trease and Evans 1989).

3.2 Cellulose Acetate

In the year 1865, cellulose acetate was first generated from the reaction of cellulose and acetic anhydride that formed a cellulose ester. Wood pulp had been widely used as the main source for cellulose acetate production. Cellulose acetate has a series of unique combination of properties that make it suitable for many applications. These properties include high transparency, good mechanical strength with excellent machinability, chemical resistance and compatibility with human skin. Different processing methods will produce cellulose acetate with different properties (Fischer et al. 2008).

In the industrial sectors, cellulose acetate is used in textiles industries, as wound dressing, membranes and filter media. Moreover, cellulose acetate fibers are used commercially in textile industries due to its characteristics that make it comfortable, breathable and easy to adsorb dye. Cellulose acetate is also used as cigarette filters. It was estimated to be the fastest growing application market for cellulose acetate. Global demand is expected to be driven by growing consumption of cigarettes in developing countries, demand for longer cigarette filters and shift towards cellulose acetate tow filters from polypropylene cigarette filters (Cellulose Acetate Market—Global Industry Analysis, Size, Share, Growth, Trends and Forecast 2013–2019).

3.3 Hydroxyethyl Cellulose

Many polymer researchers are of the opinion that polymer chemistry originated through the characterization of cellulose. Water soluble cellulose ethers, such as methylcellulose (MC), hydroxypropyl methylcellulose (HPMC), hydroxypropyl cellulose (HPC) and hydroxyethyl cellulose (HEC) are effective as hydrophilic matrix-forming components (Kamel et al. 2008). Hydroxyethyl cellulose is a gelling and thickening agent. It is widely used in cosmetics, cleaning solutions, and other household products. In solid dosage forms, they are primarily used as tablet binders, film-coating agents, and controlled release matrix formers.

Hydroxyethyl cellulose (HEC) is somewhat sensitive to water and permeable to water vapor (Mitchell et al. 1972). According to the authors, in applications where these properties are detrimental, it is customary to coat films with micro-thin layers of transparent, hydrophobic film formers such as Saran-type vinylidene and vinyl chloride copolymers or cellulose nitrate. Coated films of this type have been successful and are being produced in large quantities. Water barrier coated HEC films have a slight edge in appearance, stability and shelf life over coated cellophane films.

Hydroxyethyl cellulose film is manufactured by continuously casting a film, relatively wide stream of a caustic solution of HEC into an acidic coagulating bath which immediately neutralizes the caustic solution and coagulates a thin gel film of HEC that is continuously removed from the coagulating bath for further treatment. A typical HEC casting solution may contain 9 % of HEC and 5.6 % by weight NaOH (Mitchell et al. 1972). The HEC gel film removed from the coagulating bath is washed with water to removed residual acid and salts.

3.4 Cellulosic Ethanol

Recently, plant cellulose had been investigated as an alternative for fuel source through conversion of cellulose biofuel such as cellulosic ethanol. In this case, lignocelluloses from the plant will be utilized as a raw material for the production of biofuel. The methods to produce cellulosic ethanol include cellulolysis and gasification. Intensive studies have been conducted to develop an economical method to convert cellulose and hemicelluloses into biofuel. In 2007, starches and sugars were the main raw material for ethanol production. The trend started to change when cellulose material from wood, straw and much of the structure of plants were used for cellulosic ethanol production. Since cellulose cannot be digested by humans, the production of cellulosic ethanol does not compete with the production of food.

3.5 Cellulose and Deforestation

The use of alternative non-wood sources of fiber in preparation of pulp for industrial applications has received substantial attention. The most important reason for the continuous increase in this area is a decrease in wood availability with the increasing demand for market pulp in some rapidly developing countries in Asia, Africa, and Latin America (Frank et al. 2009). In addition, the use of cellulose crops from wood and cotton give direct impact on the Earth's carbon cycle as it used up forest resources. Therefore, bacterial cellulose or biocellulose can be used as an alternative for plant cellulose.

4 Bacterial Cellulose as Alternative for Cellulose

Bacterial cellulose or biocellulose is produced by bacteria from the species of *Aerobacter, Acetobacter, Achromobacter, Agrobacterium, Alacaligenes, Azotobacter, Pseudomonas, Rhizobium* and *Sarcina*. However, only the *Acetobacter* species produce enough cellulose to justify commercial interest. The most extensively studied member of the *Acetobacter* species is *A. xylinus*, formerly known as *A. xylinum*. Bacterial cellulose has finer structure compared to plant cellulose. This biocellulose is also relatively pure and does not contain lignin, as well as the hemicelluloses associated with wood, that need to be removed and can be grown to virtually any shape. In industries, bacterial cellulose was produced for products such as dessert, wound dressing, high strength paper and diet foods.

The industrial interest in bacterial cellulose is based on the potential use of the biopolymer in more specialized applications, in which the price per weight unit of the material is not the most critical parameter. Such applications are thus potentially possible in those cases in which the specialized requirements are correlated with the differences in the properties of bacterial cellulose and cellulose readily available from plant sources. In practice, this means that the industrialist, for instance, might take advantage of the use of bacterial cellulose in cases in which chemical purity, mechanical strength, degree of crystallinity, or adsorptive capacity are particularly important. Thus, the material in the above conditions can be employed in the production of cellulosic derivatives such as MCC due to the fact that in the production of these cellulosic derivatives, cellulose must have the highest purity possible. The production of cellulose derivatives significantly broadens the application of cellulose by modifying the polymer chemically in terms of hydrophobicity, processability, and solubility (Rafael et al. 2011).

American Chemical Society in Science Daily on February 2007 reported that biotechnology's next high-value product could be bacterial cellulose (Science Daily, 2007). The unique properties of bacterial cellulose make it suitable to be used in different fields. Many studies have been conducted which utilizes bacterial cellulose, especially in the medical field. R. Malcolm Brown Jr from Poland is one of the most famous researchers that studied about bacterial cellulose properties and used it for new purposes. Some of his significant findings are to use bacterial cellulose as wound dressings and as an electronic display paper (Shah and Brown 2005; Czaja et al. 2006). Rafael et al. (2011) have studied on the feasibility of producing MCC from a bacterial cellulose source. The results showed that the MCC produced from bacterial cellulose has the characteristics of commercial MCC produced from wood pulp (Rafael et al. 2011).

However, there are some issues that prevent larger scale commercialization such as the high price of substrates, low volumetric yields and also a lack of large-scale production capacity. The current study was conducted to enhance the production of bacterial cellulose by designing a Rotary Discs Reactor (RDR) which can also promote a sustainable and economical production of bacterial cellulose from agricultural wastes which would then encourage the reduction of global warming and the increase of forestry preservation.

5 Economic and Sustainable Production of Bacterial Cellulose from Agricultural Wastes Using Rotary Discs Reactor (RDR)

The continuous demands of plant cellulose in various uses such as paper and textile industries can lead to the depletion in the number of plants on Earth. Consequently, it can cause many environmental problems such as global warming. Thus, use of bacterial cellulose can reduce the dependency on plant cellulose. To ensure favourable production yields and controlled costs, attention needs to be given to the species and genetic modification of the bacteria used, types of medium, composition and the type of reactor for the production process (Shi et al. 2014).

5.1 Production of Bacterial Cellulose Using RDR

To date, the most usual method for production of bacterial cellulose is static fermentation. In this method, pellicles of bacterial cellulose are formed on the surface of the static culture. However, there are some issues that prevented larger scale production (Bae and Shoda 2005; Norhayati et al. 2011). Static cultivation methods also produce low volumetric yield. It requires a long fermentation period and lots of equipment as well as a large space to produce the bacterial cellulose. This situation explains why large-scale production of bacterial cellulose is lacking. Hence, an alternative method to overcome this problem by using a better production method such as Rotary Discs Reactor (RDR) is necessary.

The RDR uses the concept of rotating biological contactor that exposed bacteria to the air for better aeration. RDR consists of a series of discs that are mounted on the shaft. The shaft is connected to a driven motor that can rotate the shaft with the discs. In brief, the discs are placed in a horizontal trough that contains a biological medium in which at least a sufficient portion of the discs is submerged.

Subsequently, the discs will then alternately soak the organisms in nutrient medium and expose them to the air (Norhayati et al. 2011).

5.2 Agricultural Waste as Fermentation Medium for *Acetobacter xylinum*

Traditional sources of carbon for bacterial cellulose fermentation are sugars such as glucose, fructose and sucrose (Bae and Shoda 2005). The cultivation medium for bacterial cellulose production mainly consists of glucose and sucrose. Different carbon sources provided to the medium could lead to a different yield of bacterial cellulose production. The common medium used for bacterial cellulose production is known as defined medium, which contains a mix of chemicals and carbohydrates. These types of medium are high in cost since they utilize many types of chemicals in their preparation and will subsequently be sold at a high cost rate (Tsuchida and Yoshaniga 1997).

The effective culture medium to produce bacterial cellulose from cheaper carbon sources by *Acetobacter xylinum* was examined. More recently, unconventional feedstocks from renewable resources and waste streams have been investigated (Zeng et al. 2011). These include fruit juices (Hungund et al. 2013; Kurosumi et al. 2009), sugar cane molasses (Keshk and Sameshima 2006), sweet potato pulp (Shigematsu et al. 2005), rotten apple (Gupta et al. 2010), and maple syrup (Zeng et al. 2011). Additionally, Hungund and Gupta (2010) has studied the production of bacterial cellulose from various fruits juice, where it was concluded that fruit juice alone as a carbon source is capable of producing high yield of bacterial cellulose instead of using high cost medium. Table 2 shows the various medium used for fermentation of bacterial cellulose.

Pineapple is the leading edible manner of family *Bromeliaceae*, which embraces about 2000 species. It is known botanically as *Ananas comosus*. It is native to the southern part of Brazil, and Paraguay. It is found in tropical climate countries such

Table 2 Production of bacterial cellulose using different medium

Medium	Main carbon source	Fermentation type	pH	Yield (g/l)	Researcher
Palm oil mill effluent (POME)	Sucrose	Static fermentation	5.5	3.94	Yusof (2013)
Pineapple waste	Sucrose	Static fermentation	5.5	3.44	Junaidi and Muhammad (2012)
Maple syrup	Fructose	Rotary shaker (135 rpm)	–	2.76	Zeng et al. (2011)
Shigeru Yamanaka medium	Sucrose	Rotary disc reactor (7 rpm)	5.0	5.41	Norhayati et al. (2011)
Pineapple peel juice	Glucose	Static fermentation	3.5	2.80	Castro et al. (2011)
Sugar cane juice	Fructose	Static fermentation	3.5	2.10	Castro et al. (2011)

as Malaysia. Pineapple is eaten fresh or canned and is available as juice. It is used in desserts, salads, as a complement to meat dishes and in fruit cocktail. While sweet, it is known for its high acid content (Abdullah 2007). Krueger et al. (1992) reported that major constituents of fresh pineapple juice are glucose, fructose, sucrose, citric acid, malic acid and mineral potassium. The dominant sugar is sucrose while fructose is slightly higher than glucose.

The processing of pineapple has made the fruit well known around the world. As reported by Rosma and Ooi (2006), only about 20 % from the whole pineapple is used while 80 % from that is the waste. Pineapple wastes consist of pineapple flesh, peeled skin and pineapple crown. The pineapple wastes generated by pineapple canning industries that are located in tropical regions such as Malaysia, Thailand and Indonesia produce a large quantity of solid and liquid wastes. It was estimated that more than 1,651,672 tonnes of pineapple waste are generated by canning industries worldwide, which are 50 % liquid waste and 50 % solid waste, respectively, each year (Sasaki et al. 1991; Abdullah 2007). According to Malaysian Pineapple Industry Board, total production of pineapple fruit in 2010 showed an increase from the previous year, which is from 114,958 to 127,415 metric tons per year. This increase shows that the production items from the source of pineapple fruit are growing from year to year. If these wastes are discharged into the environment untreated, they could cause serious environmental problems.

Nevertheless, even with the many cases of pollution and hazard that result from food processing waste, these wastes (such as pineapple waste) may show potential to be recycled. Those waste can be use as valuable raw materials, adapted to useful and higher value-added products, raw material for other industries as well as to be utilized for food or feed after biological treatment (Kroyer 1991). This waste contains valuable components that are mainly sucrose, glucose, fructose and other nutrients (Sasaki et al. 1991). An attempt has been made by many researchers to utilize the waste for producing high value-added products including bacterial cellulose, single cell protein (SCP), ethanol, acetic acid, oxalic acid and methane (Bardiya et al. 1996).

6 Conclusion

The applications of bacterial cellulose are getting more demand in these recent years, especially for medical purposes. As cellulose price and the market demand increases every year, it is important to find sustainable and economical ways of cellulose production. Utilization of agricultural waste as a substrate to produce bacterial cellulose is good as it can reduce or eliminate the used of other carbon source that would then reduced the production cost. Furthermore, the use of RDR to produce bacterial cellulose will enhance the production rate of cellulose and shorten the production time. In the light of above, the biocellulose from bacteria as alternative for plant cellulose can be a good effort in reducing global warming through preservation of forest.

References

Abdullah MB (2007) Production of organic acid from local raw materials. Master degree thesis, Universiti Teknologi Malaysia, Malaysia

Ashori A, Sheykhnazari S, Tabarsa T, Shakeri A, Golalipour M (2012) Bacterial cellulose/silica nanocomposites: preparation and characterization. Carbohydr Polym 90:413–418

Bae S, Shoda M (2005) Statistical optimization of culture conditions for bacterial cellulose production using Box–Behnken design. Biotechnol Bioeng 90(1):20–28

Bardiya N, Somayaji D, Khanna N (1996) Biomethanation of banana peel and pineapple waste. Bioresour Technol 58(1):73–76

Bergen M (2011) The world's 10 most threatened forest hotspots. www.conservation.org

Castro C, Zuluaga R, Putaux JL, Caro G, Mondragon I, Ganán P (2011) Structural characterization of bacterial cellulose produced by *Glucanocetobacter swingsii* sp. from Colombian agroindustrial wastes. Carbohydr Polym 84(1):96–102

Cellulose Acetate Market—Global Industry Analysis, Size, Share, Growth, Trends and Forecast (2013–2019). https://www.ihs.com/products/cellulose-acetate-flake-chemical-economics-handbook.html

Czaja W, Krystynowicz A, Bielecki B, Brown RM Jr (2006) Microbial cellulose—the natural power to heal wounds. J Biomater 27(2):145–151

Ernsting A (2012) Sustainable biomass: a modern myth. www.biofuelwatch.org.uk

Fischer S, Katrin T, Bert V, Kay H, Ingeborg S, Klaus FM (2008) Properties and applications of cellulose acetate. Macromol Symp 262:89–96

Frank OO, Tiwalade AA, Augustine OO (2009) Processing pharmaceutical grade microcrystalline cellulose from groundnut husk: Extraction methods and characterization. Int J Green Pharm 3 (2):97–104

Gindl W, Keckes J (2004) Tensile properties of cellulose acetate butyrate composites reinforced with bacterial cellulose. Compos Sci Technol 64(15):2407–2413

Gupta G, Basavaraj S, Hungund S (2010) Production of bacterial cellulose from *Enterobacter amnigenus* GH-1 isolated from rotten apple. World J Microbiol Biotechnol 26:1823–1828

Hungund BS, Gupta SG (2010) Improved production of bacterial cellulose from *Gluconacetobacter persimmonis* GH-2. J Microb Biochem Technol 2:127

Hungund BS, Prabhu S, Shetty C, Acharya S, Prabhu V (2013) Production of bacterial cellulose from *Gluconacetobacter persimmonis* GH-2 using dual and cheaper carbon sources. J Microb Biochem Technol 5:031–033

John V (2005) Rainforest loss shocks Brazil. The Guardian, London

Junaidi Z, Muhammad AN (2012) Optimization of bacterial cellulose production from pineapple waste: effect of temperature, pH and concentration. 5th Engineering conference. 10–12th July 2012, Kuching, Sarawak

Kamel S, Ali N, Jahangir K, Shah SM, El-Gendy AA (2008) Pharmaceutical significance of cellulose: a review. Exp Polym Lett 2(11):758–778

Keshk S, Sameshima K (2006) Influence of lignosulfonate on crystal structure and productivity of bacterial cellulose in a static culture. J Enzyme Microb Technol 40:4–8

Kirk O (1993) Cellulose. Encyclopedia of Chemical Technology, Wiley, New York 4(5):476–496

Kongruang S (2008) Bacterial cellulose production by *Acetobacter xylinum* strains from agricultural waste products. Appl Biochem Biotechnol 148(1–3):245–256

Kroyer GT (1991) Food processing wastes. In: Martin AM (ed) Bioconversion of waste material to industrial product. Elviser Applied Science, London, pp 293–303

Krueger DA, Krueger RG, Maciel J (1992) Composition of pineapple juice. J Assoc Off Agric Chem 75(2):280–282

Kurosumi A, Sasaki C, Yamashita Y, Nakamura Y (2009) Utilization of various fruit juices as carbon source for production of bacterial cellulose by *Acetobacter Xylinum* NBRC 13693. Carbohydr Polym 76:333–335

Landin M, Martinez-Pacheco R, Gomez-Amoza JL, Souto C, Concheiro A, Rowe RC (1993) Effect of batch variation and source of pulp on the properties of microcrystalline cellulose. Int J Pharm 91:133–141
Liang S, Zhang L, Li Y, Xu J (2007) Fabrication and properties of cellulose hydrated membrane with unique structure. Macromol Chem Phys 208(6):594–602
Marijnissen C, Ozinga S, Richards B, Risso S (2004) Facing reality: how to halt the import of illegal timber in the EU, Technical Report (FERN, Greenpeace European Unit, WWF Policy Office, eds.):38
Matrosovich M, Tatyana M, Wolfgang G, Hans-Dieter K (2006) New low-viscosity overlay medium for viral plaque assays. Virol J (BioMed Cent) 3:63
Mitchell RL, Murphy CF, La Polla EM (1972) Heat-sealable hydroxyethyl cellulose film and process for preparing same. US Patent 3652316
Norhayati P, Khairul AZ, Ida IM (2011) Production of biopolymer from *Acetobacter xylinum* using different fermentation methods. Int J Eng Technol IJET-IJENS 11(5):90–98
Pei Y, Yang J, Liu P, Xu M, Zhang X, Zhang L (2013) Fabrication, properties and bio applications of cellulose/collagen hydrolysate composite films. Carbohydr Polym 92:1752–1760
Rafael LO, Hernane SB, Rosana MNA, Carla SM, Geandre OC, Guimes RF, Younes M, Sidney JLR (2011) Synthesis and characterization of microcrystalline cellulose produced from bacterial cellulose. J Therm Anal Calorim 106:703–709
Ron N (2006) The little green handbook: seven trends shaping the future of our planet. Picador, New York. ISBN 978-0-312-42581-4
Rosma A, Ooi KI (2006) Production of *Candida utilis* biomass and intracellular protein content: effect of agitation speed and aeration rate. Malaysi J Microbiol 2(2):15–18
Roth D (1983) Philippine forest and forestry 1565–1920. In: Tucker RP, Richards J (eds) Global deforestation and the nineteenth-century world economy. Duke Press Policy Study, Durham NC
Sasaki K, Noparatnaraphorn N, Nagai S (1991) Use of photosynthetic bacteria for the production of SCP and chemicals from agro-industrial waste. In: Martin AM (ed) Bioconversion of Waste Material to Industrial Product. Elviser Applied Science, London, pp 225–233
Shah J, Brown RM Jr (2005) Towards electronic paper displays made from microbial cellulose. J Appl Microbiol Biotechnol 66(4):352–355
Shi Z, Zhang Y, Philips GO, Yang G (2014) Utilization of bacterial cellulose in food. Food Hydrocoll 35:539–545
Shigematsu T, Takamine K, Kitazato M, Morita T, Naritomi T, Morimura S (2005) Cellulose production from glucose using a glucose dehydrogenase gene (gdh)-deficient mutant of *Gluconacetobacter xylinus* and its use for bioconversion of sweet potato pulp. J Biosci Bioeng 99(4):415–422
Sokolnicki AM, Fisher RJ, Harrah TP, Kaplan DL (2006) Permeability of bacterial cellulose membranes. J Membr Sci 272(1–2):15–27
Trease GE, Evans WC (1989) Pharmacognosy. ELBS/Bailliere Tindall, London 13:345–6, 535-6, 772-3
Tsuchida T, Yoshaniga F (1997) Production of bacterial cellulose by agitation culture system. Pure Appl Chem 69(1):2253–2458
UNECE, FAO (2007) Forest products statistics 2002–2006
UNFCCC (2007) Investment and financial flows to address climate change. United Nations Framework Convention on Climate Change, p 81
Wilson EO (2002) The future of life. ISBN 0-679-76811-4
Yusof WNHAWM (2013) Palm oil mill effluent (POME) as fermentation medium for bacterial cellulose production using static fermentation method. Bachelor Degree Thesis, Faculty of Chemical Engineering, Universiti Teknologi Malaysia, Malaysia
Zeng X, Darcy PS, Wankei W (2011) Statistical optimization of culture conditions for bacterial cellulose production by *Acetobacter xylinum* BPR 2001 from maple syrup. Carbohydr Polym 85:506–513

Author Biography

Ida Idayu Muhamad was born in Malaysia received her B. Tech. (1993), M. Sc. (1998), and in 2004 completed her Ph.D. degree in chemical engineering from the University of Manchester, United Kingdom. She is a professor at Department of Bioprocess and Polymer Engineering, Universiti Teknologi Malaysia. She published in food and biopolymer journals such as Journal of Food Process Engineering, Journal of Microencapsulation and Carbohydrate Polymers. Her current interests include particle technology, food process engineering, active packaging and biopolymer engineering. Prof. Dr. Muhamad is a Chartered Scientist, fellow of Food Science and Technology UK (FIFST) and the associate member of IChemE UK.

Norhayati Pa'e was born in Johor, Malaysia. She received the B. Eng (Chemical-Bioprocess) in 2006 and M. Eng (Bioprocess) in 2010 from Universiti Teknologi Malaysia. Currently, she is a Ph.D student in the same university. Her current interests include microbial cellulose production, functional film production from microbial cellulose, fermentation and bioreactor designing. Mrs. Pa'e is the member of Food and Biomaterial Engineering Research Group, UTM.

Khairul Azly Zahan was born in Johor, Malaysia. He was received his B. Eng (Chemical-Bioprocess) in 2010 and completed his M. Eng (Bioprocess) in 2014 from Universiti Teknologi Malaysia. He is currently working as a lecturer at Malaysian Institute of Chemical and Bioengineering Technology, Universiti Kuala Lumpur, Malaysia. His current interests include biopolymer production and characterization, bioreactor design and fabrication and phytochemicals study and product formulation. Khairul Azly is also an associate member (AMICheme) of IChemE Malaysia.

Urban Agriculture in the Manguinhos Favela of Rio de Janeiro: Laying the Groundwork for a Greener Future

Lea Rekow

Abstract

This paper presents a dramatic example of how agroecology has been used to restore chronically degraded land, create productive public space, and foster inclusive socio-economic activity inside the densely populated Manguinhos favela of Rio de Janeiro. It demonstrates how fragile, informal contexts can benefit from creating green infrastructure that is anchored to community involvement and practical action. It gives an overview of the processes involved in building such a project; observes its challenges and benefits; and in closing, outlines considerations for engaging organic agriculture to shape sustainable socio-economic activity in informal urban contexts.

Keywords

Urban agroecology · Favelas · Organic food systems

1 Introduction

This paper examines how urban agroecology can serve as a foundation to develop inclusive, socio-economic practices inside the favelas (informal settlements) of Rio de Janeiro, Brazil. It demonstrates how a whole-systems approach to producing local, organic food systems on many scales can lead to improving ecological,

L. Rekow (✉)
Arts, Education and Law Group, Griffith University, Brisbane, Australia
e-mail: learekow@gmail.com

L. Rekow
Green My Favela, 59 Franklin St, Suite 303, New York, NY 10013, USA

W. Leal Filho et al. (eds.), *Sustainable Economic Development*,
World Sustainability Series, DOI 10.1007/978-3-319-45081-0_10

economic, and social stability on many levels (Altieri 1987)—by regenerating degraded environments to improve health, generate income, multiply knowledge, and cultivate social capital (Ribeiro et al. 2015).

Specifically, this paper examines the implications of a municipal-level urban agroecology project being developed inside the Manguinhos favela of Rio de Janeiro. Until recently, Manguinhos was considered one of the city's most neglected and violent informal settlements (Calvalcanti 2015). Though not without its challenges, this organic farming project has emerged to become, both in size and ambition, the largest urban food garden in South America (O'Reilly 2014).

Characterized by multi-stakeholder participation and governance, this initiative is testimony to how agroecology can stimulate micro-economic activity and bring some level of social stability to contested urban territory. The initiative is one of approximately 30 agroecology projects developed inside the favelas and municipal public schools of Rio under the lead of *Hortas Cariocas* (HC), a favela agriculture program established by the Municipal Department for the Environment in 2006.

The paper elucidates how establishing agroecosystems inside Rio's favelas is an effective, if limited, vehicle through which to begin to address multi-pillar, informal sector needs. It develops this argument through a conceptual framework that aligns itself with ecological economics and degrowth, theoretical perspectives that offer an alternative to the notion of economic growth as the principal benefit of policy outcome.

How the Manguinhos project and HC connect to and diverge from the city's current policy framework, and why state-led interventions can fail or succeed because of their attitudes toward informality are also examined. The underpinnings of the HC project, and specifically the Manguinhos garden, is then expanded upon, specifically in relation to how the production of useful green space can foster a more inclusive connection between the built environment and the social, cultural, and economic wellbeing of Rio's informal communities.

In closing, the paper considers how municipal-scale agroecology projects can be better coproduced with the informal sector; how they can serve as a template for developing more sustainable and inclusive socio-economic policy; and how inclusive green growth within this context can shift municipal priorities and influence development plans.

2 Conceptual Framework

In Rio de Janeiro, one out of five people (1.2 million) live in urban favelas (IBGE 2010), which are growing exponentially at the rate of 6:1 compared to the growth rate of the formal city (IPP 2008–2009, IBGE 2000). The impact of living in such conditions is considerable. Undernutrition presents as the underlying cause for approximately 50 % of child deaths in developing countries worldwide (Black et al. 2003; Caulfield et al. 2004). In Rio, undernutrition is 14 % higher in favelas than in the formal city, and the mortality rate is triple for children under five years of age (Unicef 2008).

In the Manguinhos favela, the last published undernutrition study was conducted in 1987, when approximately a quarter of children in this age bracket were found to be at least mildly undernourished, according to their age-to-weight relationship (Filho et al. 1992). Undernourishment results in slow growth, physical abnormalities, and in having less energy for work. It depresses the immune system, which results in greater susceptibility to several diseases (including anemia, goiter, thyroid, diarrhea, pneumonia, mosquito-borne illnesses, and measles). Environmental degradation, a critical lack of ecosystem services, pollution, and the risk of armed conflict, also threatens residents' health and safety and exasperates instability. These conditions make clear the critical need to invest in a suite of strategies that can aid in stabilizing Rio's fragile informal sector communities.

A 2007 report compiled by the RUAF Foundation for the FAO, clearly details the sustainable economic value that urban agroecology can bring to the informal sector (Van Veenhuizen and Danso 2007). In many cities, agroecology is now being recognized as an important component of urban planning, slum upgrading, and land use redesign (Dubbeling 2011). Yet relatively little scholarly literature exists in relation to its application in the favelas of Rio de Janeiro (O'Reilly 2014; Arruda 2011).

This paper, therefore, reviews the practical benefits and challenges of establishing agroecology projects inside Rio's favelas, using the Manguinhos project as a case study. It also assesses the project's total economic value (TEV), with a focus on its role in producing social inclusion and stability, which, this paper contends, is foundational to, and shares a symbiotic relationship with generating sustainable economic activity.

There are hundreds of definitions that link sustainable development with socially inclusive, resource efficient, economic stability. A critical perspective, however, posits that the term sustainability has been hijacked by political rhetoric in order to promote a neoliberal agenda (Hornborg 2009). Assuming, however, that sustainable development means economic development conducted without depletion of natural resources, (Brundtland 1987) how does that apply in favelas (and indeed much of the world) where natural resources are already depleted or chronically contaminated, and operating far beyond carrying capacity?

The United Nations Environment Program defines a green economy "as a system of economic activities related to the production, distribution and consumption of goods and services that result in improved human well-being over the long term, while not exposing future generations to significant environmental risks and ecological scarcities" (UNEP 2011). Key concepts include a low carbon economy; green economic growth; green collar jobs; a circular economy; and an ecological economy (Bushehri 2012).

Approaches such as ecological economics are closely related to green economics, however, an ecological economy shifts away from the dominance of economic growth in terms of political focus. Ecological economics, instead, adopts a transdisciplinary approach toward cultivating sustainability. This approach also takes into account the importance of intergenerational equity, irreversible environmental change, and the unpredictability of long-term outcomes. Ecological economics is not strictly a sub-discipline of either economics or ecology, but rather functions as a bridge to offer an understanding of how these two fields are inseparably intertwined (Costanza et al. 2015), and theoretically connect to aspects of The Economics of Ecosystems and Biodiversity (TEEB).

TEEB is a structured approach to multi-pillar ecosystem valuation system that links economics to ecology, to highlight the relationship between biodiversity, ecosystem services, and human wellbeing. The attributes of TEEB confront the challenges of valuing ecosystem services; the issues related to economic discounting; the costs of environmental inaction; and the economic dimensions of ecosystem services loss and degradation (De Groot et al. 2012).

What these theoretical perspectives offer is the opportunity to examine how the informal sector can expand in terms of sustainable output and employment when linked to concepts of green activity. The informal sector of Rio—and others with considerable economic, social, and ecosystem service impacts and needs (Schneider 2012)—must be addressed if sustainability, in any form, is to be achieved.

There are several dominant perspectives that postulate how informality is connected to, and positioned in relation to the formal economy; how and why a regulatory framework should be put in place; whether or not the government should provide jobs, credit, business development services, basic infrastructure and social services to the informal sector; if and how governments should address the unequal relationship between big business, bureaucracy, and subordinated producers and workers; and if informal activities should be considered illegal because they do not conform with regularization laws (Chen 2012). Whatever the theoretical perspective, an oppositional tension between the informal and formal exists in all these perceptions.

Within this opposition lies the supposition that the informal sector is disorganized, and therefore can be discredited for its deficiencies. In contrast, it is assumed that the formal sector, because it is built on a foundation of structure and organization, is more effective. This is a critical distinction to be made within the informal/formal debate, specifically when theorizing about policy decisions, as this perspective can, and often does, lead to state interventions with catastrophic results. In recent years, theorizing has begun to move beyond this dichotomy to discuss the specific objectives of policy intervention outside of this narrow development discourse. Evidence suggests that the informally organized can be highly efficient, more so than, for example, interventions by a corrupt government (Guha-Khasnobis and Kanbur 2007).

Each region, and even favela, presents diverse and variable conditions that make generalizing about informality, and how to work in it, difficult. Rio's favelas are radically diverse in population size, topography, geography, built density, socio-economic levels, governance (gang or militia control, pacification, or none of the above), levels of human security, and infrastructure. Therefore, each possibility for intervention or collaboration presents very nuanced and complex circumstances through which to negotiate.

Understanding how Rio's informal communities, and their economies function —how they have been built independently, without the support of the state, in a climate of economic disparity, social fragmentation, and armed conflict—is critical to working with them. However, defining them in terms of lack of social and legal protection is not the only way to understand them. They are also comprised of ingenious DIY tangles of networks, economies, and social exchanges that play out in a labyrinth of spatial complexities (Fabricius 2008).

Overall, Brazil's informal economies (depending on how they are analyzed) may represent as much as 40 % of the GDP (de Holanda Barbosa Filho 2013). In Rio, favelas have been estimated to have an economic worth of US $6.1 billion (Biller and Petroff 2012). A person may prefer to work in the high end of the informal employment workforce rather than work at the minimum wage end of formal employment. This is a very different scenario than someone who works in the low end of the informal sector because of limited opportunities available to them, for example because they suffer from ill health or possess few remunerative skills.

In addition, informality that is the outcome of imposed structural constraints, including being forcibly relocated to the peri-urban periphery of the city (with up to a 6 h daily commute time); or being relocated to a mid-rise favela (where street-level business opportunities are lost); or having to absorb the ever-rising costs of public transportation (or the ever diminishing public transportation routes that the municipality offers) in order to reach employment; or the sky-rocketing cost of living increases that come with pacification (gentrification and regularization), all limit the mobility and choices of the informal sector, perpetuate poverty, and inhibit inclusive growth (Heintz 2012).

Furthermore, since 2008, Rio's pacification campaign (UPP) has used military intervention as a means to control favela territory (Neocleous 2010, 2013) and advance the city's rapid urbanization agenda, which has been accelerated by the hosting of mega-events such as the World Cup and Olympics. This agenda, thus far, has resulted in 67,000 removals of low-income and informal sector residents and put an additional 40,000 at immanent risk (Azevedo and Faulhaber 2015). Much of this plays out in what David Harvey calls a process of 'accumulation by dispossession,' (2004) whereby favelas are abandoned and devalued by the state, then strategically seized through militarized occupation. These assets are then leveraged and revalorized through a political framework of 'exceptionality urbanism' (Vannuchi and Van Criekingen 2015).

Within this context of disenfranchisement, there are a small fraction of PPP infrastructure services that have been put in place to benefit those who live inside 'pacified' territory. The Manguinhos garden is one of these examples. Connected to a larger redevelopment project launched in 2007 (Jauregui 2005). Manguinhos residents, nevertheless, have been evicted from their homes, and excluded from the decision-making processes involving the redevelopment of their community (Clarke 2013a, b).

Favela residents in general continue to be denied a voice in these processes (Braathen et al. 2013). Because HC operates independently to military pacification, however, it has more latitude to develop a structure whereby, at least on the lower rungs of the participatory ladder, decision-making and project management are built into its modus operandi. Rather than seeing informality as a separate constraint in itself, HC looks for inclusive ways to implement ground-up food security solutions to alleviate actual social and economic problems (Barros 2014).

HC is an organizational example of how integrated policy actions can be applied in favelas. Part of their solution for establishing food security involves creating new and innovative relationships, practices, and access to opportunities that, through their existence, may make living in informality a little more secure. Their working method focuses not only expanding the number of choices available to individuals but on improving the quality of those choices (Heintz 2012). Their model opens up opportunities to improve the long-term human well-being of favela constituencies by producing infrastructure and land use management practices that are managed by favela residents, and operate within carrying-capacity (Costanza et al. 2015).

The Manguinhos project is also a fitting platform with which to explore the concept of 'socially sustainable degrowth.' Degrowth challenges the 'green growth' and 'green economy' paradigms, both of which are associated with the belief that economic growth must be embedded into a desirable political agenda (Demaria et al. 2013). The concept of degrowth attempts to stimulate the debate on producing social, environmental, and economic stability rather than producing growth engines (Asara et al. 2015). In this sense, degrowth can be seen as a critique of the current hegemonic development paradigm (Rist 2014).

This development critique first began to appear in post-development theory in the 1980s and 1990s in the writings of scholars such as Arturo Escobar and Wolfgang Sachs, amongst others, (Odum and Odum 2001; Victor 2008) and has evolved into an interpretative mechanism for understanding how a diverse range of participants can engage in collective actions that mobilize the dispossessed (Della Porta and Diani 2006).

Though degrowth reaches toward an ideological social movement, (Demaria et al. 2013) the concept does offer theoretical possibilities for expanding how multi-sector partnerships can achieve practical goals (Chatterton and Pickerell 2010). In particular, it places importance on the non-use and option values of space (Plottu and Plottu 2007). Both degrowth and ecological economics offer possible paths for reassessing how to decrease human pressure on ecosystems at the inseparable junction of economics (Demaria et al. 2013). These alternative approaches may offer a constructive means for cultivating economic stability within Latin America (North 1990).

Economic degrowth is ecologically necessary, and at some point in time, likely unavoidable (Kallis et al. 2012). From this perspective, favelas provide a relevant, real-world situation in which to explore how, and in what conditions, employment and socio-economic stability can be achieved, at what public cost it can be organized, and under what socio-political conditions it can happen. Agroecology, in its broadest multidisciplinary application—as an environmentally, socially and culturally sensitive approach to agriculture—is an ideal vehicle with which to explore these concepts. The practice of agroecology focuses on cultivating sustainable, productive, equitable, and stable food (and/or forestry) systems. In the Global

South, agroecology also carries overtones of economic and social justice (Fernandes and Gotuzzo 2012).

TEV offers a multi-dimensional platform through which to assess the overall benefits of agroecology projects like Manguinhos. TEV is an evaluation method that recognizes both use and non-use values of a resource or infrastructure system:

Use value involves interaction with resources, either directly or indirectly. *Direct use value* derives from the goods and services produced. *Indirect use value* involves human interaction on both consumptive and non-consumptive levels, for example, social, recreational and educational benefits and activities.

Non-use value is associated with benefits derived simply from the knowledge that the resource or infrastructure exists. By definition, it is not associated with any tangible benefit derived from it, although users might attribute non-use value to it. Another category is *Option value,* how an individual derives benefit from ensuring that services and resources will be available for use in the future (Plottu and Plottu 2007).

Within the conceptual framework outlined above, the Manguinhos project provides a practical template for exploring agroecology as an instrument for moving sustainability forward within an informal urban context. Characterized by low-cost maintenance and a low-carbon footprint, as a generator of green jobs, and as an example that embodies the concepts of ecological economics, the Manguinhos garden represents a replicable and scalable socio-economic practice that can be implemented within the most challenging of circumstances.

3 Research Method

This report is based on multi-year qualitative field research that began in 2011, in Manguinhos, while working with the *Green My Favela* (GMF) urban restoration project in collaboration with *Hortas Cariocas*. GMF launched its first volunteer section of the HC garden as a coproduction with retired and underemployed favela residents in 2012. More recently, in 2016, it established a medicinal garden and fruit orchard with and for the children of the area. GMF continues to work with HC and favela residents in Manguinhos, and it is from within this perspective that the primary information presented in this paper is derived—including from direct observations, informal and semi-structured conversations with favela residents, participation in meetings with citizens' committees and Residents' Associations, and informal interactions with drug traffickers, UPP officers, HC, and other municipal stakeholders.

The field research has been cross-analyzed against a range of empirical data, including an examination of articles of law, statistics, development plans, architectural design documents, human rights reports, and mainstream and independent media reportage. Participation in conferences and reviews of literature produced by other researchers engaged in analyses of Rio's urbanization activities also provide a foundation for cross-referencing against the qualitative research. It must be noted, however, that obtaining reliable quantitative data inside favelas due to a range of factors, including official underreporting, renders statistics highly inaccurate.

4 Delineation of Research

This paper aims to stimulate discussion on several questions: What can a green economy look like in the context of Rio's informal communities; how can agroecology help achieve it; can it serve as a replicable, adaptable and/or scalable template for both municipal policy and in a DIY context for the informal sector; how can Rio's informal sector produce and trade to improve their own well-being; do 'green' initiatives based in informal communities require much formalization, and if not, what level of formal support do they require for them to remain viable?

To explore these questions, the Manguinhos space is examined in relation to its total economic value:

a. *Direct use value*: food, water, quality of space, income generation, consumption value, and services.
b. *Indirect use value*: human interactions—social, recreational, health and educational benefits.
c. *Non-use value*: benefits derived from the existence of the garden, specifically in relation to the therapeutic benefit it provides through stress relief, sensory perception, and aesthetic value.
d. *Option value*: benefits and degrees of hopefulness of knowing the space will be available for future use.

5 Background and Scope

The Manguinhos Complex is a large, poor North Zone cluster of ten to fifteen favela neighborhoods, located 6 km from the city center, and 4 km from the Rio's Maracaña sports stadium. It has an overall population of anywhere between 35,000 and 70,000 people. Previously under the visible armed control of the Comando Vermelho (CV) drug trafficking gang, it was pacified by the UPP in 2012. The large-scale redevelopment of Manguinhos is part of a broad suite of multi-tiered urbanization policies that have been impacting the neighborhood since 2007.

The garden has been established as part of a large redevelopment plan that connects to the city's public security interventions and 'exceptionality' policies that drive pacification, urbanization, and economic development surrounding Rio's mega-events. As such, Manguinhos is subject to the broader consequences of pacification, including forced displacement, gentrification, military occupation, and regularization, as well as the ongoing influence of the CV drug trafficking gang, which still dominates much of the social landscape of the favela.

However, though the research takes into account this broad range of shifting dimensions, it remains focused on how HC uses organic agriculture as method that can effectively function inside these complex circumstances. Manguinhos is only one of approximately fifteen HC projects operating inside several of the 38 favelas pacified throughout the city. Therefore, it is outside the scope of this research to analyze Rio's rapid urbanization policies, the UPP, or its impact on the culture of drug trafficking or gang infrastructure, other than to recognize these factors as part of an interrelated set of circumstances that must be negotiated.

The focus and intent of this research, therefore, is to offer proof-of-concept as to how the urban agroecology project in Manguinhos can function to benefit Rio's informal sector despite these problematic social, ecological, and economic conditions; how, as a strategy, it can be used as a pragmatic template to point toward how to include the informal sector in a policy framework; and moreover, how it impacts on favela residents. Finally, it explores if and how agroecology can be an effective driver for creating the underpinnings of green and sustainable socio-economic practices that promote long-term stability for Rio's informal sector while enhancing the physical environment of favelas.

6 Greening the Base of the Pyramid

Urban upgrade projects in Rio often focus on relocating informal communities to the urban perimeter where they are splintered from networks of friends and families, isolated from economic opportunities, (Perlman 2003, 2004, 2010) and subject to extended commute times of up to 6 h per day (Pereira and Schwanen 2013; Watts 2014; World Bank 2015).

However, instead of focusing on relocation, if Rio's urban planners and policy-makers look to how improvements can be made within the already existing fabric of low-rise inner-city favelas, agroecology can provide an important framework with which to lay the foundations of environmentally focused infrastructure services that may lead to increased social stability and income generating opportunities, without extending a development footprint.

In Brazil, several hundred municipalities have favela upgrade programs. Many incorporate some form of land tenure security and participatory project planning and management into their models, and many work in partnership with the private sector and other non-governmental actors.

In Rio's favelas, however, informal urbanization initiatives are effectively non-participatory, top-down PPPs that focus almost exclusively on generating private sector profits (Magalhaes and Xavier 2003). Few have aimed to increase employment opportunities, establish safe public space, restore degraded lands, or improve access to ecosystem services. The HC project is distinctive in this regard.

Though the main goal of the HC program is to provide food security (Barros 2014), the restoration of degraded land and services are necessary parts of the process. Access to healthy food and medicinal plants; a safe and consistent water supply; the mitigation of flooding; garbage removal; the restoration of the physical environment; cultivating remunerative employment opportunities; and creating educational and social value, are all part of what makes HC projects potentially so important. However, they can also suffer or collapse quickly with little warning if, for example, regular access to water is cut for extended periods; if armed conflict increases and limits physical access to space; if project funding is cut or discontinued; or if ill health befalls the custodial guardians of the projects. All these scenarios are not uncommon in favelas.

The scarcity of both physical and social infrastructure services is difficult and expensive to overcome. Policy failures compound environmental degradation and fail to provide communities with access to basic utilities and resources. Therefore, establishing any form of permanent enterprise in a favela is an incredibly difficult task. However, evidence still suggests significant socio-economic opportunities lie in the design and distribution of goods and services for poor communities.

There is a strong case emerging for governments, as well as the private sector, to invest in those living at the Base of the Pyramid (BoP) (Hart 2013). BoP theory asserts that and that integrated stakeholder partnerships can help eradicate poverty for four billion people who live on less than $5 or less a day. The purchasing power of this sector has been estimated at anywhere between $0.3 and $1.3 trillion globally (Karnani 2009). $172 billion of this potential market exists in Brazil, where the BoP comprises of 75 % of the population (Hammond et al. 2007).

So how does urban agriculture present itself as showing sustainable economic promise for Rio's BoP sector? Its value may be assessed through how it impacts on multidimensional poverty indicators such as food, nutrition and water security, health and sanitation, employment, and education, and how it reduces feelings of anxiety, scarcity, neglect, and lack (OPHI 2015).

7 The Landscape of Urban Agriculture

By 2010, almost $20 million of federal funds, benefiting 74,000 BoP recipients, had been invested in the urban and peri-urban agriculture sector of Brazil (MDS 2010). Agroecology and organic agriculture is a growing subset of this sector. The 2006 Agricultural Census (IBGE) recorded about 90,000 organic producers in total throughout the country—those not using pesticides, chemical feedstock or genetically modified seeds or organisms.

This approach to local food production has been bolstered by a three-year National Organic Production and Agroecology Plan (PLANAPO). Implemented in 2013, PLANAPO invested $4 billion into the agroecology sector, mostly through establishing credit lines. In 2013–2014, $3 billion in credit lines for agroecology yielded the Sicredi Credit Union $20 billion in assets, up 24 % from 2012. Though this figure represents less than one-tenth of federal agribusiness subsidies, it was a step toward strengthening the sector overall. Nevertheless, substantial barriers to providing micro-credit to both the informal sector, and the urban agroecology sector, still remain. The largest failure of Brazil's micro-credit system in general is that it is unable to reach the poor and informal sectors that need it most because of restrictive lending policies (Cravalho 2012).

Almost the entire state population (96.7 %) of Rio de Janeiro resides in urban areas. The municipality exceeds six million, and almost twelve million live in the larger metropolitan area (IBGE 2010). The city has one of the largest per capita GDPs in the country,[1] with the urban poor representing an economy worth $6.1 billion—33 % of the favela economy nation-wide (Instituto de Pesquisas Data Popular/ IBGE 2012). 22 % of the city's residents (1,393,314 people) live in 763 favelas—up more than 27 % in the first decade of the 21st century (IBGE 2010). Despite the urban environment, there are believed to be 3,764 family farmers active in the region (IBGE 2006).

HC launched in Rio in 2006. Its goal is to integrate urban agroecology with land use restoration to promote food security and human well-being in the favelas and the underserved municipal public school system. Under the tenacious lead of program director, Julio Barros, HC has established more than 30 food security initiatives in 28 low-income areas throughout the city. These organic agriculture projects are innovative mixes of public and private investment partnerships that work with local Residents' Associations, schools, NGOs, and individual participants to establish sustainable micro-economies and fight persistent poverty (Barros 2014). About half of the projects have been established inside or on the periphery of UPP-controlled favelas, wile the other half operate inside schools.

HC functions as an organizational tool that uses community involvement as a base for socio-economic development. The program demonstrates that urban agriculture not only has the potential to be an income generator for the informal sector, but be a driver for the larger social wellbeing of favela communities. The traction that urban agriculture is gaining in Rio has been catalyzed by many concurrent crises experienced inside the favelas. The concept that agriculture is an affordable means for building sustainable socio-economic activity is now putting poverty alleviation on the agenda of municipal action through the HC program. These actions are important in closing the gulf between the rhetoric and the reality of establishing pro-poor green initiatives, and how they can help address conditions inside fragile urban contexts.

[1]Rio de Janeiro has a per capita GDP of $ 9723.47 (2008), 57.1 % higher than the country average, and representing 5.1 % of Brazil's GDP overall.

8 Manguinhos Case Report

8.1 Context

The region known as Complexo de Manguinhos is a sprawling entanglement of government housing projects, abandoned factories that have been turned into housing occupations, and low-rise favela housing. These clusters of irregular settlements shape themselves on both public and private lands and are home to around 50,000 residents, approximately 20,000 of whom are unskilled workers with a low and unstable wage threshold (IBGE Census 2010).

The name Manguinhos denotes its origins as a mangrove ecosystem that was long ago obliterated by the dumping of waste and major urbanization. Manguinhos is now hemmed in by a series of open sewers that once flourished as rivers, a labrynth of highways congested with traffic, an elevated train track, and an above-ground metro. High voltage transmission lines cut through its middle.

It is subject to a range of environmental hazards including the flooding of its sewer/river system which is also used for recreation purposes by children on hot days, and by some residents to bathe in. Large amounts of household and industrial waste, frequently disposed of in open burns, accumulate throughout the favela. Toxins migrate from a nearby oil refinery to pollute the water and soil, and emissions from heavy traffic carry airborne toxins which blanket the favela in a thick oily film.

Numerous environmental risks afflict the residents, and in particular its youth, who are 500 times more likely to develop cancer and neurological disorders because of exposure to high levels of lead that contaminate the area (De Cássia 2009). In addition, drug trafficking influences, including gang domination of the community and the proliferation of gun violence exasperated by military counterinsurgency operations, also impacts residents.

From the late 1970s on, Manguinhos came to exemplify the impacts seen from decades of state abandonment and, consequently, drug trafficking and urban warfare. Eventually Rio's largest crackolândia—a string of shanties and armed drug selling points assembled on top of garbage—sprang up along the train track and under the transmission lines. The crackolândia was a center for more than 2000 crack addicts, armed dealers, and other forms of vice, including prostitution, drug sales, and open consumption. Counterinsurgencies, public executions, and exchanges of gunfire were not infrequent.

In 2012, Rio's special operations forces stormed the favela with 1300 troops and took military control of Manguinhos, which had been governed by the CV drug-trafficking gang since the 1980s. The UPP now occupy the favela at a ratio of approximately one officer for every 66 residents. Following the pacification of Manguinhos, the city bulldozed its massive crackolândia,[2] clearing the way for an immense organic urban food garden, the physically largest in South America.

[2]Some of the addicts were forcibly taken to a mental health facility and released a few days later to congregate further along the highway.

8.2 HC Working Procedures

At a cost of approximately US $250,000, the Manguinhos garden project is a partnership between the Municipal Department for the Environment's HC program; Brazil's second-largest private-public electric utility, *Light* (which granted the right to use the land under the high voltage lines); the Mayor's Office; and the Manguinhos Residents' Association. Another two sections of the garden—launched and supported by GMF—are managed by residents on a volunteer basis.

This undertaking has had a huge impact in the Manguinhos neighborhoods of Vila Tourismo and João Goulart. The project began with the removal of 700 truckloads of garbage (between 250 and 450). A half-meter layer of contaminated topsoil was then scraped off the top of the kilometer-long space that runs under the transmission lines. A half-meter of gravel and crushed stone was subsequently laid in order to increase drainage and prevent weeds. This has helped mitigate flooding and the build up of trash and stagnant water. More than three hundred garden beds, each ten meters long by one meter wide were built and filled with clean topsoil (Fig. 1).

Eight water tanks have been installed on site and connected to the city's water supply to provide for manual irrigation of the beds. Drip irrigation models, fashioned out of recycled PET bottles, have become a popular method of irrigation. Approximately 15–35 % less water withdrawals can be made with this more sustainable means of irrigation, and adaptive models are being ingeniously crafted by residents all the time. PET bottles are also used in several other HC gardens to create the walls of garden beds and compost areas.

Comlurb, the city's waste collection and disposal service, provides recycled garbage as compost for the garden. They also provide basic (and inadequate) trash collection for the surrounding communities (a service implemented with pacification). Two toolsheds have been built, and seeds are regularly dispensed to gardeners, along with basic tools and materials for building shade structures (Fig. 2).

Fig. 1 Partial aerial view of *Hortas Cariocas* garden under construction in Manguinhos. The garden is divided into three sections by soccer fields. (May 2013)

Fig. 2 Manguinhos garden community plot. (June 2014)

Fig. 3 HC worker with family in the Manguinhos garden. (June 2014)

Fig. 4 Children participating in GMF activities. (April 2014)

Fig. 5 GMF section of HC garden. (April 2014)

HC, in consultation with the Residents' Association, have hired more than 20 resident gardeners, none of who had any previous agricultural experience, and some of who are rehabilitating drug addicts. The gardeners receive training in agroecology and are paid a stipend of R $400 (approximately US $140) per month to cultivate a range of organic fruits, vegetables, and medicinal herbs and seasonings of their choice.

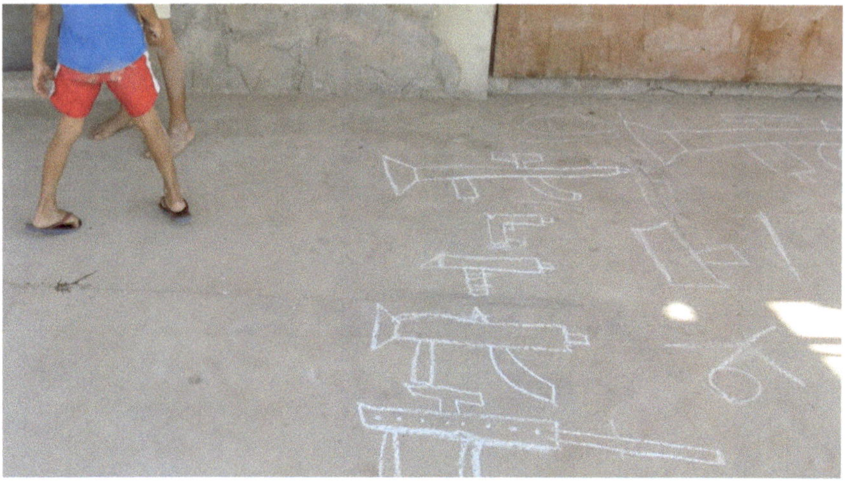

Fig. 6 Manguinhos kids making chalk drawings of guns. (June 2014)

Fig. 7 Comlurb dumpsters in the garden. Garbage collection was only introduced to Manguinhos following pacification in 2013. It remains woefully inadequate. June 2014

Because of an all-year round growing season, each gardener is able to take home two to three bags of produce a week, however, it has been difficult to accurately quantify the amount of produce harvested at Manuginhos. Though HC gardeners are technically obliged to weigh, record, and calculate the volume of food produced, production reportage remains highly inexact. According to the gardeners, however, compensation in produce has substantially increased their family's intake of

Fig. 8 Residential home gardens, Manguinhos. June 2014

Fig. 9 Kids cooling themselves at the garden, 2014

Fig. 10 Boys washing up at the toolshed in Manguinhos, April 2014

vegetables, and offset their average monthly income by approximately US $30 per month (GMF 2013/2014) (Fig. 3).

Under Hortas Cariocas guidelines, gardeners also distribute excess produce to food banks as identified by the local Residents' Association, and work with the Department of Education to give to local school lunch programs. This impact may be significant considering the the demographics in Manguinhos, a favela which sits close to the bottom of the city's Human Development Index (amongst the five lowest), where unemployment rates hover between 30 and 50 %, and where per capita income exceeds little more than US $60 per month (ENSP/Fiocruz 2012).

8.3 GMF

Gardeners in the GMF section work outside these HC guidelines, on a volunteer basis, to cultivate 48 garden beds in the section established late in 2012. They have adopted a collective approach to managing and organizing their garden. They have no formal restrictions on how they distribute or sell produce, nor any obligation to record their yields. These volunteers are made up of seven or eight retirees, the unemployed or underemployed, and their families. The garden operates at full capacity, and volunteers donate excess produce to feed additional community members, up to 100 people according to the gardeners (GMF 2014). The primary function of this area is to provide an intergenerational space for families, friends, children and neighbors to work and/or socialize in (Fig. 4).

GMF provides gardeners with ongoing practical support in the form of external volunteer labor to keep the area trash and weed free. They donate equipment and seeds, and help with other needs as they arise. Because the majority of GMF's external volunteers are foreigners living in Rio, children have a rare opportunity to experience intercultural exchanges. Other GMF undertakings include aestheticizing the space through projects such as creating murals and painting (Fig. 5).

The Manguinhos garden is open to the public at all times, and is frequented by families, children, and individuals on a daily basis. The gardeners now almost exclusively govern and manage the space, deciding what produce is grown, and how and to whom it is distributed. Gardeners have chosen to plant relatively low-value staple crops that are desirable for daily household consumption, including herbs and spices such as basil, cilantro and chili peppers; topical and non-topical biomedicinals for various ailments; fruits such as pineapple, melon, passionfruit and papaya; and a variety of vegetables, including eggplant, sweet potato, okra, tomatoes, green onions, cucumbers, and lettuce.

The ultimate economic goal for HC is to have their gardeners produce and sell enough that the project becomes fully self-sustaining. Of the 30+ gardens established overall by the program, seven have already achieved this status. In the volunteer area, where gardeners rely only on self-motivation, informal knowledge exchange, and each other to cultivate and maintain their space, the garden is already self-sustaining. This indicates that social incentive holds at least equivalent value to economic incentive under certain circumstances.

8.4 Key Challenges

HC project director, Julio Barros, would like to see community gardens formally included in future municipal planning policies, and believes they are key to promoting economic, social and ecological sustainability (2014). However, gardeners still encounter substantial barriers in regard to agricultural production, including zoning restrictions, a lack of funding for urban family farming, and an absence of designated agricultural areas in Rio's municipal master plan.

Residents' overall perception that the garden will be available for their use in the future (its option value) is low. The garden is only tenuously protected under the current PPP configuration, in a scenario that can be abandoned at any time.

There is also the probability that the UPP program will be dismantled and drug traffickers will return to armed control of the favela. Even preceding the country's multi-tiered budget crisis, this was always a possibility. With the state of Rio de Janeiro (the UPP's prinicpal funding body) dealing with a US $4.5 billion shortfall for 2016, and the future economic prognosis looking grim, the government is scrambling to sell everything from its bus terminals and land holdings, to future royalties and outstanding debt (Leal 2015).

This, together with a lack of any formal land tenure security for residents, and the lack of trust in a government who have abandoned their policies time and time again, creates the general feeling that the project may not last (GMF 2014). Given the government's track record, this is not an unrealistic conclusion to come to.

Another challenge to working in Manguinhos is the dual, contradictory forms of governance that are simultaneously in play. Though the UPP are in armed control of the favela, its goals purposefully avoid the dismantling of drug trafficking, and instead remain narrowly focused on disarming gangs and controlling territory. Moreover, the police are highly unpopular with the favela public for the lethal and non-lethal extra-judicial violence they regularly exert over individual citizens. Therefore, the undercurrent of CV drug trafficking remains strongly embedded in the community (Fig. 6).[3]

In an area with few lucrative economic opportunities other than drug trafficking, creating attractive forms of sustainable green-based employment is critical and requires long-term commitment. From an economic perspective, this presents a challenge for a country that has slid into negative growth; a state in floundering in financial crisis; and a financially overburdened municipality legally bound to absorb shortfalls from an outlandish Olympic budget. In Manguinhos, state-led community infrastructure projects—PPP ventures driven by electoral politics and public/private interests—are already being dismantled or scaled back. The Youth Reference Center, opened in 2009, has closed; the library, which opened in 2012, operates on reduced hours; the swimming pool, which opened around the same time, sits drained and abandoned; and the high school (along with 65 others throughout the state) has been abandoned by teachers whose salaries have not been paid, and subsequently occupied by students who conduct their own classes. Due to increasing fiscal pressure, the situation begs the question, how effectively will Hortas Cariocas survive the stress of an already meager municipal budget which is likely to diminish in the foreseeable future?

With HC stipends already very low, and only small crews working part-time to manage and cultivate a large production space, how long can people remain committed if the instability associated with increased economic instability

[3]This means that negotiations of any project must simultaneously satisfy dual governing bodies in order to progress. This is a role often undertaken by local Residents' Associations.

(including municipal stipend payments being delayed) and incidences of armed conflict escalating?

The small stipends offered by HC, Barros believes, encourages gardeners to make the project self-sustaining. However, with primary production focused on cultivating low-value crops, remunerative income opportunities are not high. The capacity of current yields is limited to supplying only a fraction of those living in the vicinity with an increase in their nutrition intake, and though there is the potential to intensify the yield within the existing garden footprint, it would require many more resources.

HC employees also need to have a bank account and a CPF (the Brazilian equivalent of a social security number) in order to be hired. Banking inclusion of low-income households increased between 2003 and 2012, from 70 million (40 %) to more than 120 million (63 %) by the end of 2012 (Instituto Lula 2014). More recently, the creation of special cash deposit accounts began facilitating for low-income beneficiaries of social programs and micro-entrepreneurs. However, these fee-free accounts are not available to HC beneficiaries, and opening a regular bank account remains prohibitive for those without formal identification documents. In addition, the new austerity measures being rapidly introduced by Brazil's current interim government, have—or are threatening to—roll back many social programs (including cash transfer programs), which may indirectly impact the project in a variety of ways through the further economic destabilization of families.

Another major obstacle to the project's stability is poor communication between government tiers and departments. For example, Comlurb, the city's waste collection agency, was responsible for destroying the fencing around an entire section of garden when their garbage trucks broke through it and deposited dumpsters on the beds. Comlurb took no measures to rectify the situation, and gardening in this section had to be suspended for many months until, eventually, the dumpsters were removed, the beds rebuilt, and the fence replaced. This is not the only incident of the waste collection agency breaching the garden perimeter. Indeed, it is a recurring problem in many parts of the space. Comlurb, also responsible for supplying the garden with recycled compost, has at times, delivered a product so putrid and filled with glass fragments that it has been deemed unusable (Fig. 7).

Corruption, a lack of political will, and the incapacity to fulfill infrastructure pledges has also led to the worsening of environmental crises that periodically compromises a section of the garden located next to the Faria-Timbó River. The original redevelopment plans for Manguinhos included the dredging of the river and the installation of sewerage collection infrastructure. In June 2012, leading representatives of the Caixa Econômica, Brazil's federal bank, and CEDAE, the state's water and sewerage company, stood next to State Governor Sergio Cabral to announce R $176 million in financing to build the sewage collection system (Clarke 2013a, b). Not only was the promise not fulfilled, but the money quickly disappeared.

A second guarantee, repeated only four months later, was accompanied by another R $250 million pledge, with assurances the construction would begin by

December the same year (Jornal do Brasil 2012). Again, the project was never actualized. In June the following year, another round of infrastructure funding pledged the less ambitious goal of reducing river waste (Clarke 2013a, b). Not only has the promise not been fulfilled, but the little work that has been done has exasperated flooding. Now during periods of intense rain, the river (which functions as an open sewer) overflows to contaminate this corner of the garden.

Produce is not tested for any possible contamination that may result from floodwaters or the buildup or migration of bioaccumulators (in particular lead) that may potentially contaminate soil or collect on the surface of produce. This also presents possible health risks.

There is a gross absence of civic participation in the decision-making and top-down planning and redevelopment of favela communities, despite government rhetoric espousing inclusive policies (La Rocque and Shelton-Zumpano 2014). In Manguinhos, attempts by residents to formally organize and assert their voices into the community's redevelopment processes have not only been ignored, but swiftly pronounced illegal and shutdown by the city (Braathen et al. 2013).

In general, a lack of clearly defined and secure property rights, policy goals, environmental policies, compounded by institutionalized corruption, forced displacement, and police brutality, leave favela residents with little confidence or trust in the government (Da Cunha and Mello 2011; Freeman 2012).

In addition, results of such projects remain dependent on the specific social, cultural, geographic and economic contexts involving relevant stakeholders, fluctuating levels of armed conflict, and the ongoing fight for territorial control. Yet, despite these setbacks and flaws, Manguinhos shows itself as having the potential to overcome many of the challenges it faces.

HC, as demonstrated by the dozens of gardens the program has produced, has shown itself to be a replicable, adaptable and scalable model—both on a municipal level, and within in a DIY or third sector context. These projects suggest that such green growth initiatives only require minimal structural formalization, economic support, and social stability for them to remain viable.

8.5 Key Benefits

Confronting the complex conditions in Rio's favelas with innovative approaches offered by programs such as *Hortas Cariocas* and *Green My Favela*, points to how favelas can be reimagined as places where creative environmental practice, when linked to engaged action, can encourage mobilizing efforts that lead to reclaiming space for productive use.

The Manguinhos project offers evidence that even in the most fragile of contexts, the informal sector can adapt and evolve quickly into semi-autonomous, productive self-management units, if given just minimal social support by the state. Not only is the garden flourishing but some residents that live adjacent to it are beginning to green their own homes (Fig. 8).

The garden has shown itself to provide direct use value in terms of yielding year-round fresh produce to augment the tables of up to 400 people (GMF 2014). In every conversation with adults and children that frequent the garden, their vegetable intake has substantially increased as a consequence. And because produce is organic, it does not contribute to environmental contamination or consumptive pesticide uptake.

An increase in the amount of fresh nutritious food put on household tables at mealtimes is especially important for young children under five years of age who suffer from undernutrition. Though the majority of cases reported in this age group only suffer from mild malnutrition, almost half of those living on a quarter of the minimum wage or less still experience moderate to severe food insecurity. Those in this income bracket spend 29.8 % of their money on food, (CAISAN 2011) on diets that consist almost exclusively of bread, margarine, coffee, beans and rice. Half of all these households go on average for three weeks at a time without consuming any fresh vegetables or meats (Monteiro and de Mendonça 2004). The garden can alleviate economic stress on this income bracket by approximately 20 % per month.

The HC program demonstrates that Rio's informal sector is capable of forming internal circuits for distribution that improves community well-being. In addition, participating in local market chains that breach informal/formal sector barriers cultivate basic, remunerative income generation. Though production in Manguinhos has yet to reach a point where gardeners are regularly selling at organic markets, this is happening in other HC gardens, most notably the garden located in the nearby favela of Formiga. In Manguinhos, because the venture is so new, it will likely take more time to develop formal revenue streams.

Cash stipends provided for HC employees, though low, when coupled with the income offset that regular take-home produce provides, increases socio-economic stability not only the for direct recipients, but for food bank recipients; the children who benefit from school lunches; and the individuals who procure small amounts of free food from the garden on a daily basis. There is very little theft of food, despite the kilometer long space being open and accessible 24/7.

The garden registers high in terms of its indirect use value. The formal training provided for HC employees, and the agricultural knowledge exchanged by the volunteers, has built up a range of useful skills for participants, through hands on experiential learning techniques. This is particularly relevant in an environment with a 10 % + national illiteracy rate for those over fifteen years of age (IPP/ IBGE 2010). Educational activities are also provided for school and NGO groups that visit the space to learn about agroecology and the importance of environmental stewardship (O'Reilly 2014).

In addition, the garden provides better drainage and a comparatively trash and debris free environment that can help mitigate vermin infestations and the spread of diseases such as dengue fever and leptospirosis during flood periods. Safe irrigation water, provided in the form of regulated access to city supplies ensures the project's success and provides respite for children who like to drink from the hoses and spray themselves and each other on hot days as alternative to diving in the polluted river. Because of improved drainage, water is less likely to pool and stagnate.

This reduces the risk of further elevating incidences of dengue, which is almost five times higher in Manguinhos than it is in the affluent neighborhood of Leblon (Municipal Secretary for Health 2008) (Fig. 9).

Neighbors who do not participate directly in gardening voluntarily safeguard the space to protect the community resources it provides. The physical and social transformation of the neighborhood provides some feeling of security and freedom for residents, especially children (GMF 2014). In addition, the stench of garbage and the presence of vermin no longer overwhelm the physical environment. In this sense, the garden's tremendous non-use value is derived from the simple existence of the space. This is especially important for children in Manguinhos, where more than 30 % of the population is aged fourteen years or younger; 15 % of girls between the ages of fifteen and seventeen are already mothers (ENSP/Fiocruz 2012); and boys still come under intense pressure to join the CV drug gang (Fig.10).

The main value the space brings to the community as a whole, and gardeners in particular, is the therapeutic benefit it provides through stress relief. Gardeners say working in the project has greatly reduced their hypertension. Recreational interactions and activities also provide families and friends with a cleaner and more aesthetically pleasing social space in which to stroll through, and children with a safer space in which to play (GMF 2014).

In large part, the sense of pride that the community takes in the garden has helped build self-esteem and transformed the social and aesthetic face of the neighborhood. Access to this sort of public space is a new, complex and unfamiliar concept in favelas. Yet, it is foundational to providing the necessary enabling conditions on which sustainable green enterprise can be built.

9 Conclusion

The removal of Rio's largest crackolândia and its replacement with a large-scale community food garden is, in itself, incalculably valuable on multiple dimensions. The remediated physical infrastructure is helping to rebuild social as well as natural capital, though it is not feasible to expect that the presence of the garden can address more than a limited reversal of the chronic economic impoverishment, social complexities, or environmental damage suffered in Manguinhos. Overall, however, the benefits the garden does bring helps to produce the enabling conditions for building the foundations for degrowth.

First, it shows how urban agriculture can potentially help develop infrastructure for food production and distribution systems that can also potentially strengthen citizen security, improve the delivery of and access to basic goods and services, restore livelihoods, and promote community solidarity. This may also lead to future micro-financing opportunities and/or market chains that link to the formal economy.

Second, the project makes an economic case for shifting PPP investments toward the informal societal sector by demonstrating how transforming abused lands into productive space advances the integrated political, social, economic, and environmental components necessary for the informal sector to establish key building blocks for progressing more sustainably and achieving semi-autonomous stability.

Manguinhos also indicates how integrated stakeholder partnerships may help communities begin to fight persistent poverty by addressing inequities in access to social services and building socio-economic opportunities. Advancing green policies that are targeted to meet the actual needs and vulnerabilities of the people living in favelas through committed municipal, local, and third sector engagement, also points to how it may be possible to strengthen the long-term resilience of informal communities and help them meet current and future challenges.

Policy-makers can be assisted in better comprehending the benefits of incorporating favela agriculture into policy by being conveyed information that demonstrates, through applied proof-of-concept:

- multi-pillar benefits that are derived from transforming degraded space into functional spaces for organic agroecological production;
- the interface between agroecology, sustainability, degrowth, and socio-economic planning;
- the social, economic, and ecosystem services that are directly or indirectly provided or influenced by this sort of activity;
- the wider costs and impacts involved when ecosystem services are degraded in comparison to when they are (at least partially) restored; and
- how these benefits can be distributed among different recipient groups and stakeholders.

Informal sector urban agroecological projects also offer an opportunity to explore alternative economic perspectives. The gift economy is a model of exchange where produce is not traded or sold, but rather given without an explicit, immediate, or future expectation of rewards. The circular economy utilizes recycled or upcycled resources in the production of space, and keeps them in use for as long as possible.

From a traditional economic perspective, where agricultural practices shape the way goods and services are produced and distributed, the remunerative advantages of cultivating high-value crops (Ebert 2014) may be considered as a way to augment subsistence output and create prospects for trade. From an ecological economic viewpoint, increased crop diversification can also attract biodiversity, create green corridors, mitigate erosion, reduce CO_2 buildup, and encourage surplus trade. Aspects of all these models are evident, at least to some degree, in Manguinhos.

Manguinhos also offers an opportunity to better understand the complex relationship between favela rehabilitation, urban sustainability, and conflicts over land. The negative effects on economic, social, spatial and ecological development brought about in the favelas due to armed conflict; simultaneous dual forms of governance; weak institutional support; an unregulated real estate market; and the dominance of illicit economies, are widespread. The sorts of territorial conflicts endured in favelas have disastrous effects not only for those individuals living in these conflict situations, but impact across the formal city as a whole.

Land tenure security, a basic canon of urban upgrade policy (UN 2004), is also an issue that requires scrutiny. In Rio, obtaining a formal land title inside a favela is a minimum five-year, heavily bureaucratized process (David 2014). Formalization linked to rapid urbanization activities, including the regularization of utilities, and real estate development and speculation, has led to massive cost of living spikes in favelas (Watts 2013), resulting in a culture of displacement and gentrification that has pushed as many as 800,000 residents from their homes (Simões 2015). In order to implement land tenure security without gentrification, urbanization processes must be slowed down (Williamson 2015).

The active involvement of citizens, especially at a neighborhood level, in the actions surrounding comprehensive land use and its collective management, may have the capacity to benefit both illicit and institutional stakeholders. Taking direction from favela residents on how to meet their needs, and turning this into coproduced achievements, may enable people to feel more committed to protecting their social assets as a consequence (Wehrmann 2008).

Where citizenry rights over material assets are clearly recognized, and people feel secure both in and outside of their homes, conflicting interests for control of land may be alleviated to a degree. In the context of the favelas this not only depends on the recognition of informal rights, but means the introduction of innovative forms of cooperative land tenure security (Lloyd-Jones and Rakodi 2002; Payne and Durand-Lasserve 2013). In this respect, the garden offers the potential to anchor principles of social equity, the protection of common property, and the social obligations of community ownership in policy and law (Dekker 2006).

The market for the trade of organics continues to grow worldwide. 30 % more jobs are generated by the organic production sector compared to conventional agriculture, and CO_2 emissions are reduced by 48–68 % (Bushehri 2012). Yet as far as recognizing the benefits of consolidating agricultural initiatives into Rio's municipal master plan, the city has a long way to go. Further policies could be implemented to create the enabling conditions necessary for the informal sector to transition toward a green economy. Public security policies, regulatory instruments, public-private partnerships, and voluntary initiatives that collectively shape projects such as Manguinhos are flawed and in need of reform, but as prototypes, they also demonstrate the potential of agroecology as a viable means with which to regenerate fragile urban contexts.

In a city where the social divide is great, and access to capital is limited for the poor, public investments in green economic initiatives are particularly important. Rio can lead by example through urban agriculture efforts that stimulate and fill the demand for organic products and services, and in doing so, potentially indicate how to better help bridge the gap between the city's formal and informal sectors.

Initiatives such as HC, that demonstrate a positive benefit stream from specific green investments and policies, if scaled up and integrated into a comprehensive municipal strategy, could offer an alternative pathway to pro-degrowth, pro-jobs, and pro-poor.

In order for agroecology to become an effective engine for driving green degrowth in favelas, it must not focus on efficiency, independence and profit, but rather on social equity, economic stability, and cooperative interdependence. This may be achievable by promoting innovative forms of land tenure reform and security that include building communal systems of land tenure that also leave room for individual proprietorship; by providing access to solidarity financing for the informal sector; by strengthening and supporting the social and economic capacity of informal stakeholders as they build productivity; by creating an inclusive and transparent legal framework that facilitates green economic activity; and by generating local, informal-to-formal production and distribution chains.

These changes need to be considered carefully and implemented over time. They require long-term commitment that combines human and economic resources with social intelligence. At best, this is tremendously challenging. Yet, if cities such as Rio hope to enhance urban experience and wellbeing, and achieve sustainable productivity, they must enable informal communities to stabilize and contribute as valued assets of an inclusive city.

References

Altieri MA (1987) Agroecology: the scientific basis of alternative agriculture. Westview Press, Boulder, p 196

Arruda J (2011) Agricultura urbana na Região Metropolitana do Rio de Janeiro: sustentabilidade e repercussões na reprodução das famílias, Thesis for Doctorate of Sciences in the Area of Development, Agriculture and Society, UFRJ. http://r1.ufrrj.br/cpda/wp-content/uploads/2012/07/Tese_Juliana_Arruda_20111.pdf

Asara V, Otero I, Demaria F, Corbera E (2015) Socially sustainable degrowth as a social–ecological transformation: repoliticizing sustainability. Sustain Sci 10(3):375–384

Azevedo L, Faulhaber L (2015) SMH 2016: remoçoes no Rio de Janeiro Olympico. Morula Press, Rio de Janeiro

Barros J (2014) Interview with towards the human city project. https://www.youtube.com/watch?v=ZL7M5ELV2S8

Biller D, Petroff K (2012) In Brazil's Favelas, a Middle Class Arises, Bloomberg Business News, 20 December. http://www.bloomberg.com/bw/articles/2012-12-20/in-brazils-favelas-a-middle-class-arises

Black RE, Morris SS, Bryce J (2003) Where and why are 10 million children dying every year? Lancet 361(9376):2226–2234

Braathen E, Sørbøe C, Bartholl T, Christovão A, Pinheiro V (2013) Rio de Janeiro: Favela policies and recent social mobilizations, Working Paper 110, Norwegian Institute for Urban and Regional Research: Oslo, p 42. http://www.nibr.no/filer/2013-110.pdf

Brundtland GH (1987) Brundtland Commission, Report of the World Commission on Environment and Development: Our Common Future. Oxford University Press, Oxford

Bushehri F (2012) Green economy in the Arab region: TEEB capacity-BUILDING workshop for MENA Region. UNEP, Beirut

Calvalcanti M (2015) Waiting in the ruins: the aesthetics and politics of favela urbanization in "PACification" Rio de Janeiro. In: Graham S, McFarlane C (eds) Infrastructural lives: urban infrastructure in context. Routledge, New York, p 97

Câmara Interministerial de Segurança Alimentar e Nutricional (2011) Plano Nacional de Segurança Alimentar e Nutricional: 2012/2015, CAISAN: Brasília, p 19. www.mds.gov.br/segurancaalimentar/LIVRO_PLANO_NACIONAL_CAISAN_FINAL.pdf.pagespeed.ce.NSQXeyLv0S.pdf

Caulfield LE, de Onis M, Blössner M, Black R (2004) Undernutrition as an underlying cause of child deaths associated with diarrhea, pneumonia, malaria, and measles. Am J Clin Nutr 80 (1):193–198

Chen M (2012) The informal economy: definitions, theories and policies, women in informal employment globalizing and organizing, WIEGO Working Paper 1, Cambridge, pp 5–6. http://wiego.org/sites/wiego.org/files/publications/files/Chen_WIEGO_WP1.pdf

Chatterton P, Pickerell J (2010) Everyday activism and transitions towards post- capitalist worlds. Trans Inst Br Geogr 35:475–490

Clarke F (2013a) A Visita do Papa Poderá (finalmente) Desencadear Melhorias no Saneamento de Manguinhos? Rio on Watch, 25 July. http://rioonwatch.org.br/?p=7099

Clarke F (2013b) Demolition in Beira Rio, Manguinhos: Resistance Amidst the Rubble, Rio On Watch, 13 March. http://www.rioonwatch.org/?p=7664

Costanza R, Cumberland J, Daly H, Goodland R, Norgaard R, Kubiszewski I, Franco C (2015) An Introduction to Ecological Economics, 2nd edn. Taylor & Francis CRC Press, Boca Raton

Cravalho de Franca Filho et alii G (2012) Policies of Microcredit in Brazil: an analysis from the case of Community Development Banks (CDBs), Journal du Mauss, March. http://www.journaldumauss.net/?Policies-of-Microcredit-in-Brazil

Da Cunha NV, Mello M (2011) Novos conflitos na cidade: a UPP e o processoda urbanicacão. DILEMAS: Revista de Estudos de Conflito e Controle Social 4(3):371–401. http://revistadil.dominiotemporario.com/doc/DILEMAS-4-3-Art1.pdf

David F (2014) Programa de Regularização Fundiária beneficia 68 mil domicílios em áreas do Morar Carioca, Prefeitura da Cidade do Rio de Janeiro website. http://www.rio.rj.gov.br/web/guest/exibeconteudo?id=4691721

De Cássia Oliveira da Costa Mattos R (2009) Avaliação dos fatores de risco relacionados à exposição ao chumbo em crianças e adolescentes do Rio de Janeiro. Centro de Estudos da Saúde do Trabalhador e Ecologia Humana, Escola Nacional de Saúde Pública, Fiocruz. Ciência & Saúde coletiva, 14(6) Rio de Janeiro, December. http://www.scielo.br/scielo.php?pid=S1413-81232009000600011&script=sci_arttext

De Groot R, Fisher B, Christie M, Aronson J, Braat L, Gowdy J, Haines-Young R, Maltby E, Neuville A, Polasky S, Portela R, Ring I (2012) The Economics of Ecosystems and Biodiversity: The Ecological and Economic Foundations. Routledge, Abington, p 23

de Holanda Barbosa Filho F (2013) An estimation of the underground economy in Brazil, FGV/IBRE | FGV, January. http://portalibre.fgv.br/main.jsp?lumPageId=8A7C8233253AEA0A01253B17F4990DE9&contentId=8A7C82C53CF3C398013D03AB713C0A50

Demaria F, Schneider F, Sekulova F, Martinez-Alier J (2013) What is degrowth? From an activist slogan to a social movement, Environmental values, White Horse Press, 22:191–215. http://www.jnu.ac.in/sss/cssp/What%20is%20degrowth.pdf

Dekker H (2006) In Pursuit of Land Tenure Security. Pallas Publications, Amsterdam, p 128

Della Porta D, Diani M (2006) Social movements: an introduction, 2nd edn. Blackwell, Malden, pp 226–249. http://www.hse.ru/data/2012/11/03/1249193172/Donatella_Della_Porta_Mario_Diani_Social_Mov.pdf

Dubbeling M (2011) Integrating urban agriculture in the urban landscape, Urban Agriculture, RUAF Foundation 25:43–46. http://www.ruaf.org/sites/default/files/UAM%2025-Integrating%20urban%20agriculture%2043-46.pdf

Ebert AW (2014) Potential of underutilized traditional vegetables and legume crops to contribute to food and nutritional security, income and more sustainable production systems. Sustainability 6:319–335

Escola Nacional de Saúde Pública, Fundaçao Oswaldo Cruz (2012)

Fabricius D (2008) Resisting representation: the informal geographies of Rio de Janeiro, Harvard Design Magazine, Spring/Summer 28:5–6. http://wenku.baidu.com/view/3c5c0a030740be1c650e9a5b.html

Fernandes LA, Gotuzzo G (2012) Agroecology and solidarity economics: a sustainable symbiosis, Presented at ISEE conference: ecological economics and Rio + 20: challenges and contributions for a green economy, Rio de Janeiro. http://www.isecoeco.org/conferences/isee2012-versao3/pdf/765.pdf

Filho B, Silva M, Silva DO, Sousa H (1992) Desnutrição em crianças de áreas faveladas: Manguinhos, Rio de Janeiro. Cad Saúde Pública 8(1):69–76

Freeman J (2012) Neoliberal accumulation strategies and the visible hand of police pacification in Rio de Janeiro, REU, Sorocaba, SP, June 38(1):95–126. http://spectrum.library.concordia.ca/977282/1/Freeman_REU_UPP.pdf

Green My Favela (2014) Documentation of conversations with adult and adolescent residents and gardeners of Manguinhos. https://soundcloud.com/learekow/gmf-at-manguinhos-with-helen-and-tyson, https://soundcloud.com/learekow/manguinhos-garden-conversation

Guha-Khasnobis B, Kanbur R (2007) Linking the formal and informal economy: concepts and policies. Oxford University Press, New York, pp 3–6

Hammond AL, Kramer WJ, Katz RS, Tran JT, Walker C (2007) World Resources Institute, International Finance Corporation, World Bank Group, The next 4 billion, market size and business strategy at the base of the pyramid, p 26. http://www.ifc.org/wps/wcm/connect/3c2787004cc75e6094d7b59ec86113d5/Pub_009_The%2BNext%2B4%2BBillion.pdf?MOD=AJPERES

Hart S, Cañeque F, Shpak A, Dasgupta P (2013) Raising the base of the pyramid through enterprise: innovative case studies of BoP ventures and initiatives. Barcelona: GlobalCAD, p 8. http://www.bopglobalnetwork.org.br/labnetwork_web.pdf

Harvey D (2004) The 'new' imperialism: accumulation by dispossession. In: Panitch L, Leys C (eds) Socialist register 2004: the new imperial challenge. Fernwood Publishing, Black Point, pp 63–87

Heintz J (2012) Informality, inclusiveness, and economic growth: an overview of key issues, International Development Research Centre (IDRC), Canada. SIG Working Paper 2012/2, Political Economy Research Institute, Supporting Inclusive Growth program, University of Massachusetts, Amherst, pp 16–17. http://www.idrc.ca/EN/Documents/SIG-WP2-Informality.pdf

Hornborg A (2009) Zero-sum world: challenges in conceptualizing environmental load displacement and ecologically unequal exchange in the world-system. Int J Comp Sociol (50) 3–4:237–262. Sage http://cos.sagepub.com/content/50/3-4/237.abstract

IBGE (Instituto Brasileiro de Geografia e Estatística) (2000, 2004, 2008, 2010) Census; IBGE Agricultural Census (2006)

Instituto Lula (2014) Banking inclusion statistical information. http://www.brasildamudanca.com.br/en/microeconomia/simplified-accounts-facilitate-access-banking-beneficiaries-social-programs

IPP (Instituto Pereira Passos), Prefeitura do Rio de Janeiro, Rio Negócios (2008–2009) DataRio: the city in numbers, p 32. http://rio-negocios.com/en/uploads/2011/04/Relatorio-Data-Rio-Ingles.pdf
Instituto de Pesquisas Data Popular (2012), http://www.datapopular.com.br/
Jauregui J (2005) Public space: broken city. In: Benites C, Lyster C, Abruzzo E, Briseno A, Solomon J (eds) 30 60 90 09: Regarding Public Space, 1st edn. Princeton Architectural Press, Princeton, pp 32–37
Jornal do Brasil (2012) Complexos de Manguinhos e Jacarezinho ganharão obras de saneamento, 22 October. http://www.jb.com.br/rio/noticias/2012/10/22/complexos-de-manguinhos-ejacarezinho-ganharao-obras-de-saneamento/
Kallis G, Kerschner C, Martinez-Alier J (2012) The economics of degrowth. J Ecol Econ, Elsevier, August. http://degrowth.org/wp-content/uploads/2012/11/Kallis_2012_The-economics-of-degrowth.pdf
Karnani A (2009) United Nations Department of Economic and Social Affairs Working Paper 80, The bottom of the pyramid strategy for reducing poverty: A failed promise, p 2. http://www.un.org/esa/desa/papers/2009/wp80_2009.pdf
La Rocque E, Shelton-Zumpano P (2014) The Sustainable Development Strategy of the Municipal Government of Rio de Janeiro, Woodrow Wilson International Center for Scholars, Washington, DC. https://www.wilsoncenter.org/sites/default/files/Paper_%20La%20Rocque%20and%20Shelton-Zumpano_2014_0.pdf
Leal L (2015), Estado precisará de R$ 14 bilhões extras para cobrir déficit em 2016, Estato Rio blog post, 1 October
Lloyd-Jones T, Rakodi C (2002) Urban livelihoods: a people-centred approach to reducing poverty. Taylor & Francis, New York
Magalhaes F, Xavier HN (2003) The case of Rio de Janeiro, Brazil, Understanding slums: Case Studies for the Global Report on Human Settlements Urban Slums Report. http://www.ucl.ac.uk/dpu-projects/Global_Report/pdfs/Rio.pdf
Monteiro D, de Mendonça M (2004) Quintais na cidade: a experiência de moradores da periferia do Rio de Janeiro. Agriculturas 1:29–30
Municipal Secretary for the Environment, Hortas Cariocas webpage. http://www.rio.rj.gov.br/web/smac/hortas-cariocas
Neocleous M (2013) The dream of pacification: accumulation, class war, and the hunt. Socialist Studies/Études Socialistes: J Soc Social Stud 9(2):7–30
Neocleous M (2010) War as peace, peace as pacification. Radic Philos 159:8–17
North D (1990) Institutions institutional change and economic performance. Cambridge University Press, Cambridge, p 101
Odum HT, Odum EC (2001) A prosperous way down: principles and policies. University Press of Colorado, Boulder
O'Reilly E (2014) Agricultura Urbana—Um Estudo de Caso do Projeto Hortas Cariocas em Manguinhos, Rio de Janeiro. Environmental Engineering thesis, UFRJ, p 58. http://monografias.poli.ufrj.br/monografias/monopoli10009377.pdf
Oxford Poverty and Human Development Initiative (2015) Multidimensional Poverty Index, Department of International Development, University of Oxford. http://www.ophi.org.uk/wp-content/uploads/Multidimensional-Poverty-Index-2015-2-March-2015.pdf?0a8fd7
Payne G, Durand-Lasserve A (2013) Holding on: security of tenure-types, policies, practices and challenges, Background paper UN OHCHR, January. http://www.ohchr.org/Documents/Issues/Housing/SecurityTenure/Payne-Durand-Lasserve-BackgroundPaper-JAN2013.pdf
Pereira R, Schwanen T (2013) Commute time in Brazil (1992–2009) Discussion Paper 192, March, Brazilia: Institute for Applied Economic Research. http://repositorio.ipea.gov.br/bitstream/11058/5140/1/DiscussionPaper_192.pdf

Perlman J (2010) Favela: four decades of living on the edge in Rio de Janeiro, Oxford University Press, New York, p XX, 62, 85, 271, 277
Perlman J (2004) Marginality: from myth to reality in the Favelas of Rio de Janeiro, 1969–2002. In: Roy A, AlSayyad N (eds) Urban informality: transnational perspectives from the Middle East, Latin America, and South Asia (Transnational perspectives on space and place), Lexington Books, Lanham, pp 105–146
Perlman J (2003) The chronic poor in Rio de Janeiro: what has changed in 30 Years? Conference on chronic poverty in Manchester, April
Plottu E, Plottu B (2007) The concept of total economic value of environment: a reconsideration within a hierarchical rationality. Ecol Econ 61(1):52–61
Ribeiro SM, Bógus C, Watanabe H (2015) Agroecological urban agriculture from the perspective of health promotion. Saúde e Sociedade, São Paulo 24(2):730–743
Rist G (2014) The history of development: from western origins to global faith, 4th edn. Zed Books, London
Rio de Janeiro Municipal Secretary for Health (2008) A Dengue e a Desigualdaade Na Cidade Maravilhosa. http://www.conhecendomanguinhos.fiocruz.br/files/T&Cdenguebiblioteca BETA_0.pdf
Schneider F (2012) The shadow economy and work in the shadow: what de we (not) know? Discussion Paper 6423, Institute for the Study of Labor, Bonn
Simões G (2015) National coordinator for the Homeless Workers' Movement, quoted by Tavener B, in The Olympics Are Screwing Rio de Janeiro's Poorest Citizens Out of Housing, Vice News, 16 April. https://news.vice.com/article/the-olympics-are-screwing-rio-de-janeiros-poorest-citizens-out-of-housing
Unicef (2008) The State of Latin American and Caribbean Children, Unicef, p 15. http://www.unicef.org/publications/index_45087.html
UNEP (United Nations Environment Program) (2011) Towards a green economy: pathways to sustainable development and poverty eradication—a synthesis for policy makers, p 16, 35. http://www.unep.org/greeneconomy/Portals/88/documents/ger/ger_final_dec_2011/Green%20EconomyReport_Final_Dec2011.pdf
Van Veenhuizen R, Danso G (2007) Profitability and sustainability of urban and peri-urban agriculture. FAO Agricultural Management, Marketing and Finance Occasional Paper No 19. Rome: FAO
Vannuchi L,Van Criekingen M (2015) Transforming Rio de Janeiro for the Olympics: another path to accumulation by dispossession? Articulo—J Urban Res, Special Issue 7. https://articulo.revues.org/2813
Victor PA (2008) Managing without growth: slower by design, not disaster. Advances in ecological economics Series, Cheltenham, UK; Northampton, MA: Edward Elgar publications
Watts J (2014) Fury and frustration in Brazil as fares rise and transport projects flounder, The Guardian, 6 February. http://www.theguardian.com/world/2014/feb/06/brazil-bus-chaos-fare-rise
Watts J (2013) The Rio favela transformed into prime real estate, The Guardian, 23 January. http://www.theguardian.com/world/2013/jan/23/rio-favela-real-estate
Wehrmann B (2008) Land conflicts: a practical guide to dealing with land disputes. GTZ, Eschborn. https://www.giz.de/fachexpertise/downloads/Fachexpertise/giz2008-en-land-conflicts.pdf
Williamson T (2015) A new threat to favelas: gentrification. Arch Rev, 30 May. http://www.architectural-review.com/opinion/a-new-threat-to-favelas-gentrification/8682967.article
World Bank (2015) The state of social safety nets 2015, The World Bank, Washington DC, p 70. http://repositorio.ipea.gov.br/bitstream/11058/5140/1/DiscussionPaper_192.pdf

Author Biography

Lea Rekow (Ph.D.) is director of Green My Favela, an urban restoration project and applied research network based in the favelas of Rio de Janeiro. Her current research focuses on transdisciplinary practices for reclaiming degraded space in areas where people are living under extreme socio-economic and environmental stress. Lea is an advising consultant to the sustainable development action network GlobalCAD, an advisor to the Integrated Media program at Cal Arts, and a member of the Institute for Australian Geographers and of New York Women in Film and Television. Previously based in New York, Lea has held positions as director of Gigantic ArtSpace, media director at the former Center for Peace and Human Security, adjunct professor of media and communications at Pratt Institute, and cultural advisor for Advance. She has also held positions as executive director of the Center for Contemporary Art in Santa Fe, and research fellow at the Center for Art and the Environment. Lea has sat on numerous advisory panels for Urban IxD, NYFA, SVA, Parsons, Amnesty International, the MacArthur award, and the LMCC.

Sustainable Finance Role in Creating Conditions for Sustainable Economic Growth and Development

Magdalena Ziolo, Filip Fidanoski, Kiril Simeonovski, Vladimir Filipovski and Katerina Jovanovska

Abstract

The sustainability issue is the crucial one in distressed, socio-economic environment and in the new and old social risks era. Additionally, the welfare state crisis and the regulations failure which has been revealed after 2008 are significant factors determining the need for searching a new ways and solutions for stabilising of national economies and creating conditions for sustainable economic development. Social and financial exclusion, increasing income disparities, inefficient redistribution system, and negative economic externalities (including environment protection problems) are selected challenges facing states and local governments. This chapter will provide an introduction to these and other broad issues and point out the role of sustainable finance as a toll supporting sustainable economic development. Our empirical research examines the OECD countries after the crisis 2008. We found that soundness of banks, quality of educational

M. Ziolo (✉)
Department of Banking and Comparative Finance, University of Szczecin, Szczecin, Poland
e-mail: magdalena.ziolo@wzieu.pl; magda.ziolo@wp.pl

F. Fidanoski · V. Filipovski
Ss. Cyril and Methodius University in Skopje, Skopje, Republic of Macedonia
e-mail: filipfidanoski@gmail.com

V. Filipovski
e-mail: vladfil@eccf.ukim.edu.mk

K. Simeonovski
Ministry of Finance, Skopje, Republic of Macedonia
e-mail: kiril.simeonovski@gmail.com

K. Jovanovska
University of Ljubljana, Ljubljana, Slovenia
e-mail: katerina.jovanovska@ef.uni-lj.si

W. Leal Filho et al. (eds.), *Sustainable Economic Development*,
World Sustainability Series, DOI 10.1007/978-3-319-45081-0_11

system and burden of government system have a positive impact to the nominal GDP per capita. On the other side, the misery index and ease of access to loans negatively affected nominal GDP per capita in the period of our observation. Hence, as a policy recommendations, it is very important to construct a model which will reflect these and other similar results in order to accelerate the economic development. That is the main challenge for people, governments, financial organisations and companies worldwide.

Keywords
Economic development · Financial inclusion · Sustainable development · Sustainable finance

1 Introduction

The concept of sustainable development has evolved largely and the term sustainable development has received much attention in many spheres: in scientific discourses, in daily life practices, in international negotiations, in local policy measures, in marketing, in business. In other words, the use of the term *sustainability* has become almost inflationary in both science and society (Enders and Remig 2015).

Sustainable development as a concept promises many things to many people. Aspects of government policy, business strategy and even lifestyle decisions have been shaped around the concept. Sustainable development is a complex subject which is difficult to encapsulate in one paper. Authors could only include some of the pieces of sustainable development theory and practice (Mawhinney 2002). Sustainable development means different things to different people.[1] The concept is used in many locations and contexts, by people from varying cultural backgrounds and disciplinary schools of thought, and for different purposes (Lawn 2006). The second element of the term, *-able*, of course means *capable of* or *in a position to* do something. The verb *to sustain*, on the other hand, is more complex. The 1961 edition of the Oxford English Dictionary dedicates several columns to it, and traces it back to the Middle Ages. Of the various definitions, the fourth gives the meaning that appears most interesting to us: it is to *keep in being*, in other words to maintain the existence of something. The next definition is *to cause to continue in a certain state*, and the one after that, *to keep or maintain at the proper level or standard*, and finally, *to preserve the state of*. Thus, sustainable means to be able to *maintain*, to *preserve*, or to *bear, as a load*. It stems from the Latin *sustinere*, which various Latin–English dictionaries render as *support*, *sustain*, or *hold back* (Grober 2015).

The World Commission on Environment and Development (WCED 1987) in the Brundtland Report inaugurated and defines sustainable development in the following way: *sustainable development is a development that meets the needs of the*

[1]Jagdish Bhagwati (2004) very cynically pointed that *even God does not know what sustainable development means.*

present without compromising the ability of future generations to meet their own needs. In other words, according to Asheim (1994), sustainability is a *requirement for our generation to manage the resource base such that the average quality of life we ensure ourselves can potentially be shared by all future generations.*[2]

Considering the role that financial institutions and banks play in society, they can significantly influence the sustainable development achievement. A lot of banks in the United States and Europe, started to adapt their financial policies for investments in environmentally sustainable business (e.g. fiscal green funds) (Jeucken 2001). However, in some old societies, the emphasis was on economic materialism, neglecting the immaterial aspects and the environmental consciousness. Hence, sustainability requires financing and investing activities with long-term capacity for improvement (Hargreaves and Fink 2004). Yet, to say than to do. Starting form early education, the mankind has been preconditioned to focus on the short-term results rather than long-term results; complemented with the wealth maximisation concept, societies are pushed to maximise short-term pay offs at the expense of long-terms' (Fatemi and Fooladi 2013). *Mutatis mutandis*! This calls for a systematic change in our paradigm; shifting from the traditional shareholder wealth maximisation concept to more sustainable and value creative one. Consequently, *homo oeconomicus* and *mammonism* evolve to *moral economic man* and firms from traditional to sustainable, multi-attribute optimisers (profit, people and planet), thus generating financial, social and environmental returns (Emerson 2003; Soppe 2004).

Sustainable investment funds evaluate firms by their attitude towards environment and social sustainability stance. Following multilateral funds for sustainability, such as Global Environment Facility (GEF), Renewable Energy and Energy Efficiency Fund (REEF), SDG (Solar Development Group), and Prototype Carbon Fund (PCF) were established to address global environmental problems in developing countries (Jeucken 2001). Some evidence provides support that sustainability does pay off (Epstein and Roy 2001; Jeucken 2001). Yet, there are no reliable and unified standards for measuring and evaluating sustainability risks and returns-compared to financial risks and returns, neglecting the blended returns related to the natural capital preservation and socioeconomic well-being. More general theory of finance needs to be developed that will incorporate *the socio-ecologically embedded nature of finance* (Fullwiler 2015, p. 18).

---

[2]This statement can be interpreted in many different ways. Some would ridiculously interpret the sustainability requirement as a requirement that all options are preserved, which would imply the preservation of all kinds of resources (we would never use iron, oil or other exhaustible resources). This enforces us to find a clearer definition of sustainable development. So, the economic development in a specified area (region, nation, the globe) is sustainable if the total stock of resources—human capital, physical reproducible capital, environmental resources, and exhaustible resources—does not decrease over time (Bojö et al. 1992). Invest all profits or rents from exhaustible resources in reproducible capital such as machines. This injunction seems to solve the ethical problem of the current generation shortchanging future generations by *overconsuming* the current product, partly ascribable to current use of exhaustible resources. Under such a program, the current generation converts exhaustible resources into machines and lives off current flows from machines and labour (Hartwick 1977).

The rest of this paper proceeds as follows. In Sect. 2, we noted the importance of sustainable finance in the contemporary economic theory and practice. Further, we noted the importance of financial sector in the contemporary economic theory and practice (Sect. 3). Section 4 is dedicated on the connection of sustainable finance from the one side, and economic development, financial inclusion, and stability from the other side. Readers who are interested in our empirical results only can directly go to Sects. 5, 6 where we define the core variables and present the empirical findings from our research. At last, Sect. 7 contains concluding remarks.

2 What Makes the Concept of Sustainable Finance Important?

The basic economic model of private-profit maximisation may be too narrow to incorporate the analysis of maximisation of social returns, or of social value-added in activities with strong externalities or spillovers affecting many sections of the society (intra-temporal perspective) as well as many generations (inter-temporal perspective). In such activities, the problem of insufficient investment may arise due to difficulties in adequately incorporating the external benefits and costs into the private-profit maximisation calculations. Moreover, the *private returns only* approach may be considered as rather narrow from the moral point of view; by being seemingly ethical value-neutral, it may miss to address some wider aspects of human well-being. In this context, the concept of sustainability of economic development is meant to overcome the narrowness of the mainstream economic models and to incorporate wider aspects of economic welfare[3] related to environmental and social *values*. That may mean channeling the investment in such a way that will put the economy on different growth trajectory, one which would achieve improvement in living standard by resolving the problems of negative and positive externalities which may hinder achievement of higher quality of social welfare. Investment that helps achieve sustainable growth trajectory may be considered as socially responsible investment (SRI).[4] Investment Leaders Group (ILG 2014, p. 7) defines the SRI as *investment that creates long-term social, environmental and*

[3]People differ substantially in their economic well-being and welfare, both among countries and among families within a given country (Becker 1962). Maximisation of welfare of every individual, in economic terms, can mean maximisation of the total volume of consumption in the life-time of any individual consistent with the similar maximisation of any other individual (Horvat 1958).

[4]SRI has grown dramatically over the past decade. SRI is a booming market globally. As of 2008 it had a worth of about $2.71 trillion in the United States alone, and nearly one out of every $10 under professional management in the United States belonged to socially responsible investment funds like Calvert, Trillium, and Portfolio 21. Celent, a financial consultancy, predicts that the SRI market in the United States will grow beyond $3 trillion within the next few years. These trends are not are not just confined to the United States; as the European SRI market grew by $892 billion in the last two years (Wilhelm 2013).

economic (sustainable) value, investment that combines financial and non-financial value creation, or investment that correctly prices social, environmental and economic risk. And accordingly, the concept of sustainable finance would basically aim at providing the financial instruments and market structures that would allocate capital in a way which will maximise overall social returns adjusted for the financial as well as non-financial risks.

The implosion of the housing, equity, and debt markets exposed the need for increased regulation, transparency, and proper governance. Therefore, the debt, socially responsible investment, and venture capital sectors have all begun to pressure businesses to factor social, environmental, corporate governance, and climate issues into their financial statements, policies, disclosures, credit assessments, and lending guidelines. The good guide for best practices for financial institutions for dealing with the environment are Equator Principles and the United Nations (UN) Principles for Responsible Investment. The Equator Principles are a comprehensive set of socially and environmentally responsible practices that over 63 of the world's major financial institutions have voluntarily adopted. Borrowers that want loans from these signatories must categorise and fully disclose all risks associated with social and environmental performance and provide a mitigation plan for managing these risks. The UN Principles for Responsible Investment is another voluntary set of guidelines that provides investors with a range of options for not only fulfilling their fiduciary duty but also giving consideration to the Environmental, Social and Corporate Governance (ESG) issues of the companies they choose (Wilhelm 2013).

From a more institutional point of view, sustainable finance may be viewed as creating institutional arrangements in the financial system that can contribute to achieving sustainable development of the economy (see more in Pisano et al. 2012). In this context, sustainable finance would include financial instruments, institutions and markets which would take into consideration the issues like environmental and social responsibility of entities that seek to get finance. So, it can be seen that sustainability and responsibility in finance and investment requires an ethics. By ethics is not meant ethical investment but a telos of awareness of the social interconnections forged by financing activity (Haigh 2012). The ethical (or *green*) financial instruments would properly involve provisions related to the purposes which financial funds provided by their issuance and may be used for advanced financial operations. The ethical (or socially responsible) financial institutions would have their investment policy aims include financing entities which pay due respect to sustainability issues like: environmental protection, social responsibility, and high quality governance standards; and would also take into account the environmental, social and governance risks when designing their risk management processes. The ethical (or socially responsible) financial markets would create market structures which enable environmentally/socially/governance responsible firms to get lower cost finance due to their higher standards for transparency and engagement in sustainability issues.

One overriding principle of sustainable finance could be formulated in the following manner: when choosing their investment portfolio goals and strategies, asset owners (the principles) and asset managers (the agents) should strive to accomplish a two-fold objective—to pursue the objectives of diversification of risk and an adequate level of liquidity of the portfolio (and hence adequate flexibility), but at the same time to pursue a more long-term (sustainable) strategy which would enable dynamic optimisation of investment portfolio from the perspective of achieving societal values—i.e. maximising long-term social returns adjusted for environmental/social/governance risks.

3 The Nexus Between Financial Sector and Economic Growth and Development

The relationship between financial development and economic growth has received a great deal of attention throughout the modern history of economics. An early and intellectual development came from Bagehot, in his classic *Lombard Street*, where he emphasised the critical importance of the banking system in economic growth and highlighted circumstances when banks could actively spur innovation and future growth by identifying and funding productive investments (Arestis 2006). We can learn from the economic history that the main cause of early U.S. economic growth (1780–1850) was the development of the financial sector.[5] Adam Smith, more than 200 years ago, recognised the crucial role of finance in economic growth in his memorable book on wealth of the nations (Smith 1776/2008). Only when the financial system is sufficiently advanced significant numbers of major, wealth-creating projects be profitably undertaken (Wright 2002). Larger and more efficient financial markets help economic agents hedge, trade, and pool risk, raising investment and economic growth (Rioja and Valev 2004).

It is increasingly recognised that the financial system plays a crucial role in the process of economic development.[6] In principle, both technological and financial innovations have driven modern economic growth (Todaro and Smith 2012). The precise mechanism by which financial development spurs economic growth, though

[5]Buchanan also argued that the major difference between the two nations was their respective financial sectors (Wright 2002).

[6]It is obviously clear that macroeconomic policy can have important impacts on the environment in a different and many ways. For example, easier access to the financial services can help the poor avoid intensive degradation or mining of the environment (Taylor 1996). Moreover, sustainable development has to do with inter-temporal resource allocation. It is therefore very natural to expect a close connection between the interest rate and sustainability (Bojö et al. 1992). Without financial sector it is impossible to reach and attain sustainable development. Expanding access to financial services holds the promise to help reduce poverty and spur development. Economists have long linked the expansion of financial markets to the spread of broader economic activity. By the same token, economists have focused on ways that barriers to financial markets undermine economic efficiency. In other words, the dominant part of the economic thought is dedicated to ensure a soundly financial environment as a prerequisite for development (Karlan and Morduch 2010).

not accepted by all economists,[7] has long been understood.[8] Basically, most trace the idea to Joseph Schumpeter, who academically argued that the services that financial without intermediation, visionary projects that would have created wealth wither and die (Wright 2002). The cornerstone and central focus of his theory is the concept of a circular flow of production and consumption. Importantly, credit is the key instrument for ensuring of entrepreneurial success and for break of the circular flow through innovation (Schumpeter 1961). Consequently, financial arrangements that lower transaction costs can promote specialisation, technological innovation and growth (Levine 2005). Better financial systems improve the probability of successful innovation and thereby accelerate economic growth and vice versa. Similarly, financial sector distortions reduce the rate of economic growth by reducing the rate of innovation (King and Levine 1993b).

Growth requires investment, so poor agents without access to financial markets will not have the resources to invest (Acemoğlu et al. 2005). In a narrower view, all financial systems in market economies, whether they are developed or developing, perform two basic functions[9]: administering the country's payments mechanism and intermediating between savers and investors (Fry 2000). To bridge the gap between saving and investment, which widens with increasing specialisation, further financial development is needed to create those assets which can be used to channel surplus funds into investment use. In short, financial deepening does contribute to capital accumulation and economic development. That said, the greater the extent to which financial deepening has occurred, the greater is the extent to which investment is undertaken (Lin 1981).

[7]Robinson (1952) stressed that *where enterprise leads, finance only follows*. The possible interpretation is that there is an *unholy marriage* between finance and growth. Bisarrely, it is ironic that Lucas (1988) also dismisses financial matters as a very badly over-stressed determinant of growth in popular and even much professional discussion.

[8]Keynes (1930) also recognised the important role of banks for the growth. He argued that the credit is the pavement along which production travels; and the bankers, if they knew their duty, would provide the transport facilities to just the extent that is required in order that the productive powers of the community can be employed to their full capacity.

[9]The primary function of any financial system is to facilitate the allocation and deployment of economic resources, both across borders and across time, in an uncertain environment. From the most aggregated level of the single primary function of resource allocation, we can further distinguish six basic functions performed by the financial system: (i) a financial system provides ways of clearing and settling payments to facilitate the exchange of goods, services, and assets; (ii) a financial system provides a mechanism for the pooling of funds to undertake large-scale indivisible enterprise or for the subdividing of shares in enterprises to facilitate diversification; (iii) a financial system provides ways to transfer economic resources through time, across geographic regions, and among industries; (iv) a well-functioning financial system facilitates the efficient allocation of risk-bearing; (v) a financial system provides price information that helps coordinate decentralised decision-making in various sectors of the economy; (vi) a financial system provides ways to deal with the incentive problems when one party to a financial transaction has information that the other party does not, or when one party is an agent for another (Merton and Bodie 1995). The basic functions of a financial system are essentially the same in all economies-past and present, East and West (Merton 1995).

It is important to express that the role of financial sector in economic development was either neglected or relegated to a secondary or accommodating position, while primary emphasis was placed on the role of real factors, such as physical and human capital (Lin 1981). We believe that physical, human capital, together with the finance, are the parts of the same coin. Their joint synergistic effect is *sine qua non* for growth and development. From other perspective, first authors and antecedents of the contemporary and substantial analysis of finance-growth were Goldsmith (1969) and McKinnon (1973), who drew attention to the contribution of financial activities to economic growth and development. In the next period, the financial sector and its role in the process of economic development has attracted notable attention since the early 1990s (Asian financial crisis). Commercial banks provide an intermediation service that brings savers and investors together. Neoclassical growth models tell us that an increase in the efficient investment of savings in new and innovative projects is one of the main engines and factors of economic growth (Armenta 2007).

In the spirit of recent time and works, many theoretical and empirical researches have shown that a sound and effective financial system is critical for economic development and growth. The financial system, however, is also subject to boom and bust cycles and fragility, with negative repercussions for the real economy (Beck 2011).

In the literature, there is insightful debate *about whether financial intermediaries*[10] *and services affect long-run growth*? While the basic idea that finance affects growth is not new and can be traced back at least to Schumpeter's *Theory of Economic Development* (1969), it is fair to say that until recently most economists looked with scepticism at the proposition that financial conditions could explain part of the cross-country differences in levels of development and rates of growth. Nevertheless, many countries have witnessed a resurgence of interest in the study of how financial intermediaries and services affect long-run growth (Galetovic 1994). In market economies, financial intermediaries develop during the early stages of industrialisation. Galetovic (1996) argues that it occurs because as firms specialise the number of transactions involving credit increases. Famously, we can conclude that banks and other financial intermediaries are the main source of external funds to firms in the all stages of firm development (Diamond 1996). Countries with more banking facilities per capita in 1830 experienced greater rates of economic growth up

[10]In a Walrasian model, firms and individuals use markets to borrow and to diversify idiosyncratic risks. But because firms can issue perfectly divisible securities, information is symmetric, and complete contracts can be written and enforced at no cost, there is no need for financial intermediaries, and exchange is organised through direct and impersonal markets. As it is well known, under such assumptions an efficient risk allocation obtains whenever markets are complete; but financial markets are quite uninteresting: they work like any other commodity market, their efficiency is a straightforward implication of the technological environment, and financial arrangements are irrelevant (Modigliani-Miller). One explanation of financial intermediaries starts with the observations that the securities issued by individual firms are not perfectly divisible and there are scale economies in transaction technologies—the ideas of Gurley and Shaw (Galetovic 1994).

to 1860 than states with less-developed banking sectors (Bodenhorn 2003). Therefore, we will agree with Alexander Hamilton, one of the founding fathers of the United States, who argued that *banks were the happiest engines that ever were invented for spurring economic growth* (Beck 2011).

It will be convenient to assess the degree of *how important is financial intermediation and development for economic development*? Increasing evidence and a substantial body of empirical work assesses the impact of the operation of the financial system on economic growth suggests that the components of the financial system, e.g. banks and stock markets, may play a more direct or active role in the development process (Lin 1981; Levine 2005). The contemporary theoretical and empirical literature supports the proposition that financial development plays a significant and positive role for higher economic growth (Filipovski 2004). But the *financial development* is a too generic term; to gauge the impact on growth, one must specify the particular financial market concerned (Pagano 1993). Financial development is the accumulation of financial assets at a rate that is more rapid than the rate of non-financial assets accumulation. Economic growth is the evolution of gross domestic product (GDP) in the short, medium and long terms. From the collected data observed on 21 Sub-Saharan African (SSA) countries during the period 2000–2014 and by using the dynamic panel GMM technique, Ngongang (2015) showed that there exists a positive link between financial development and economic growth. In addition, Aghion et al. (2005) documented that all countries above some critical level of financial development should converge in growth rate, and that in such countries financial development has a positive but eventually vanishing effect on steady-state GDP.

Mainly given the renewed attention of finance issues, there is a sizable amount of theoretical literature on the relationship between financial development and economic growth. Arestis et al. (2001) showed that while stock markets may be able to contribute to long-term output growth, their influence is, at best, a small fraction of that of the banking system. Also, the analysis of Greenwood et al. (2013) suggests that financial intermediation is important for economic development. Significantly, Hasan et al. (2009) tested the quality finance-growth nexus for a comprehensive sample of more than 100 countries during 1996–2005. They reported that the relation between volume-based measures of financial development and economic growth is weak.

One important study is prepared by Xu (2000), who used a multivariate vector-autoregressive approach to examine the effects of permanent financial development on domestic investment and output in 41 countries between 1960 and 1993. The results reject the widely accepted theoretical proposition that *financial development simply follows economic growth* and has very little effect on it. Additionally, there are also other papers that confirmed the positive relationship between finance and growth and the beneficial effects of financial development (King and Levine 1993a; Rajan and Zingales 1998; Khan and Senhadji 2000;

Rioja and Valev 2004[11]; Loayza and Rancière 2006[12]). Particularly, Christopoulos and Tsionas (2004) have combined cross-sectional and time series data to examine the relationship between financial development and growth in ten developing countries. Their empirical results provide a clear support for the hypothesis that there is a single equilibrium relation between financial depth, growth and ancillary variables (investment share and inflation), and that the only cointegrating relation implies unidirectional causality from financial depth to growth.

More carefully, Rousseau and Wachtel (2005) found that the finance–growth relationship in not as strong with more recent data as it was in the original studies with data for the period from 1960 to 1989. They offered two possible explanations. First, financial depth may have had greater value as a shock absorber in the 1970s and 1980s, decades characterised by worldwide nominal shocks. Second, the spread of financial liberalisation in the 1980s may have led to increasing financial depth in countries that lacked the legal or regulatory infrastructure to successfully exploit financial development.

Other authors have wider points of view on finance-growth hypothesis. Miller (1998) defends a position directly counter to the anti-financial market views. At last, it is important to stress that financial structure also matters for the growth. Financial intermediation promotes growth because it allows a higher rate of return to be earned on capital, and growth in turn provides the means to implement costly financial structures[13] (Greenwood and Jovanovic 1990). The organisation of financial activities within a country affects economic growth through its impact on how corporations raise and manage funds. Financial structure can hinder or promote financial development, though. With a poor financial structure, the cost of capital is too high, so that it is difficult for entrepreneurs to create firms and for these firms to invest efficiently. Financial structure has to be designed so that both financial intermediaries and capital markets can play their role effectively (Stulz 2001). Growth provided the wherewithal to develop financial structure, while financial structure in turn allowed for higher growth since investment could be more efficiently undertaken. In maturity, an economy has a fully developed financial structure, attains a stable distribution of income across people, and has a higher growth rate than in its infancy (Greenwood and Jovanovic 1990). In the similar

[11]In a panel of 74 countries during 1961–1995, authors found that finance has a strong positive influence on productivity growth primarily in more developed economies. In less developed economies, the effect of finance on output growth occurs primarily through capital accumulation.

[12]They suggested that a positive long-run relationship between financial intermediation and output growth co-exists with a mostly negative short-run relationship. Authors found that financially fragile countries, namely those that experience banking crises or suffer high financial volatility, tend to present significantly negative short-run effects of intermediation on growth. For more stable countries, this effect is in average nil.

[13]Financial development and the component defined by the legal protection of outside investors explain long-term cross-country growth rates. Financial structure, namely, the distinction between market- and bank-based financial systems, does not offer any additional information. These results presented by Beck et al. (2001) are robust to the use of different indicators of financial development and structure and different conditioning information sets.

story, by using of advanced and theoretically sound econometric methodology, Arestis et al. (2004) reported that financial structure exerts significant effects on the level of output per capita in all but one country (the Philippines). Yet, the magnitude of the long-run effects (cointegrating parameter) of financial structure on per capita output is extremely heterogeneous across countries.

4 Sustainable Finance for Development, Inclusion, Stability and Equity

Sustainability is a well-established concept in the field of environmental economics (Soppe 2004). There are numerous aspects of sustainability, including sustainable development, sustainable economics, sustainable growth, sustainable finance, etc. Soppe (2004) pointed out that there is one key point to understanding sustainability as a whole: the core concept of sustainability explicates that a connection should be made between present and future generations. In summary, sustainability is a concept that seeks to find balance among social, economic, and environmental issues and is a crucial requirement to secure a good quality of life for future generations. This balance is especially important in terms of the economics of inequality (Piketty 2015), with particular attention given to increasing income disparities, an aging society, social problems, the crisis of capitalism, and unprecedented environmental challenges that will become millstones. The scope of the aforementioned problems creates social and economic costs and consequences, all of which impact negatively on sustainable development and economic growth. These problems must be addressed, because they strongly affect quality of life, economic performance, and fundamental societal rights and values such as liberty, equality, justice, and fairness. In such conditions, sustainable development and sustainable growth are in high demand. One significant part of sustainability is sustainable finance. The concept of sustainable finance has received a great deal of attention in recent decades, parallel to the rise of sustainable development as an idea.

The general approach to sustainable finance is to encompass the economic, environmental, and social sustainability of finance (Wilson 2010). As finance and economics are strictly bounded and coherent, sustainable finance may be an instrument for ensuring and supporting the process of building the framework for sustainable development and sustainable growth; on the other hand, it could be one of the sources of distress in a given economy. The phenomenon of financialisation is the best proof that finance and the financial system may be dangerous for economy and society without efficient control; a *black swan* scenario is a real threat to sustainability and the stability of markets (Taleb 2008). The crisis of 2008 was definitely a global financial crisis that spilled over from financial markets to the public finance space and strongly affected every sphere of economic and social life. The costs as a result of the crisis have been especially problematic to mitigate. Today, global leaders are looking forward and seek for policies which will ensure a better and more sustainable development.

There is strong evidence in the literature that finance is an efficient tool that can be used to overwhelm the exclusion problem in an economy. The so-called financial inclusion process results in poverty reduction and lower income inequality. Importantly, Sarma (2008) defined financial inclusion as a *process that ensures the easy access, availability, and usage of a formal financial system for all members of an economy*. In his paper, Honohan (2008) showed that higher financial access significantly reduces income inequality, as reflected by the Gini coefficient (Park and Mercado 2015). Furthermore, Burgess and Pande (2003), based on research carried out in 2005 in India, found that the -led expansion of rural bank branches was one of the crucial factors determining poverty reduction. The presence of banks in an economy and in the financial system is a significant stimulus for creating a sustainable finance framework. Many banks have recently implemented policies and procedures that address the environmental and sustainability impacts of their operations. The term *green banking* reflects this kind of approach in the banking sector, which requires a strong policy framework, transparency of implementation, adapting an environmental and social management system, and exercising leadership in sustainable finance.

Research concerning sustainable finance (especially green banking) has provided details regarding the sorts of policies implemented by different banks in order to meet the expectations of sustainable finance, particularly environmental or social concerns such as human rights, labour rights, indigenous people, climate and energy, dams, biodiversity, forests, fisheries, extractive industries, sustainable agriculture, chemicals, transparency, and reporting done by clients and environmental and social management systems (WWF-UK and BankTrack 2006). Projects covering the fields of interest mentioned above are important for financing and cooperation in the analysed banks. The allocation of money (lending policy) within selected scopes of interest in the social and environmental dimensions allows for the stimulation of sustainable development and growth. One of the best practices is to address financial regulations while taking into account their impact on environmental and social risk (ESR). In the literature, the number of studies and reports regarding the inclusion of environmental and social aspects in financial market regulation are systematically increasing for different groups of countries. The efficient management of ESR requires discussion and consideration related to: (1) the method of regulation (quality of regulations, level of restrictions); (2) legacy portfolios; and (3) the method of attracting green investments.

Financial regulations are important and stand as a leading factor that determines the success of sustainable finance goals. That is why the quality of such regulations matters. The more regulated the financial system in terms of financial sources, the more difficult it is to achieve financing for green projects. Similarly, Dewatripont et al. (2010) suggested that regulation is too often designed to *fight the previous crisis* rather than the next one, and is typically one step behind new market trends. Further, it is important to consider that an additional regulatory burden may encourage the shift of financing activity to the shadow banking sector, which is less regulated and better for the provision of loans and other financial services such as project finance. The crowding out effect of shadow banking is one possible way to achieve easier financing in the overregulated financial sector. In his speech at the 2012 Financial Markets Conference, sponsored by Federal Reserve Bank of Atlanta, Ben Bernanke stressed the importance of regulation of the shadow banking, as an action for prevention of potential channels for the propagation of shocks through the financial system and the economy. There is a strong evidence that the greater risk-taking capacity of the shadow banking system leads to an increased demand for new assets to fill the expanding balance sheets and an increase in leverage. The picture is of an inflating balloon which fills up with new assets. As the balloon expands, the banks search for new assets to fill the balloon and on that way they hurt and distort the financial markets (Shin 2009).

However, the global financial regulatory landscape and mosaic is more likely to resemble a *Japanese garden*, with new details and new perspectives emerging at each step, rather than a centralised and symmetrical *jardin à la française*. Consistency will not be uniformly achieved, but as political philosopher Francis Fukuyama put it in a lecture in 2005 at Yale University: *creating new institutions that will better balance the requirements of legitimacy and effectiveness will be the prime task for the coming generation* (Rottier and Véron 2010).

Legacy portfolios are the second concern, because in order to avoid the lock-in effects of investments with environmental and social risks in highly polluted industries, quick moves must be made in the loan books of banks and other regulated lenders. Therefore, in this case, shifting from the banking sector to shadow banking may be slower since it is difficult to quickly make changes and revise the portfolio of loans issued in the past. One of the suggestions to support sustainable finance is to not punish but rather attract and stimulate green investors.

Sustainable finance, in some circumstances, is an effective tool for ensuring development, inclusion, stability, and equality. The first condition for that is creating and sustaining a framework for sound banking practices, risk management, and adequate consumer protection due to the over-indebtedness risk that occurs in inclusive financial systems (bin Ibrahim speech 2010). Other requirements linked to building efficient mechanisms for sustainable finance are strictly connected with implementing (WWF-China 2015):

- international guidelines and collaboration (cross-sectoral agreements, sector-specific agreements, international exchange and coordination);
- the development of finance institution standards (IFC Sustainability Framework and Performance Standards, the World Bank standards, etc.);
- national policies and regulation (environmental regulation, such as pollution limits, mandatory environmental and social provisions in risk assessments, etc.); and
- the voluntary commitment of financial institutions (for example, in banks, environmental and social standards to better manage their operations, Global Reporting Initiative).

The quality of legal regulation and legislative framework is crucial to ensure the successful implementation of a sustainable finance concept.[14] Gerster (2011) points out that cost savings, increased revenues, reduced risks, the development of human capital, and the improvement of access to capital are the potential positives listed by providers of financial services in the context of outcomes created by sustainable practices. According to Gerster (2011), the sustainability approach to banking (and finance) can be expressed in the business area in a comprehensive way through products (including project financing, funds), outreach (retail, including SMEs, microfinance), private banking, investment banking, processes, and logistics. All of these financial services implicate and support the process of sustainable development and reduce ESR risks.

Sustainability is a leading criterion in the decision-making process in sustainable finance with respect to financial markets and private entities as well as in public finance. State and self-governing entities play a key role in implementing and supporting the sustainable finance concept. There is some empirical evidence that a positive association exists between higher public revenue (as a % GDP) and income inequality reduction, as well as the key role of public social spending on human development. This is not the only factor determining the importance of a redistributive policy as an instrument supporting sustainability. A redistributive policy also determines the assets and capital allocation between the private and public sectors for equity and sustainable development in order to promote financial transparency, which could facilitate tackling harmful tax competition and the taxation of mobile capital and financial wealth (Kohler 2015). Sustainability in finance is also expressed by effective debt management, which means that state authorities need to control the public deficit and debt level effectively since excessive deficits and debt create negative effects and costs for society.

5 Definition of Variables

Peace and security, freedom, development and the environment remain prominent aspirations today, and it has been increasingly acknowledged that they are closely interconnected. Insufficient level of development can significantly threaten the peace and security, and vice versa. Development provides the capacity to sustain nature's life-support systems, but can also threaten them, in turn setting back development. Strong interdependencies are now recognised among the economic, social and

[14]Given the inter-temporal character of financial transactions and the high degree of asymmetric information and the resulting agency problems, legal institutions play an especially important role in the financial sector. Among the institutions that financial economists have focused on are those governing agency relationships, such as the rights of secured and unsecured creditors vis-a-vis borrowers in- and outside bankruptcy and the rights of minority shareholders vis-a-vis management and blockholders, as well as institutions that help overcome information asymmetries, including the quality of accounting and auditing frameworks and systems of credit information sharing (Beck 2010).

environmental dimensions of sustainable development. In its original version, sustainable development objectives have been widely defined along three dimensions: *economic*, *environmental* and *social* or *ecology*, *economy* and *equity* (3E) (UN DESA 2014). Following the consideration of sustainable development in the context of a three-dimensional model and its proposed extension with a fourth pillar (UN DESA 2014), the variables in this research can be properly classified into four groups with each representing the economic, social, environmental or institutional dimension. The variables selected for the purpose of the research are explained in turn.

The economic group includes *the real GDP per capita*, *ease of access to loans*, *soundness of banks* and *misery index*. Real GDP per capita (Y) is used as a proxy for economic development and represents the dependent variable in the model. The values for this variable are calculated using constant prices in 2005 and for calculation purposes are included as logarithms. Ease of access to loans (EAL) is an indicator developed by the World Economic Forum for the Global Competitiveness Index to signify how easy businesses can get bank loan only with a good business plan and no collateral. Countries are ranked on a scale from 1 to 7, with 1 denoting most difficult and 7 least difficult conditions. In general, one would logically expect that greater ease in getting bank boosts the private sector and supports higher economic development; the Global financial crisis, however, confirmed that an extremely high ease of extending loans may end up with disastrous effects on the economy. Soundness of banks (SB) is another constituent indicator of the Global Competiteveness Index, which shows the level of soundness of banks in terms of the necessity for their recapitalisation. The indicator ranks countrics on a scale from 1 to 7, where 1 indicates an extremely low soundness with high necessity of recapitalisation and 7 indicates healthy banks with sound balance sheets. Banks with sounder balance sheets are of key importance for ensuring financial stability. The higher financial stability, the higher prospects for ensuring long-term economic development. The last variable in this group the misery index (MI), first developed by Arthur Okun, sums up the unemployment rate with the inflation rate and assumes that higher values of these rates both increase the economic and social costs, putting downward pressures on economic development.

The social group includes the *quality of educational system* (EDU), also part of the Global Competitiveness Index, which shows how well the educational system confronts the needs of a competitive and market economy. Similarly to the ease of access to loans and soundness of banks, this indicator assigns values on a scale from 1 to 7, with 1 indicating a very low quality of educational system and 7 indicating a very high quality of educational system. It is a pure convenience to assume that higher quality of the educational system increases the productivity of the labour force, improves the innovation capacity and positively affects the economic development.

The environmental group includes the *energy intensity* (ENI), calculated as total primary energy supply per capita, as a proxy for the level of energy supplied in the economy. Higher levels of this variable point out to a higher energy-intensified production and may have positive effects on economic development; however, this assumption cannot be taken into account so strong, as its values also depend on other factors (e.g. effects of climate change, population growth, and population density). De Cian et al. (2013) have more intensively studied the energy intensity by decomposing the aggregate energy intensity into two components: (i) changes in industrial activity composition (structural effect) and (ii) changes in sectoral energy intensity (technological effect).

The institutional group includes the *burden of government regulation* (BGR), included in the Global Competitiveness Index, which is associated with bureaucracy red tape and shows how burdensome is for businesses to comply with governmental administrative requirements in their country. On a scale from 1 to 7, a score of 1 denotes highest burden, while a score of 7 lowest.

A complete review of the selected variables with their description, notation and value can be found in Table 1.

Table 1 Definition of variables

Group	Variable	Description	Notation
Economic	Real GDP per capita	Logged value of GDP per capita at constant prices in 2005	Y
	Ease of access to loans	Score on a scale from 1 to 7 (1 = most difficult; 7 = easiest)	EAL
	Soundness of banks	Score on a scale from 1 to 7 (1 = least sound; 7 = soundest)	SB
	Misery index	Absolute value of the sum of unemployment rate and inflation rate	MI
Social	Quality of educational system	Score on a scale from 1 to 7 (1 = very low quality; 7 = very high quality)	EDU
Environmental	Energy intensity	Total primary energy supply per capita	ENI
Institutional	BGR	Score on a scale from 1 to 7 (1 = most burdensome; 7 = least burdensome)	BGR
Other	Year dummy for 2010	Dummy variable capturing the year-specific effects in 2010	D2010
	Year dummy for 2011	Dummy variable capturing the year-specific effects in 2011	D2011
	Year dummy for 2012	Dummy variable capturing the year-specific effects in 2012	D2012
	Year dummy for 2013	Dummy variable capturing the year-specific effects in 2013	D2013

Sources IMF World Economic Outlook Database, OECD Data and World Economic Forum Global Competitiveness Reports

6 Empirical Research and Results

This section includes an empirical research on the role of sustainable finance for the ensuring of sustainable economic development across the OECD members[15] in the 2008–2013 period. As the data obtained for the research are cross-sectional, covering 34 countries for 6 years, the study employs a panel regression analysis. In its simplest form, a static panel regression model takes the following form:

$$y_{i,t} = \alpha_i + \beta_i x_{i,t} + v_{i,t}, \tag{1}$$

where $y_{i,t}$ is the dependent variable; α_i is the time-invariant individual effect; $x_{i,t}$ is a time-invariant matrix of explanatory variables; β_i is a vector of regression coefficients; and $v_{i,t} = u_{i,t} + \varepsilon_{i,t}$ is the error term, incorporating the unobserved country-specific effects, i.e. fixed effects, $v_{i,t}$ and the idiosyncratic error $\varepsilon_{i,t}$.

The assumption of a static model, however, does not usually hold in practice, as tendency of persistent values of the dependent variable may exist. This leads to the necessity of adopting a dynamic panel regression model, which includes lagged values of the dependent variable as regressors. Assuming that the lag length is one period, the dynamic panel regression model would be:

$$y_{i,t} = \alpha_i + \theta y_{i,t-1} + \beta_i x_{i,t} + v_{i,t}, \tag{2}$$

where $y_{i,t-1}$ is the one-period lagged value of the dependent variable $y_{i,t}$; γ is the convergence rate towards the equilibrium level; and anything else remains the same as in Eq. (1).

The specification of Eq. (2), albeit capturing the tendency of persistent values of the dependent variable over time, may still suffer from other econometric problems. One such problem that usually occurs is endogeneity of the explanatory variables, implying correlation between regressors of the explanatory variables, i.e. the endogenous variables, and the error term. In fact, causation does not always stem from the explanatory variables but it may run in both directions. For instance, soundness of banks or quality of educational system can lead to changes in real GDP per capita but it is worth taking into account that changes in these two variables can be strongly affected by the overall conditions in the economy. Another problem may be the existence of correlation between the time-invariant country characteristics, i.e. the fixed effects, and the error term. A solution to these problems was first proposed by Anderson and Hsiao (1981), who developed instrumental variables (IV) estimation by taking the first difference of the regression equation to eliminate the fixed effects and utilising higher lag-order values of the

[15]The empirical research includes all OECD members. These are: Australia, Austria, Belgium, Canada, Chile, Czech Republic, Denmark, Estonia, Finland, France, Germany, Greece, Hungary, Iceland, Ireland, Israel, Italy, Japan, South Korea, Luxembourg, Mexico, Netherlands, New Zealand, Norway, Poland, Portugal, Slovakia, Slovenia, Spain, Sweden, Switzerland, Turkey, the United Kingdom and the United States.

dependent variables as instruments in the model, leading to increased asymptotic efficiency. Arellano and Bond (1991) made improvements of the Anderson–Hsaio estimator by implementing a generalised method of moments (GMM) estimation rather than IV estimation to the same set of instruments and thereby achieving higher asymptotic efficiency with lower variance. Another important property of the Arellano–Bond GMM estimator is its usefulness in panel data analyses with short time dimension and large country dimension.

Because of the properties mentioned above, the study in this paper employs a balanced dynamic panel regression analysis using the Arellano–Bond GMM estimator of the following form:

$$\Delta y_{i,t} = \Delta\theta y_{i,t-1} + \Delta\beta_i x_{i,t} + \delta_i d_t + \Delta\varepsilon_{i,t}, \tag{3}$$

where Δ is a sign for the difference of the regression equation; d_t is a matrix of year dummy variables; δ_i is a vector of regression coefficients; and anything else remains the same as in Eqs. (1) and (2). It should be noted that a full set of year dummy variables were introduced in the model to control for the unobserved macroeconomic effects that may cause significant changes in the dependent variable over the course of the years. Also, Eq. (3) includes only the idiosyncratic error $\varepsilon_{i,t}$, as the time-invariant fixed effects captured in $u_{i,t}$ have been eliminated by taking first difference:

$$\Delta v_{i,t} = (u_t - u_t) + \left(\varepsilon_{i,t} - \varepsilon_{i,t-1}\right) = \Delta\varepsilon_{i,t}.$$

The results of the panel regression analysis are reported in Table 2. Statistically significant coefficients were estimated for the lagged real GDP per capita, soundness of banks, misery index and energy intensity, while the estimated coefficients for the ease of access to loans, quality of educational system and burden of

Table 2 Estimation results

Variable	Coefficient	Standard error	t-statistic
logY(−1)	0.491810***	0.023521	20.90917
EAL	−0.006622	0.005979	−1.107469
SB	0.012883***	0.003258	3.954073
MI	−0.004357***	0.000373	−11.69531
EDU	−0.009723	0.005875	−1.654872
ENI	0.040462***	0.007313	5.532481
BGR	0.003442	0.005916	0.581803
D2010	0.044473***	0.002288	19.44117
D2011	0.008203***	0.002002	4.097943
D2012	−0.005496***	0.001640	−3.350975
D2013	0.003470	0.002134	1.625769

***, ** and * denote statistical significance at the level of 1, 5 and 10 %, respectively governmental regulations are statistically insignificant. Additionally, all statistically significant coefficients exhibit strong significance at the level of 1 %.

The positive coefficient of the lagged real GDP per capita shows that the initial level of real GDP per capita positively affects its future movement. The absolute convergence rate of 0.492 points out that convergence occurs at a high speed and that the real GDP per capita would increase by about 0.48 % had its initial level increased by 1 %.[16] This confirms that the economic development across the OECD members can be properly considered a dynamic process, in which the initial level of real GDP per capita significantly impacts its future movement.

Positive effects were estimated for the soundness of banks, energy intensity and burden of government regulations. Following the Global financial crisis, largely triggered by the developments in the financial sector and banks' aggressive behaviour primarily in the OECD members, the banking regulators have been looking for policies that enjoin more prudent behaviour in order to ensure greater financial stability. Such policies usually require recapitalisation towards greater proportion of equity hold and thereby increasing balance sheet's soundness. The positive impact of the soundness of banks on the real economic growth shows that the policies supporting more prudent bank behaviour, albeit reducing banks' credit potential, are supportive in the assurance of economic development. In connection with the magnitude of this impact, the estimated coefficient of 0.013 implies that an increase in the score by 1 would increase the real GDP per capita by 1.3 %.[17] The finding about the positive impact of the energy intensity is consistent with the assumption that higher energy-intensified production may have positive effects on economic development. Technological improvements have always been common for OECD members, bringing the sectoral energy intensity on a higher level. Since changes in technology are generally seen as significant contributor to the level of economic development, their role in increasing energy intensity can be properly considered a channel to higher economic development. The magnitude of the positive effects through the estimated coefficient of 0.040 implies that an increase in energy intensity by 10 % would be associated with an increase of the real GDP per capita by about 0.41 %. The positive impact of the burden of government regulations would mean that any decrease in the number of regulations and the associated cost that the businesses have to bear brings up the real GDP per capital and increases the level of economic development; but the impact is statistically insignificant and therefore the demonstrated relationship cannot be considered as reliable.

[16]Since the values of both the real GDP per capita as dependent variable and its lagged value as independent variable are expressed as logarithms, the increase in real GDP of 4.8 % given the increase in its lagged value by 10 % and the estimated coefficient of 0.492 is calculated as follows: $0.48 \approx (1{,}01^{0.492} - 1) \times 100$.

[17]Since the values of the real GDP per capita as dependent variable are expressed as logarithms and those of the independent variables except the lagged value of real GDP per capita are expressed as real numbers, the increase in real GDP of 1.3 % given the increase in the score for soundness of banks by 1 and the estimated coefficient of 0.013 is calculated as follows: $1.3 \approx (e^{0.013} - 1) \times 100$. The changes in the real GDP per capita are calculated in the same way for the other independent variables as well.

The estimated coefficients for the ease of access to loans, misery index and quality of educational system show that all these variables have negative impact on the real GDP per capita. A negative impact of the ease of access to loans can be conveniently linked to the Global financial crisis and its aftermath in increasing banks' prudent behaviour and posing more stringent bank regulations. In fact, the rising difficulty in getting bank loans after the crisis can be perceived as a way to primarily avoid adverse selection and moral hazard in order to ensure financial stability. However, this finding is statistically insignificant and cannot be taken into account with much credit. In concordance with the theory, the misery index negatively affects the real GDP per capita at a rate of 0.004, which suggests that an increase of the index by 1 % would bring down real GDP per capita by 0.4 %. As the effects from the Global financial crisis resulted in increased unemployment rates across OECD members and low interest rates that eventually lead to lower inflation rates, many countries faced difficulties in bringing back their economies on the growth path and experienced relatively modest economic growth. Also, the negative impact of the misery index on the real GDP per capita is most likely driven by the increased levels of unemployment, which implies that policies aimed at decreasing the unemployment to the pre-crisis levels are welcome in boosting economic growth and increasing the level of economic development. Finally, the estimation results point out to negative effects of the quality of educational system on real GDP per capita, which is a bit illogical and ambiguous, but can be ruled out as irrelevant because of the statistical insignificance of the coefficient.

Despite the statistically significant effects of the explanatory variables, it is to be noted that the results also show significant effects of all year dummy variables except for 2013, meaning that changes in the nominal GDP per capita may be caused by other factors not captured in the model. Specifically, the estimated positive coefficients of 0.044 in 2010 and 0.008 in 2011 point out that effects of these factors would increase real GDP per capita by 4.5 % in 2010 and 0.8 % in 2011, respectively; the negative coefficient of −0.005 in 2012 indicates to a decline in real GDP per capita by 0.5 % as a result of the year-specific effects.

7 Concluding Remarks

The role of sustainable finance in creating conditions for sustainable economic development have become an important topic of interest in the recent period, as many economists are trying to estimate the effect that the social, environmental, institutional and the financial inclusion policies on ensuring sustainable economic development. To that extent, many international economic institutions and regulatory bodies have expanded their focus of interest towards the development of indicators that capture the effects of these policies.

The study in this paper contributes to the literature by employing variables that proxy for the economic, social, environmental and institutional policies. The results generally support the economic theory and confirm that policies directed towards more stringent bank regulation, technological improvements and higher energy-intensified production create a positive base for sustainable economic development.

However, the analysis shows that year-specific effects are present and significantly affect the economic development, implying that it is very difficult to develop a model which will include exactly those variables that contribute the most. Other limitations of the study are the unavailability of data and the relatively short time series for performing one such analysis.

References

Acemoğlu D, Johnson S, Robinson AJ (2005) Institutions as a fundamental cause of long-run growth. In: Aghion P, Durlauf SN (eds) Handbook of economic growth, vol 1A. Elsevier, Amsterdam, pp 385–472

Aghion P, Howitt P, Mayer-Foulkes D (2005) The effect of financial development on convergence: theory and evidence. Q J Econ 120(1):173–222

Anderson TW, Hsiao C (1981) Estimation of dynamic models with error components. J Am Stat Assoc 76(375):598–606

Arellano M, Bond S (1991) Some tests of specification for panel data: Monte Carlo evidence and an application to employment equations. Rev Econ Stud 58(2):277–297

Arestis P (2006) Financial liberalization and the relationship between finance and growth. In: Arestis P, Sawyer M (eds) A handbook of alternative monetary economics. Edward Elgar Publishing, Cheltenham, pp 346–364

Arestis P, Demetriades PO, Luintel KB (2001) Financial development and economic growth: the role of stock markets. J Money Credit Bank 33(1):16–41

Arestis P, Luintel AD, Luintel KB (2004) Does financial structure matter? Working Paper No. 399, The Levy Economics Institute of Bard College, New York

Armenta MW (2007) The financial sector and economic development: Banking on the role of human capital. J Public Int Affairs 18(Spring):188–203

Asheim GB (1994) Net national product as an indicator of sustainability. Scand J Econ 96(2):257–265

Beck T (2011) The role of finance in economic development: benefits, risks, and politics. Discussion Paper No. 2011-038, European Banking Center, Tilburg University, Tilburg

Beck T, Demirgüç-Kunt A, Levine R, Maksimovic V (2001) Financial structure and economic development: firm, industry, and country evidence. In: Demirgüç-Kunt A, Levine R (eds) Financial structure and economic growth: a cross-country comparison of banks, markets, and development. The MIT Press, Cambridge and London pp 189–241

Beck THL (2010) Legal institutions and economic development. Discussion Paper No. 2010-94, Center for Economic Research, Tilburg University

Becker GS (1962) Investment in human capital: a theoretical analysis. J Polit Econ 70(5):9–49

Bhagwati J (2004) In defense of globalization. Oxford University Press, New York

Bodenhorn H (2003) State banking in early America: a new economic history. Oxford University Press, New York

Bojö J, Mäler K-G, Unemo L (1992) Environment and development: an economic approach. Springer, Dordrecht

Burgess R, Pande R (2003) Do rural banks matter? Evidence from the Indian social banking experiment. Bureau for Research in Economic Analysis of Development, BREAD Working Paper: No. 037

Christopoulos DK, Tsionas EG (2004) Financial development and economic growth: evidence from panel unit root and cointegration tests. J Dev Econ 73(1):55–74

De Cian E, Schymura M, Verdolini E, Voigt S (2013) Energy intensity developments in 40 major economies: structural change or technology improvement? Center for European Economic Research, ZEW Discussion Papers No. 13-052

Dewatripont M, Rochet J-C, Tirole J (2010) Balancing the banks: Global lessons from the financial crisis. Princeton University Press, New Jersey

Diamond DW (1996) Financial intermediation as delegated monitoring: A simple example. FRB Richmond Econ Q 82(3):51–66

Emerson J (2003) The blended value proposition: integrating social and financial returns. Calif Manag Rev 45(4):35–51

Enders JC, Remig M (2015) Theories of sustainable development: an introduction. In: Enders JC, Remig M (eds) Theories of sustainable development. Routledge, London, pp 1–5

Epstein MJ, Roy M-J (2001) Sustainability in action: Identifying and measuring the key performance drivers. Long Range Plan 34(5):585–604

Fatemi AM, Fooladi IJ (2013) Sustainable finance: a new paradigm. Glob Finance J 24(2):101–113

Filipovski V (2004) The financial sector and the economic development in the Republic of Macedonia (in Macedonian). In: Bogoev K, Fiti T, Petreski G (eds) Open challenges to the Macedonian economy, proceedings. Macedonian Academy of Sciences and Arts, Skopje, pp 357–379

Fry MJ (2000) Financial markets, financial flows, and economic growth in LDCs. In: Oosterbaan MS, van Steveninck TdR, van der Windt N (eds) The determinants of economic growth. Springer, New York, pp 107–161

Fullwiler ST (2015) Sustainable finance: building a more general theory of finance. Binzagr Institute for Sustainable Prosperity, Working Paper No. 106

Galetovic A (1994) Finance and growth: a synthesis and interpretation of the evidence. International Finance Discussion Papers No. 477, Board of Governors of the Federal Reserve System, Washington, DC

Galetovic A (1996) Specialization, intermediation and growth. J Monet Econ 38(3):549–559

Gerster R (2011) Sustainable finance: achievements, challenges, outlook. Gerster Consulting, Richterswil

Goldsmith RW (1969) Financial structure and development. Yale University Press, New Haven

Greenwood J, Jovanovic B (1990) Financial development, growth, and the distribution of income. J Polit Econ 98(5):1076–1107

Greenwood J, Sanchez JM, Wang C (2013) Quantifying the impact of financial development on economic development. Rev Econ Dyn 16(1):194–215

Grober U (2015) The discovery of sustainability: the genealogy of a term. In: Enders JC, Remig M (eds) Theories of sustainable development. Routledge, London, pp 6–15

Haigh M (2012) Publishing and defining sustainable finance and investment. J Sustain Finance Invest 2(2):88–94

Hargreaves A, Fink D (2004) The seven principles of sustainable leadership. Educ Leadersh 61 (7):8–13

Hartwick JM (1977) Intergenerational equity and the investing of rents from exhaustible resources. Am Econ Rev 67(5):972–974

Hasan I, Koetter M, Lensink R, Meesters A (2009) Bank efficiency, financial depth, and economic growth. Available at SSRN (September 18, 2009): http://ssrn.com/abstract=1475415)

Honohan P (2008) Cross-country variation in household access to financial services. J Bank Finance 32(11):2493–2500

Horvat B (1958) The optimum rate of investment. Econ J 58(272):747–767

Investment Leaders Group (ILG) (2014) The value of responsible investment: the moral, financial and economic case for action. Institute for Sustainability Leadership, University of Cambridge, Camridge

Jeucken M (2001) Sustainable finance and banking: the financial sector and the future of the planet. Earthscan Publications Ltd., London

Karlan D, Morduch J (2010) Access to finance. In: Rodrik D, Rosenzweig MR (eds) Handbook of development economics, vol 5. Elsevier, Amsterdam, pp 4703–4784
Keynes JM (1930) A treatise on money. Harcourt, Brace and Company, New York
Khan MS, Senhadji AS (2000) Financial development and economic growth: an overview. Working Paper No. 00-209, International Monetary Fund, Washington
King RG, Levine R (1993a) Finance and growth: Schumpeter might be right. Q J Econ 108 (3):717–737
King RG, Levine R (1993b) Finance, entrepreneurship, and growth: theory and evidence. J Monet Econ 32(3):513–542
Kohler P (2015) Redistributive policies for sustainable development: looking at the role of assets and equity. DESA Working Paper No. 139
Lawn P (2006) Sustainable development: concept and indicators. In: Lawn P (ed) Sustainable development indicators in ecological economics, current issues in ecological economics. Edward Elgar Publishing, Cheltenham, pp 13–51
Levine R (2005) Finance and growth: theory and evidence. In: Aghion P, Durlauf SN (eds) Handbook of economic growth, vol 1A. Elsevier, Amsterdam, pp 865–934
Lin VL (1981) The impact of the financial sector on economic development: some cross-section evidence. Stud Comp Int Dev 16(1):44–52
Loayza NV, Rancière R (2006) Financial development financial fragility and growth. J Money Credit Bank 38(4):1051–1076
Lucas RE (1988) On the mechanics of economic development. J Monet Econ 22(1):3–42
Mawhinney M (2002) Sustainable development: Understanding the green debates. Blackwell, Oxford
McKinnon RI (1973) Money and capital in economic development. Brookings Institution Press, Washington
Merton RC (1995) A functional perspective of financial intermediation. Financ Manage 24(2):23–41
Merton RC, Bodie Z (1995) A conceptual framework for analyzing the financial system. In: Froot KA, Mason SP, Perold AF, Merton RC, Bodie Z, Sirri ER, Tufano P (eds) The global financial system: a functional perspective. Harvard Business School Press, Boston, pp 3–31
Miller MH (1998) Financial markets and economic growth. J Appl Corp Finance 11(3):8–15
Ngongang E (2015) Financial development and economic growth in Sub Saharan Africa: a dynamic panel data analysis. Eur J Sustain Dev 4(2):369–378
Pagano M (1993) Financial markets and growth: an overview. Eur Econ Rev 37(2):613–622
Park C-Y, Mercado Jr, R (2015) Financial inclusion, poverty, and income inequality in developing Asia. ADB Economics Working Paper No. 426
Piketty T (2015) The economics of inequality. The Belknap Press of Harvard University Press, Cambridge
Pisano U, Martinuzzi A, Bruckner B (2012) The financial sector and sustainable development: logics, principles and actors. European Sustainable Development Network, Quarterly Report No. 27
Rajan R, Zingales L (1998) Financial dependence and growth. Am Econ Rev 88(3):559–586
Rioja F, Valev N (2004) Finance and the sources of growth at various stages of economic development. Econ Inq 42(1):127–140
Robinson J (1952) The generalization of the general theory. In: Robinson J (ed) The rate of interest and other essays. MacMillan, London, pp 67–142
Rottier S, Véron N (2010) Not all financial regulation is global. Peterson Institute for International Economics Policy Brief, No. PB10-22
Rousseau PL, Wachtel P (2005) Economic growth and financial depth: Is the relationship extinct already? Discussion Paper No. 2005/10, World Institute for Development Economics, Helsinki
Sarma M (2008) Index of financial inclusion. ICRIER Working Paper No. 215
Schumpeter JA (1961) The theory of economic development. Oxford University Press, New York
Shin HS (2009) Securitisation and financial stability. Econ J 119(536):309–332
Smith A (1776/2008) The wealth of nations. Wilder Publications, Radford

Soppe A (2004) Sustainable corporate finance. J Bus Ethics 53(1–2):213–224
Stulz R (2001) Does financial structure matter for economic growth? A corporate finance perspective. In: Demirgüç-Kunt A, Levine R (eds) Financial structure and economic growth: a cross-country comparison of banks, markets, and development. MIT Press, Cambridge, pp 143–188
Taleb NN (2008) The black swan. The impact of highly impossible. Penguin Books Ltd, London
Taylor L (1996) Sustainable development: an introduction. World Dev 24(2):215–225
Todaro MP, Smith SC (2012) Economic development. Pearson Education, Boston
United Nations Department of Economic and Social Affairs (UN DESA) (2014) Prototype Global sustainable development report. New York: United Nations Department of Economic and Social Affairs, Division for Sustainable Development
Wilhelm K (2013) Return on sustainability: how business can increase profitability and address climate change in an uncertain economy. FT Press, New Jersey
Wilson C (2010) Why should sustainable finance be given priority? Lessons from pollution and biodiversity degradation. Account Res J 23(3):267–280
World Commission on Environment and Development (WCED) (1987) Our common future. Oxford University Press, Oxford
World Wide Fund (WWF-China) (2015) Financial market regulation for sustainable development in the BRICS countries. World Wide Fund-China, Beijing
World Wide Fund (WWF-UK) and BankTrack (2006) Shaping the future of sustainable finance: Moving the banking sector from promises to performance. Woking, and Nijmegen: World Wide Fund-UK, and BankTrack
Wright RE (2002) The wealth of nations rediscovered: integration and expansion in American financial markets, 1780–1850. Cambridge University Press, Cambridge
Xu Z (2000) Financial development, investment, and economic growth. Econ Inq 38(2):331–344

Author Biographies

Magdalena Ziolo Ph.D., hab is Associate Professor, Head of Department of Banking and Comparative Finance at University of Szczecin, Faculty of Management and Economics of Services, Poland. Her main research interests include finance and banking, public finance and regionalism.

Filip Fidanoski is one of the leading Macedonian economists and scholars from the young generation. His *alma mater* is Ss. Cyril and Methodius University in Skopje. His major subject of career is professionally dedicated to banking, financial economics, economic history, entrepreneurial economics, institutional economics, macroeconomics, through consultancy and scientific research. He is a reviewer, author and coauthor of a many articles, papers and chapters published in a valuable journals. In addition, his texts are appeared in influential mediums in Republic of Macedonia. Also, he actively participates in organization of different scientific and professional events from his fields of interest. At last, he is also a poet and translator.

Kiril Simeonovski is an economist, independent scholar, open knowledge activist, public speaker and author. He holds undergraduate and postgraduate degrees from the Faculty of Economics—Skopje. His fields of interest are diverse and include macroeconomics, financial regulation, corporate governance, strategic planning, entrepreneurship, experimental economics, information economics, online marketing and others. He has also been actively involved in project and event management and has participated in numerous international conferences both as attendee and speaker.

Vladimir Filipovski Ph.D., is Full Time Professor at Ss. Cyril and Methodius University, Faculty of Economics in Skopje, Republic of Macedonia. His main research interests include financial markets, banking, international finance and macroeconomics.

Katerina Jovanovska MEE, MBS is a Teaching Assistant and Young Researcher at the Faculty of Economics University of Ljubljana. She is a member of the Unit for Management and organization where she is collaborating within undergraduate and graduate courses Learning and Knowledge Management and Organization and Management in Slovene and English language. Her research has been published in Review of Financial Markets and International Journal of Productivity and Performance Management.

The Role of Social and Environmental Information in Assessing the Overall Performance of the Enterprise

Camelia Catalina Mihalciuc and Anisoara Niculina Apetri

Abstract

The entities, that relate increasingly more to the concept of sustainable development and who engage in sustainable development actions, communicate and transmit to markets a more transparent image, through the use of extra-financial information, but without forgetting their economic role, fixing the ethical rules and applying operating principles that take into account the interests of all partners. In terms of performance of an enterprise, it can not only appreciate from the financial perspective, as necessary the use of extra-accounting information, especially those environmental and social. The objective of this paper is to present on the one hand, the annual reports types that may be published by economic entities, with respect to their social and environmental impacts, on the other hand to show that, increasing the performance of organizations involves combining a sustainable strategy that does not neglect any moment the purposes of all the stakeholders and above all to respect social and environmental principles. To respond to new information needs created as a result of the awareness of the economic entities responsibility towards stakeholders, in the

C.C. Mihalciuc (✉) · A.N. Apetri
Stefan Cel Mare University of Suceava, Str. Universitatii no. 13, 720229 Suceava, Romania
e-mail: cameliam@seap.usv.ro; cameliamihalciuc@yahoo.com

A.N. Apetri
e-mail: anisoarad@seap.usv.ro; anisoara_1978@yahoo.com

C.C. Mihalciuc · A.N. Apetri
Department of Accounting and Finance, Faculty of Economics and Public Administration,
Stefan Cel Mare University of Suceava, Suceava, Romania

W. Leal Filho et al. (eds.), *Sustainable Economic Development*,
World Sustainability Series, DOI 10.1007/978-3-319-45081-0_12

theory and practice of financial reporting appeared reporting models that are based on stakeholder requirements, for ethical corporate information, environmental and social information.

Keywords
Sustainable performance (global performance) · Corporate social responsibility (CSR) · Non/financial reporting · Social and environmental balance-sheet · Triple bottom line (TBL)

1 Introduction

Through this paper, the authors want to highlight that a coherent CSR—Corporate Social Responsibility strategy, based on integrity, strong values and a long-term approach generates obvious benefits to the company that administers and contributes positively to the welfare of society.

Financial performance encompasses all aspects of the entity's economic interactions, including traditional indicators used in financial accounting, while sustainability performance includes, besides the financial performance also the financial non-performance, such as environmental and social performance. The approach of this work aims to present first the concept of sustainable performance and to achieve this performance by involving organizations in CSR acts, as well as presenting the integrated reporting of these organizations. In this way, one shall take into account the process of highlighting the importance of non-financial reporting of the economic entities, as this form of reporting is increasingly worldwide used, providing a large range of benefits, and companies realize this way that a company's success is not measured only through financial results but also through their engagement in acts of protecting the environment and society.

2 Aspects of the Concept of Sustainable Performance (Global Performance)

To say that something is performing one will always make a comparison between results and a benchmark, and when we refer to economic entities the benchmarks can be multiple: competition, situations in the past, objectives.

In the economic field, the performance can be defined according to various criteria:

- based on the level of achievement of the objectives set;
- based on value creation;
- based on the productivity and efficiency of the enterprise.

The performance of a company is measured and evaluated using the accounting tool, because accounting represents the main source of information for interested users in measuring performance (Tabara 2006).

Depending on the interests of users of the financial accounting information, performance is defined and measured differently. Shareholders and potential investors are interested in the profitability of investments, managers aim at the world-wide performance of the company, creditors and suppliers are interested in company's solvency and stability. As we can see, performance measurement is based on the objective of interested users.

In general terms, financial performance includes all aspects of the entity's economic interactions, including traditional indicators used in financial accounting, and also intangible elements that does not typically appear in the financial statements. Financial indicators relate primarily to the profitability of an entity and are intended to provide information for both the entity's management and its shareholders. But in the context of sustainability reporting, economic indicators are more focused on how an entity affecting economic stakeholders that interact directly or indirectly. Therefore, the main purpose of economic performance measurement is to detect changes in the economic status of stakeholders, as a result of the activities of the entity, not only recording changes in the financial condition of the entity. In some cases, the existing financial indicators can directly assess these changes, but in other cases, different measures may be required such as the reformulation of traditional financial information to emphasize the impact on stakeholders.

It often happens that a company's performance to be analyzed just through profit or loss. This analysis does not provide comprehensive information on the overall performance; it only provides an image of the financial performance because users can not know how it was obtained that result, which were the "flows of income and expense that explain this result" (Tabara 2006).

Company's performance does not simplistically refer only to the higher accounting results, respectively maximum profitability, stable financial balance, ability to generate cash necessary for the operation and expansion into perspective, but it should cover all aspects of non-financial and financial aspects of its business.

Nowadays, the concept of performance has evolved to a comprehensive approach including financial aspects and non-financial aspects, which are primarily concerned with elements of social responsibility. To improve their performance, the entities are turning to certain activities that meet social and environmental safety and reliability requirements, namely the environmental and social performance. Thereby, short-term programs may include aspects of risk improvement, implementation of new control standards and business integrity management. The long term programs aim at improving the risk management and quality performance.

The environmental aspect of the performance of a company refers to the impact on natural systems, including ecosystems, soil, air and water. It is particularly important to provide information on the environmental performance both in absolute and relative values. The absolute figures give an idea of the size, scope or impact, which allows the user to analyze performance in the context of wider systems. The figures expressed in percentage illustrate the effectiveness of the company and allow comparisons between entities of different sizes. Companies are encouraged to report individual performances to broader ecological systems in which they operate. For example, companies could report the amount of environmental pollution capacity (local, regional or global) to absorb emissions. Environmental performance includes information characterizing the quantity of energy consumed and its origin, treatments applied to resource, management of emissions, waste, residues, etc., how to land exploitation and management of the environment in which the entity exists and operates.

Social performance can be measured by analyzing the impact of the company on stakeholder at local, national and global level. In some cases, the indicators of social performance influence the intangible elements of a company, such as human capital and reputation. Social performance focuses on the interactions between company, human resource engaged and any other individuals with whom contacts are established. Social Performance Indicators include: relations between employees, labor health and safety and human resources, the report wage/cost of living, elimination of discrimination, respecting the rights of local communities, impact on community development, customer satisfaction etc.

There is a broad consensus on social performance indicators in comparison with indicators of environmental performance. These indicators identify specific aspects of performance on labor practices and human rights, and broader issues that have an impact on consumers, communities and other stakeholders in society. Some social performance indicators differ considerably in nature from the economic and environmental performance indicators. Many of the social issues that are the subject of performance indicators are not easy to quantify, so a number of social indicators are, in fact, qualitative measures of the systems and activities of a company, such as policies, procedures and management practices. These indicators do not refer to the general policies, but specific social issues, narrowly set, such as forced or required labor and freedom of association.

Corporate social performance is a concept on which literature has not yet accepted a universally definition. For the first time, this term was created by the american author S. Prakash Sethi (1975), enriched by Carrol (1979) and redefined by Wartick and şi Cochran (1985). These authors have traced the evolution of corporate social performance model, focusing on three challenges of the concept of social responsibility: economic responsibility, public responsibility and social perceptiveness. At the base of, the idea related to corporate social performance is "recognition of ethical obligations of organizations and the needs to respond pragmatic to social pressures" (Cochran 2007).

Also, for Wartick and şi Cochran (1985), corporate social performance requires "Integrating social responsibility principles (economic, legal, ethical and discretionary), processes of social responsiveness and of the programs and policies aimed to fulfilling social objectives (management objectives)".

According to other authors (Zeller et al. 2003), "social performance transposing into "the organization's relations with its customers and other stakeholders". Social performance may represent also "the process of response of business to social relations through policies and projects enterprise" (Zhang and Ma 2011).

Social performance is related to "social problems that are relating to the environment, strategic governance, labor relations and stakeholder management" (Nin Ho et al. 2012).

For certain industry sectors or companies, discussions of sustainability performance become practical when the social or environmental issues affect the ability of companies to expand operations or harm the reputation or brand. New codes of corporate governance bring to the fore the debate on sustainability risk. New methodologies are needed to connect the economic performance of social and environmental. Like the standard tools of economic analysis, basic concepts should focus on individual industries so as to provide performance comparable and understandable models.

Items like safety, society and global performance are addressed by the new management systems that companies currently implementing them. The aim is to cover a large range of information on compliance, risk management, governance and performance evaluation, to ensure both present information and clues to the future and be a basis for responsible reporting.

3 Corporate Social Responsibility (CSR)—Profitable Business Strategy

A strategy, by which companies gain from an economic point of view, by improving social and ecological environment in which they operate, becomes certainly a good and fair approach. Thus, corporate social responsibility is a profitable business strategy, positively correlated with corporate financial performance, as long as they remain with the economic objective as their primary scope of activity.

In order to constitute a benefit both for society and for company, corporate social responsibility must always maintain a balance between costs and benefits of social involvement induced by taking on social initiatives. (Iamandi and Filip 2008). Finding this balance point is, in fact, the current issue of world-wide companies, given that they were already aware of the need to involve in the problems of society.

The concept of Corporate Social Responsibility (CSR) has occurred since the early 1970s, consisting of continuous voluntary commitment of a company to behave ethically and contribute to economic development while improving the quality of life of the workforce and their families, and the local community and society in general.

Corporate social responsibility (CSR) is a concept whereby organizations serve the interests of society by taking responsibility for the impacts of their activities on customers, employees, shareholders, communities and the environment in all aspects of their operations (Muresan 2010).

Socially responsible companies take into account their impact on communities and the environment when making decisions, balancing stakeholder needs with their need to make profits.

Among the advantages of applying the Corporate Social Responsibility are as follows (CODECS 2010):

- CSR can be an advantage in recruiting and retaining staff. CSR contributes to improving the company's image in the eyes of employees and the population, especially if people are convinced to help each individual level, paying money from his own pocket to raise funds intended for good causes, or when directly involved in volunteering in their community.
- RSC helps to anticipate and manage the risks. A reputation built carefully in years can collapse overnight because of corruption scandals or because of environmental disasters, which attract unwanted attention from legislative and judiciary bodies,—or from the media. Creating a corporate culture of accountability for their own actions can help protect reputation in case of scandals or accidents.
- CSR can contribute to winning customer's loyalty by creating a distinct image of the organization, based on ethical values. Every company struggles to stand out from among its competitors to differentiate in the eyes of consumers, offering them the so-called "unique offer for sale". CSR can fulfill an important role in attracting customer loyalty, demonstrating that the organization shares the distinctive set of ethical values. They benefit from such an advantage not only organizations providing products to market, but also those that provide services to other companies, if they build a reputation of integrity, best practice and environmental responsibility.
- CSR can help protect the company from political interference. Organizations are striving hard to avoid distortions caused by tax and regulatory systems. By that takes one voluntarily, accountability measures, a company can convince regulators, and the general public, it takes very seriously things like health and safety, protecting diversity and the environment thus avoiding the intervention from outside. This also applies to companies who need to justify higher profit rates and high levels of rewarding their members. Finally, companies that are far away from their home country can ensure good corporate citizens if they prove that respects local laws and standards on labor or environmental protection.

There are four dimensions of corporate social responsibility, namely economic responsibility, legal responsibility, ethical responsibility, discretionary responsibility (Coelho 2011; Carroll 1979).

Economic Responsibility—First of all, companies represent the basic economic unit in society. As such, they have a responsibility to produce goods and services that society needs and then selling them for a profit. Companies must be managed in a way that can be economically sustainable. They must fulfill their contracts with suppliers on time; they must pay their employees properly, etc., which can be difficult when dealing with financial problems.

Legal Responsibility—refers to the hope that companies will comply with the established laws of society to regulate competition in the market. Companies have thousands of legal responsibilities governing almost every aspect of their operations and products including consumer laws, laws on the environment, on employment, etc.

Ethical Responsibility—refers to society's expectations to go beyond the law, such as the hope that companies will carry out their business in a fair and equitable way. This means that companies are expected to do more than comply with the law, also make proactive efforts to anticipate and meet society's norms, even if those rules are formally adopted in law.

Discretionary Responsibility—refers to society's expectations to be good citizens. This may involve things like philanthropic programs to benefit a community or a nation.

In 2008, Carroll and Buchholz authors have introduced in their writings the concept of overall social responsibility, as follows: "Economic Responsabilities + Legal Responsibilities + Ethical Responsibilities + Philanthropic Responsibilities = total CSR" (Carroll and Buchholtz 2008). Also, there have been identified three dimensions of social performance: "Social responsibility, social perceptiveness and the solving of social problems of stakeholder" (Carroll and Buchholtz 2008).

CSR is characterized as firm obligation of a company to act beyond the legal obligations or restrictions imposed by economic and pursue long-term goals to benefit the community. That company is held responsible not only with the owners (shareholders), but also with customers, suppliers, employees, government bodies, creditors, local communities, public opinion.

In this context, a manager of a socially responsible company has several categories of ethical obligations. These obligations branch in four directions to four groups interested in that business, namely (Muresan 2010):

- to shareholders—these obligations relate to the a proper management, loyalty, information, transparency, confidentiality;
- to employees—these obligations relate to fair remuneration, professional development, respect for privacy, respect the right of petition;
- to customer—these obligations relate to product /service quality, ensure safety, the information;
- to the community—these obligations relate to environmental protection, contribution to solving social issues, respect for cultural diversity.

The literature shows a chronological summary of the three approaches to CSR, each including an approach and going beyond coverage as the previous one: (Iamandi and Filip 2008).

(a) The approach of corporate responsibility as an obligation only to shareholders (Shareholder approach): This is the classic approach of corporate social responsibility. This approach assumes that the basic objective of the company is to make profits for shareholders and social responsibility actions not related to the company's obligation.

(b) The approach corporate responsibility as an obligation to all stakeholders (Stakeholder approach): This means that companies are accountable to their shareholders and to the stakeholders alike.
(c) Addressing corporate responsibility to society as a whole obligation (Societal approach): This is the most comprehensive approach to corporate social responsibility which requires that companies are responsible to society as a whole because companies are components of the social environment.

4 Non-financial Performance Reporting Tools of a Company

In order to report the non-financial performance, companies present in the financial reports the following types of reports: balance social, environmental balance, Virtue Matrix (Matrix Morality) and other multidimensional reports, covering practical way integrated reporting companies.

4.1 Standardization of Non-financial Reporting

Due to more frequent use of environmental and social balances there was an attempt to standardize them, of their content and reporting process, in order to provide transparent and comparable information.

Process Standard—represents a set of coordinates designed to improve the economic and social reporting in a company. This helps businesses to design an efficient reporting.

Content standard—involves content identification and normalization of the document, assuming that this document is the result of a process based on dialogue with stakeholders.

The balance sheet reporting was standardized by AA1000, Q-RES, and Copenhagen Charter. The content of a balance sheet is covered by GRI (Global Reporting Initiative), GBS (Italy), LBG.

The advantages of these standards are as follows: it can be adopted in a way that is quick and easy, allows a comparison of the outcomes due to the standardization document. And a disadvantage may be that it can happen from year to year only figures to change the document, as in the case of the balance sheet.

Global Reporting Initiative (GRI) is a non-profit organization that works towards a sustainable global economy by providing guidance in implementing sustainability reports. The headquarters of the Organization is in Amsterdam promoting the view that corporate reporting on economic, social and environment performance should be a routine comparable to financial reporting type (GRI 2014).

GRI has initiated and developed a comprehensive framework of sustainable reporting that is widely used worldwide. The framework enables organizations to measure and report all economic performance, environmental, social and governance—the four key domains of sustainability (GRI 2014).

The framework of reporting—which includes guidelines on sustainable reporting, sector guidelines and other resources—allows a more transparent organization on economic, environmental, social and governance performance. This transparency and responsibility make stakeholders have trust in organizations, and can lead to many other benefits. Thousands of organizations of all sizes and in all sectors use the GRI, in order to understand and communicate their sustainability performance.

Over the years there have been several versions of the Sustainable Reporting Guidelines, penultimate version was G3.1. G3.1 which included expanded guidance for reporting on human rights, local community impacts, and G3.1 was launched in March 2011. Also, the latest version is G4, the fourth generation of the Guidelines, which was launched in May 2013, being the most comprehensive sustainability reporting guide available today (GRI 2014).

It consists of two parts: the first part contains guidelines on how to make reporting and the second part contains instructions on what should report including information on the profile and the company's strategy, performance indicators and the approached management.

Performance indicators are organized into three categories: Economic, Social and Environment. The social category is broken down into three sub-categories Labour, Human rights, and Society and product responsibility.

Performance indicators are organized into three categories: Economic, Social and Environment. The social category is broken down into three sub-categories Labour, Human Rights, Society and Product Responsibility.

The principles for defining the contents of the report are intended to help organizations provide transparent and designed to identify information to be reported.

There are four principles defining the content of the report, namely:

1. Include stakeholders—the organization must provide information on how to identify stakeholders, and explain how they have responded to their reasonable expectations and interests a.
2. Background of sustainability—the report should present the company's performance in the context of sustainability.
3. Materiality—the report should cover topics and indicators that: reflect the economic, social and environmental significance of the organization; substantially influence the assessment and decisions of stakeholders.
4. Completeness—Coverage of material topics and indicators and define the boundaries ratio should be sufficient to reflect the significant economic, environmental, and social impact and permit to interested parties to evaluate the performance reported by the organization in the reporting period.

The principles for defining the quality of the report are designed to ensure the quality of the reported information, including fair presentation and they are:

1. Balance—report should reflect both the positive and negative aspects of the organization's performance to enable a reasoned assessment of overall performance;
2. Accuracy—reported information should be sufficiently accurate and detailed for stakeholders to assess the reporting organization's performance;
3. Actuality—reporting is done by an established timetable and information becomes available in time for stakeholders in order to make informed choices;
4. Comparability—The issues and information should be selected and reported consistently. reported information should be presented in a manner that enables stakeholders to analyze changes in the organization's performance over time and can make comparisons with other companies;
5. Clarity—the information should be written in an easily understandable way and accessible to all persons using the report;
6. Trueness (Veracity)—Information and processes used in the preparation of a report should be gathered, recorded, compiled, analyzed and presented in a way to allow examination and determining of their quality.

After they finished up the report, companies must have applied to the self-declared GRI Reporting Framework. There are three levels of application A, B and C, depending on its complexity (whether they have passed through all the stages GRI stages). It is recommended that organizations use also an external confirmation of the level of implementation. Plus status—"+"—can be added to an application level, when the sustainability report was externally assured (GRI 2014).

4.2 Social Balance Sheet

The best tool to give visibility to the questions and needs for information and transparency of the target audience is social balance. That use of a model for reporting on the quantity and quality of the relationship between society and reference groups represented by the entire community seeks to establish a homogeneous, timely, complete and transparent framework of the complex interdependence of economic, intrinsic and socio-political factors and resulted from choices taken. (bilanciosociale.it 2000).

Social balance sheet represents the main tool for sustainability reporting as a tool for non financial reporting, whereby companies that practice social responsibility communicate transparently with its stakeholders. So far, the social balance sheet has been described as a final document containing the results, targets and actions taken to emphasize the socially responsible behavior of an undertaking. The balance must contain both the positive and negative issues facing society. It should be written in plain language, clear and transparent for all stakeholders to be understood.

The first types of social balance sheets were drawn up in the late 60s early 70s. In the early years of the advent of social balance, they have spread widely in the world (Europe and the United States).

Social balance sheet development stages are:

1. 80s—a stage of stagnation;
2. 90s—s stage when the social balance spread most of all;
3. from 2000 to present—a stage of institutionalizing the balance (try their standardization)

At an international level, the social balance by different names such as: Social Audit, Social Accounting, Social Balance Sheet and Intellectual Capital.

The social balance sheet can be used as an external communication means, means for improving internal management organization, means of developing social strategies and as a means of identifying social mission.

Social balance is a very important tool, a fundamental means for developing public relation activities to enhance social and industrial relations.

The objectives of a social balance sheet are (bilanciosociale.it 2000):

- enable the company to know their role in society;
- be a tool for comparing the results with existing needs, providing information about the social objectives;
- provides information on the social objectives;
- demonstrate that the company goal is not only to make profit but to provide added value to the community;
- help develop democracy and transparency in the company's activities;
- to become a reporting tool if the undertaking social actions have consequences in terms of usability, legitimacy and effectiveness.

4.3 Environmental Balance Sheet

A growing attention of public opinion and governments on environmental issues related to production activities led companies to use communication tools purely environmental.

Environmental balance sheet is a document that describes the main relationship between company and environment. This document is published voluntarily aimed at direct communication between the company and the public interested in environmental issues (bilanciosociale.it 2000).

Environmental balance sheet can contain various indicators such as: (bilanciosociale.it 2000):

- environmental management indicators that assess efforts in controlling the environmental aspects;
- absolute environmental indicators, which measure impact factors generated by the company;
- environmental performance indicators, to assess the ecological efficiency by decoupling from fluctuations in production;

- potential effect, which assess possible damages that the company could produce as a result of its work;
- effect on the environment, which assess real change of environment as a result of company's work.

Environmental audit can be divided into two main categories related to size of the related company:

- environmental position balance sheet, where data and information relates to individual plants or factories;
- corporate environmental balance sheet, the information and data refer to a company with several fields of activity.

The main benefits that the company receives from the preparation of such documents consists in monitoring all environmental aspects identified in the following situations: analysis of environmental costs and waste; considering investments in technologies with reduced environmental impact; evaluating the level of protection and improving the area where it operates; identifying the critical factors related to the activities and measures taken to improve the environment; building a monitoring network that allows to develop an environmental management system (EMAS or ISO14000); developing effective communication environmental tools with the local population and public administration.

4.4 Virtue Matrix

In order to assess their performance and develop a CSR strategy, companies can use the Virtue Matrix. This instrument consists of four quadrants (top side comprises: Strategic Frontier and structural Frontier and bottom side includes the following quadrants: Civil Foundation Choice and Civil Foundation Compliance), which provides a framework for assessing applications for civilian, government and stakeholders, while encouraging business leaders to be as innovative in enriching society as in increasing the welfare of their shares. This concept was introduced by those from Hardvard Business Review magazine in 2002 (Martin 2003).

The bottom side of the quadrant is the Civil Compliance Choice Foundation, which is a description of the historical evolution of laws, rules, regulations and behaviors that underpin civil society. Civil right quadrant Foundation—Imposed (Compliance) includes laws and regulations that companies must follow. In the lower left quadrant, the Civil Foundation—Volunteer (Choice), includes habits and rules that corporations tend to adhere to the choice (Corporate Social Responsibility Review 2006).

The top left quadrant, Strategic Frontier includes activities for which motivation becomes intrinsic and whose impact on shareholder value, when the business is launched, is ambiguous. In this case, the activity once established, produces sufficient positive feedback from customers, employees and other stakeholders as to make a positive contribution to creating value to shareholders. Structural Frontier from upper right quadrant includes activities that are so intrinsically motivated but also contrary to the interests of shareholders. The benefits of this quadrant primarily revert to local communities.

Virtue Matrix is a dynamic model, therefore Strategic Frontier actions tend to migrate to Civil Foundation over time. Successful Strategic Frontier actions tend to be copied by competitors and become the norm in the concerned industry and therefore part of Civil Foundation—Volunteers. Structural actions tend to be clearly established in laws and regulations, becoming part of Civil Foundation—Imposed.

A company with a great commitment to social responsibility Virtue Matrix will have a strategy that has the following four characteristics: (Corporate Social Responsibility Review 2006).

1. Construction of mechanisms to ensure that it is in strict compliance with all laws and regulations of Civil Foundation—Imposed;
2. Initiative in adoption of its industry norms and conventions in Civil Foundation—Volunteers;
3. Existing ongoing Strategic Frontier projects are of particular importance to stakeholders in the spectrum of CSR fields;
4. Implement Structural type projects, which are actively building coalitions for change and overcome external associations with these new initiatives.

This matrix can be a "SWOT analysis of shareholders'. The strengths can be compared with the Strategic Frontier actions, because these actions bring benefit to shareholders thus representing strengths for shareholders. Structural actions may be weaknesses for shareholders, because shareholders such type have no benefit from these actions, but the contrary—is an extra cost to them. Threats can be compared with required actions as may complicate activity and opportunities can be compared with voluntary actions.

Performance indicators are organized into three main categories: Economic, Social and Environmental. The social category is broken down into three sub-categories: Labor, Human Rights, Society and Product Responsibility.

Part I of the sustainability reporting guidelines includes principles for defining the content of the report and principles for defining the quality of the report. (GRI 2014).

The principles for defining the contents of the report are intended to help organizations provide transparent information and designed to identify information to be reported.

The four main principles in defining the contents of the report were presented at point 4.1 Standardization of non-financial reporting.

After companies finished up the report they must have applied to the self-declared GRI Reporting Framework. There are three levels of application A, B and C, depending on its complexity (whether they passed through all the stages of GRI Reporting Framework). It is recommended that companies shall use also an external confirmation of the level of implementation. Plus status—"+"—can be added to an application level, when the sustainability report was externally assured. (GRI 2014).

At the beginning only large companies resorted to such reporting, but this trend has extended to SMEs. This growth among SMEs may be because people at GRI encourages and provides consultations and special packages for small and medium companies. In 2011, 2810 companies were chosen as model of sustainability reporting GRI.

4.5 Non-financial Dimension by Reporting Triple Bottom Line

The idea of TBL Triple Bottom Line was put into circulation by Elkington (1994), according to which an organization should not be obliged to submit only financial performance, but also those on its impact on the environment and society—namely to show which is the total cost of its business (Elkington 1994).

For a sustainable development of a company, issues like strategy and value creation can not be analyzed in purely financial terms. Companies must apply the principle of balanced development, relying on various aspects. A well-known example is the concept of multidimensional performance evaluation focused on the theory of Triple Bottom Line (TBL). This theory underlies a balanced development of the economic entity have terms of environmental, social and economic capital (Savitz and Weber 2006).

The major objective of an economic entity is not simple a profit. Increasingly more companies began to approach, for a sustainable development of their activity, a TBL concept—introducing performance at three levels: environmental, economic and social. (Morschett and Zentes 2010). If they ignore the social and environmental aspects, large companies risks to lose in international market share, to sustain the costs on greening of the activity areas, and to allocate large sums for damage control to regain consumer confidence.

In the broadest sense, TBL is a philosophy that guides corporate performance. In a narrow sense, the concept refers to the tactic measurement, management and business performance reporting beyond financial dimension and perspective of an integrated vision of the entity, with clear evidence of the economic, social and environmental (Caraiani et al. 2010).

Triple Bottom Line designate a concept designed to potentiate the non-financial dimensions, difficult to assess the performance and accountability of a company. The concept of Triple Bottom Line (TBL) focuses entities not only on economic added value, but also the social and environmental values that creates or conversely destroy them. The strict sense of TBL is used as a framework for measuring and reporting the performance of a company on three dimensions: social, environmental and economic (Caraiani et al. 2010).

The social dimension is given by the impact on employees and the local economy by providing details of employees (gender, location, wages, holiday bonuses, dividends or interest paid) or information about payments made for goods, services received and their place of birth (national or local economy).

The dimension of environement received much more attention than the social dimension in recent years, although many of the reports were not designed to help the public. In order to get the reader's interest about the environmental impact of business entity reports, it should contain comparative data to allow the public entity to extract information provided on legislative requirements and achieved the industry averages.

The economic dimension. Traditional accounting reports represent still the most important part of TBL reporting. However, we must stress that although much improved lately, accounting and financial reporting does not provide a lot of information about particular assessment. Values are still linked to single digits without using strings of values which, multidimensional aspect, could provide more information, but would nevertheless be a higher subjectivity. The contribution of transforming the information and accounting reporting under the concept of TBL is to express a strategy in simple terms: a company adds value in the short, medium and long term by, the effective management of economic, social and environmental performance, and also by creating best opportunities with low risk; company develops not only value, but by opening three dimensions creates great benefits to the community and the area in which it exists.

TBL performance requires equal involvement in all three reporting areas of action: economic performance, social performance and environmental performance.

This classification is based on the conventional model of sustainable development. It is possible that by limiting performance indicators to these three categories may not be able to play a complete picture of the performance of a company, for several reasons (Caraiani et al. 2010):

- changes suffered to an aspect of economic, environmental and social performance may often lead to other aspects of sustainability;
- sustainable development strategies often use one area of sustainability as a reference point for defining the objectives of another domain;
- the progress of sustainable development entails coordinated action across the range of performance indicators, rather than random improvement to a whole range of measurements.

Therefore, in addition to the economic, environmental and social dimension of the information, it is necessary to add a fourth dimension, namely the integrated performance.

Given the unique relationship of each company to the economic, environmental and social system they operate in, we could not identify a standard set of integrated performance indicators. Based on the reality of worldwide performance reporting, the integrated indicators were classified into two categories: systemic indicators and integrated indicators (Caraiani et al. 2010).

Systemic indicators show the activity of a company in relation to the economic systems and social environment of which it forms part. For example, a company may describe her performance against an entire system or relative to a standard, such as the percentage rate of the total number of accidents at work in a sector of a particular country, or a certain company can provide the actual number of created jobs as a proportion of the total number of jobs created in a particular region.

Absolute systemic indicators describe a company's performance in relation to limit or capacity of the system of which it is part of. An example would be the amount of emission of some sort as a proportion of total permissible emissions by a public authority in a particular region. In general, systemic indicators allow us to understand the extent to which performance of a company can influence the performance of a larger system. Such indicators are useful for companies operating in a relatively limited geographical area.

Integrated indicators directly connect two or more dimensions of economic, environmental and social performance as a proportion. Eco-efficiency measurements (e.g., the amount of emissions per unit of product or the monetary unit of turnover) are the best known examples. Many companies have proposed standardized sets of environmental efficiency indicators that measure various modes of resource use or pollution issues in relation to an economic or productivity indicator. Integrated indicators effectively demonstrate the size of the positive or negative impact of any changes produced in a variable.

Literature and practice of the countries that have adopted the method of TBI reporting, revealed behaviors and attitudes essential for companies to understand and adhere to a management and reporting system, according to the concept TBL (Caraiani et al. 2010):

(a) Responsibility acceptance—TBL is based on the assumption that companies are responsible not only to shareholders, in order to obtain profit, but also to stakeholders (customers, suppliers, government agencies, employees' families, special interest groups) to help according to their interests and potential in sustainable development;
(b) Transparency—according to a mercantilist conception, companies are required to be open-minded about their activities and their impact beyond the strictly financial performance connotation. Accepting the legitimate right of stakeholders to know and recognize the multidimensional acknowledgement of the results represents the central idea of TBL, that most often it can be founded in business philosophy of the company, in communication practice with stakeholders and in the content of the reporting;
(c) Integrating operational plan—so that a company contribute to economic prosperity, quality environment improvement and social welfare is necessary to reflect these dimensions in the strategic plan, operational management systems and motivational schemes. In other words, the integration of the economic, social and environmental inside that defines business processes, represents a prerequisite performance evaluation and reporting in the context of TBL;
(d) Loyalty to the commitment of stakeholders—cooperation with internal and external stakeholders is a process for communicating business goals,

developed after a thorough investigation of the dimensions involved. The prospects stakeholder loyalty and strategy developement are being considered actions that add value;

(e) The multidimensional assessment and reporting—the systematic analysis and certification of economic, social and environmental performance, correlated with a formalized presentation of the results, represents the main mechanism that creates a clear picture about a company's objectives, how they act and how they honor their commitments.

Reporting of the financial performance can be considered limited in accuracy compared with market performance. Therefore, there is a growing need for non-financial reporting. In recent years, TBL reporting has also become a tool for presenting information, based on that performance by monitoring social, economic and environmental performance, and economic entities can better prepare for future opportunities and requirements.

The TBL approach of the accounting, we should remove the unilateral representation and analysis of the outputs. Account TBL access is not looking for a new system of performance indicators for expression, but rather a shift towards relocation management and performance in order to strengthen the importance and interdependence of economic, social and environmental aspects. TBL concept must be analyzed and understood in relation to other concepts on the role and responsibilities of a company and its performance. In this context, it is required a transdisciplinary conceptual framework that intensifying understanding of multidimensional corporate role and responsibilities. In a first approach appears indispensable the concepts of sustainable development, social and environmental responsibility and stakeholder engagement.

The emphasis of this transdisciplinary and multidimensional approach focuses on opportunities that emerge from the reporting the non-financial aspects and how they can be exploited.

5 Conclusions

Depending on the interest of users of financial accounting information, the term of performance is differently defined and measured. Shareholders and potential investors are interested in the profitability of investments, aimed at the overall performance of company; managers, creditors, suppliers are interested in company solvency and stability. As we can see, performance measurement is based on the objective of interested users.

In order to improve their performance, companies are turning to activities that meets social and environmental safety and reliability requirements, namely to the environmental performance and social performance.

At a worldwide level, the major objective of companies is no longer simply a profit, observing that increasingly more entities began to approach for a sustainable development of the business concept of reporting multidimensional introducing performance at three levels: environmental, economic and social, which enhances non-financial dimensions.

As shown in the content of this paper, the best known and used standard for sustainability is GRI (Global Reporting Initiative), which has developed a comprehensive framework for sustainability reporting and is widely used, allowing companies to measure and report all economic, environmental, social and governance performance—the four key areas of sustainability. There are over 3000 companies, most of them multinationals who adopts such a reporting because they understand that a company's success is measured not only in financial results but also in engaging in actions to protect the environment and society.

In conclusion, we note that, non-financial reporting should be used by all companies not only large ones, as is currently the case, because the companies that make also a non-financial reporting enjoys greater attention from stakeholders. Such reporting permits companies to offer transparent and comparable information about its activity, thereby leading to increased company credibility among stakeholders.

The research will be continued in a future paper by an empirical study based on a specific case or a systemic assessment for Romania in the field of social and environmental information use.

Acknowledgment This paper has been financially supported within the project entitled ***"SOCERT. Knowledge society, dynamism through research"***, contract number POSDRU/159/1.5/S/132406. This project is co-financed by European Social Fund through Sectoral Operational Programme for Human Resources Development 2007–2013. "**Investing in people!**"

References

Caraiani C et al (2010) Green accounting—transdisciplinary strategy towards social and environmental accounting, studies and research. Publishing House ASE, Bucharest, pp 19–35

Carroll A (1979) A three-dimensional conceptual model of corporate performance. Acad Manag 4 (4):499

Carroll AB, Buchholtz AK (2008) Business and society: ethics and stakeholder management, 7th edn. South-Western Cengage Learning, Ohio, p 46

Cochran PL (2007) The evolution of corporate social responsibility. Bus Horiz 50:450

CODECS (Centre for distance education) (2010) The power of knowledge, BZT628 Management 1: Organizations and People, Course guide for the Professional Certificate in Management, Business School, First edition in 2009, the first edition in roumanian language in 2010, The Open University Business School. Publishing House Codecs, Bucharest, 24 – 37. http://www.codecs.ro, http://www.oubs.open.ac.uk

Coelho A (2011) Corporate social responsibility—strategies and tools for stakeholder engagement—the handbook. http://www.sairdacasca.com/pdfs/TheHandbook.pdf. Accessed 3 Dec 2014

Elkington J (1994) Towards the sustainable corporation: win-win-win business strategies for sustainable development. Calif Manag Rev 36(2):90–100

GRI (2014) What is GRI? https://www.globalreporting.org/information/about-gri/what-is-GRI/Pages/default.aspx. Accessed 20 Dec 2014

Iamandi I, Filip R (2008) Ethics and corporate social responsibility in international affairs. Publishing House Economica, Bucharest, p 342

Martin R (2003) The virtue matrix. A tool for understanding corporate social responsibility. Rotman Management, Rotman School of Management University of Toronto, Canada, pp 7–11. Available on http://www.rotman.utoronto.ca/rogermartin/Virtue_Matrix.pdf. Accessed 13 Nov 2014

Morschett D, Zentes J (2010) Strategic international management: text and cases, vol. 2. Publishing House Gabler, Wiesbaden, p 222

Muresan I (2010) Ethics and social responsibility in marketing—economic and legal aspects. Publishing House All beck, Bucharest, p 53

Nin Ho F, Wang H-MD, Vitell SJ (2012) A global analysis of corporate social performance: the effects of cultural and geographic environments. J Bus Ethics 107:424. [online] Available at: http://www.springerlink.com/content/b871964073634n3j/fulltext.pdf. Accessed 10 Dec 2014

Reporting RP (2000) Available on http://www.bilanciosociale.it/bilancio_sociale.html, http://www.bilanciosociale.it/obiettivi.html, http://www.bilanciosociale.it/bilancioambientale.html. Accessed 5 Nov 2014

Savitz AW, Weber K (2006) The triple bottom line: how today's best run companies are achieving economic, social and environmental success—and how you can too. Jossey-Bass/Wiley, JB Printing, USA, San Francisco

Sethi SP (1975) Dimensions of corporate social performance: an analytical framework. Calif Manag Rev 17(3):58–64

Tabara N (2006) Modernizing accounting and management control. Publishing House Tipo Moldova, Iaşi, p 146

The Conference Board of Canada (2006) Corporate social responsibility review. pp 3–16. Available on http://penny.greatappsfast.com/web/uploads/CR_Review.pdf. Accessed 15 Nov 2014

Wartick SL, şi Cochran PL (1985) The evolution of the corporate social performance model. Acad Manag Rev 10(4):758

Zeller M, Lapenu C, Greeley M (2003) Measuring social performance of micro-finance institutions: a proposal Social Performance Indicators Initiative (SPI) Final Report, Argidius Foundation and Consultative Group to Assist the Poorest

Zhang Y, Ma R (2011) Comprehensive benefits assessment of demand response considers corporate social performance. In: International conference on electrical and control engineering, pp 4416–4418

Authors Biography

Camelia Catalina Mihalciuc Ph.D. in economic doctoral studies, sub domain accounting from "Alexandru Iona Cuza" University, Faculty of Economy and Business Administration, Iaşi, from 2008; is Associate Professor at "Ştefan cel Mare" University Suceava, Faculty of Economic Science and Public Administration, Department of Accounting and Finance, Romania, where she teaches Financial accounting and Managerial accounting; her main research interests include financial reporting and integrate reporting.

Anisoara Niculina Apetri Ph.D. in economic doctoral studies, sub domain finance from "Alexandru Iona Cuza" University, Faculty of Economy and Business Administration, Iaşi, from 2012; is Associate Professor at "Ştefan cel Mare" University of Suceava, Faculty of Economic Science and Public Administration, Department of Accounting and Finance, Romania, where she teaches General Finance, Public finance, Budget and public treasury, Currency and credit; her main research interests include Central Bank, National Bank and Finance.

Greening the Economic Growth in Romania: the Environmental Footprint Approach

Florian Marcel Nuță and Alina Cristina Nuță

Abstract

The modern world is on the verge to abandon the "growth no matter what" approach in favor of a more responsible behavior. The environment is gaining a more important position in the enterprises' and states' strategies and both researchers and decision makers are assessing new, green models of maintaining and improving the economic welfare without affecting the future generations' opportunities. The aim of our paper is to determine the correlation between the growth variables and the quality of the environment. We seek to investigate the marginal loss of economic growth for the targeted environmental protection. We consider the case of Romania interesting regarding the matter of greening the economic growth for the fact that as any other emerging economy needed structural changes affecting the peoples' welfare. In such situations the tendency is too neglect the less represented parts of the economic cycle. One of these is the environment. This fact is creating imbalances between the biological capacity of a region and its development impact upon the natural resources and environment as a whole. Once Romania had to adapt its regulations and environmental behavior as a result of EU integration, we need to evaluate how the gap between the economic development and the environmental protection is filled and compromise is made.

Keywords

Environment · Economic growth · Ecological footprint · Pollution · CO_2

F.M. Nuţă (✉) · A.C. Nuţă
Danubius University from Galați, Galați, Romania
e-mail: floriann@univ-danubius.ro

A.C. Nuţă
e-mail: alinanuta@univ-danubius.ro

W. Leal Filho et al. (eds.), *Sustainable Economic Development*,
World Sustainability Series, DOI 10.1007/978-3-319-45081-0_13

1 Introduction

The concept of sustainable development plays an important role in the evolution of a nation because, unless there is population fair and reasonable means of subsidence and development is not possible to ask the conservation and proper use of natural resources.

It expresses the emergence of a new optimistic model, to record and preserve the environment, respond to an increasing regulations or to consumers of political issues, but also of natural maturities. The planet is an ecosystem where much of the natural resources are not renewable. This reality, recognized for a long time, through the understanding of the need for extensive management of resources and a global approach. Remember that the true stakes of tomorrow unrelated to the value of the action and environmental performance, but with his great themes that climate change has become a reality, access to water resources management and inadequate funding demographic development of Asia. Sustainable development is a framework and not a use possible to allow a rational development which will prevent degradation of the global eco-system (Baek et al. 2008; Borucke et al. 2013; Haung and Cai 1994; Lafferty and Meadowcroft 2000; Nuţă 2011; O'Faircheallaigh 2010).

Globalization-environment debate largely operates around the multinationals and industrial relocation. These businesses are often major polluters and/or users of natural resources, but insufficient rules and ethics in international competition does not facilitate a process harmless for the environment and profitable for all nations of the Earth.

This competition requires more ethical and calls the precautionary principle. It needs rules and corrective actions in order to provide more information and transparency for investors and consumers. This includes reconciling economic competitiveness with the need to protect the environment. In these circumstances the imperative of a new governance is marked by institutional adjustments, new instruments for measurement as environmental accounting.

It is well known the fact that the main advantages of an emerging economy are the low labor cost, cheap natural resources and low environmental regulation. It is also a fact that the developing countries are less involved in environmental stewardship (Jayanti and Gowda 2014). Moreover there are still discussions whether is financially efficient to have a green behavior or not (Margolis and Walsh 2003; Etzion 2007). Generally for developing countries pollution control and environmental monitoring it is a challenging task due to lack of know-how, poor funding, old technology, high levels of corruption, etc. (Kathuria 2007).

Though some authors consider that greening the growth and development mean both new constraints and new opportunities (Collier and Venables 2012) the tendency is to ignore the opportunities and to complain about the constraints and find ways of avoid it even if it means relocating industrial capacities to areas less regulated.

Romania is one of the ex-communist highly industrialized countries that had to overcome the shock of exist outside the communist economic block and find new ways of becoming competitive by itself. The old industrial technologies generating high level of pollution and high levels of energy consumption had to be technologically renewed, closed or replaced. Most of the cases this generated unemployment and development gaps. Other studies identified evident trade-offs between the ecological balance and the economic performance using agriculture conditions as case studies. It was stated that both economic and environmental sustainability can be obtained as a result of the proper combination of high profitability crops (Blasi et al. 2016). This results sustain a wider approach about the appropriate combination between nature and economy, as the high developed countries concentrate on high profitability activities and externalize dirty industries.

In literature we can find also different combinations for describing the ecological impact. For instance the carbon footprint is used together with Emergy (which is a type of energy used in direct or indirect processes needed to realize a product or service) (Jamali-Zghal et al. 2013) and more recent Emergy is used together with the ecological footprint (Mikulcic et al. 2016). These is a evidence of the complexity of the ecological and sustainability issues that cannot be described alone by one indicator no matter how developed it is.

The accounting and generally the economic activity neglected the ecology and biodiversity (Vackár 2012). However recent researches included it and ecosystem accounts (related to land, ecosystem resources and water) were created for better explanation and assessment of such an important issue (Weber 2007).

In the Romanian biodiversity structure, the forest is losing spaces in the last decades in the favor of agriculture and artificial areas (tourism construction, etc.). This is a sign of imbalances between the ecological footprint and the biological capacity of the region, under the economic and development pressures (Fig. 1).

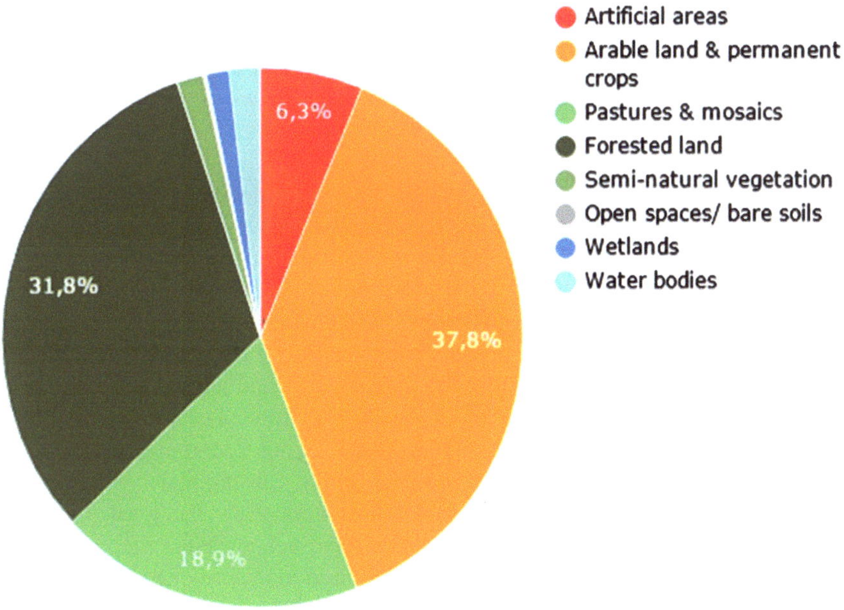

Fig. 1 The structure of the biodiversity—Romania. *Source* http://biodiversity.europa.eu/

The ecological footprint developed by Wackernagel and Rees (1996) is one of the environmental indicators trying to explain the planet biological capacity of sustaining life at a given level of resources consumption. The ecological footprint is based on 6 main land categories (cropland, grazing, forest, fishing ground, built up and land for carbon uptake). The biocapacity measures the potential productive zones available for human use (Wackernagel et al. 2002). It is known that only about a quarter of the countries analyzed still have reserves of ecological footprint at their present levels of consumption (WWF 2012).

Lately the ecological footprint was subject of discussion between prominent researchers (Galli et al. 2016) for clarifying some methodological and practical aspects. One of the things clarified was that this indicator alone cannot describe the sustainable development as a whole but only some of its attributes.

Given the macroeconomic conditions there are questions whether the countries respect their biological stock limits or influence the depletion of other countries by externalizing their high pollution industries. Moreover if the country is a poor or developing one, receiving industrial capacities and foreign capital is understood as a way of improving the economic growth. At the other hand if a country is highly developed, it seeks to relocate the heavy industries and environmental unfriendly production capacities to more permissive regions and seeking for comparative advantages (as cheaper work labor). The correlation between the income growth (economic wealth) and national environmental quality (different ways of describing environmental degradation) is generally known in the literature as the Environmental Kuznets Curve (EKC), as a inverse "U" shape showing how the environmental degradation rises as the income increases up to a point where it starts to decline (Aşici and Acar 2016).

2 The Modeling

Although we accept that the environmental policies (including taxation as a major regulating tool) may spread its beneficial effects upon not only the environmental welfare, the main benefit is keeping the environment damages and pollution at lowest levels as possible. We consider the environmental welfare as being the sum of all natural resources (water, wild life, etc.) and environmental conditions (mainly related to public health such as the air) unaffected by the human activity. The CO_2 emissions are one the most important air contaminant and are associated to urban agglomerations and combustion engines. The Kyoto protocol established limits and the automotive industry developed new standards trying to diminish the pollution. Although the carbon dioxide is widely recognized as causing major impact on the ozone layer, the variety of environmental indicators show the great interest of academics and decision makers to understand and acknowledge the critical sustainability issues of our time. Nevertheless it is hard to choose one single indicator or a category of ecological variables for assessing the environmental issues. The complexity of environmental issues itself is the first big impediment for it. The

Table 1 Correlation analysis between the environmental integrity and economic growth

Dependent variable: CO_2				
Method: least squares				
Variable	Coefficient	Std. Error	t-Statistic	Prob.
R-squared	0.724238	Mean dependent var		55.05769
Adjusted R-squared	0.699168	S.D. dependent var		4.221540
S.E. of regression	2.315435	Akaike info criterion		4.657711
Sum squared resid	58.97365	Schwarz criterion		4.744626
Log likelihood	−28.27512	Hannan–Quinn criter		4.639846
F-statistic	28.88943	Durbin–Watson stat		1.718676
Prob(F-statistic)	0.000225			

human activity affects the quality of air, water, damages the ozone layer, contribute to natural resources depletion and so on. That is why a single complex index is hard to establish and use to describe all the damage we provoke to the environment (Table 1).

To determine whether is a correlation between the economic potential and the environmental quality we assess the CO_2 emissions trend and the GDP growth rate in the same period of time (Table 2).

The analysis shows a direct correlation between the economic development described by the GDP growth rate and the environmental damage described by the CO_2 emissions. The correlation is also statistically significant. The equation describing the correlation model is:

$$CO_2\,emissions = 0.855200 * GDP\_growth\_rate + 51.93292$$

The findings confirm us the connection between the economic growth of a developing country and the environmental damage caused by the process. This is because the growth potential is based on low cost labor, poor technological endowment and lack of environmental and social regulation (Fig. 2).

The graph also shows the same trend for the both variables. The interesting fact in the graph is that the low levels of economic growth common for the crisis years determine low levels of CO_2 emissions and so less environmental harm.

The specific conditions of the Romanian economy contributed to the environmental issues we are trying to explain and solve today. After being a highly industrialized economy having great industrial platforms based mainly of high energy consumption the economic profile of the country loss many of these production capacities. This should be seen as a benefit for the environment and for the sustainable development targets. However parts of the environmental impact was transferred to the households given the fact that consumption (of all kinds) grew at this level. Moreover foreign investments brought dirty industries and natural resources exploitation contributing to environmental depletion and pollution. In this conditions Romania's development still needs more than its own biological capacity (Fig. 3).

Table 2 The model coefficients

Dependent variable: CO_2				
Method: least squares				
Variable	Coefficient	Std. Error	t-Statistic	Prob.
C	51.93292	0.866249	59.95149	0.0000
GDP_GROWTH	0.855200	0.159110	5.374889	0.0002

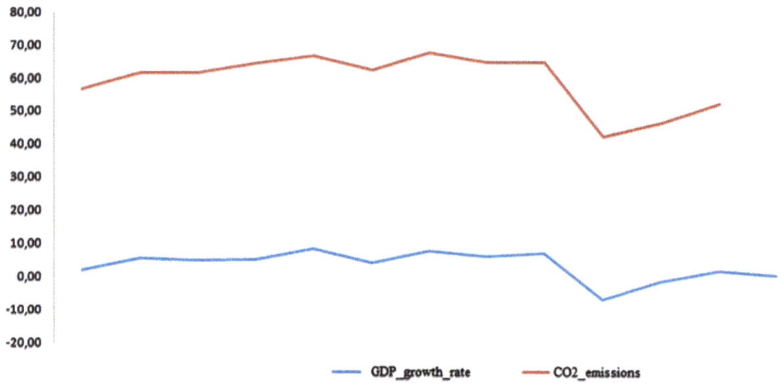

Fig. 2 The economic growth trend versus pollution

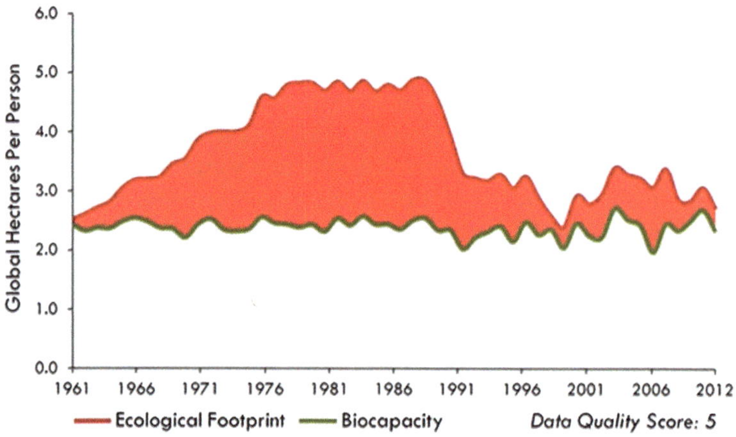

Fig. 3 The ecological footprint—Romania 1961–2012. *Source* http://www.footprintnetwork.org/en/index.php/GFN/page/trends/romania/

3 Conclusions

It is still a challenge for all the socio-economic actors to find a way to keep economic growth going and at the same time to diminish the environmental damage. Also we still have to manage our way of life into a more environmental friendly behavior. This because is a fact that less developed countries have a smaller ecological footprint than the developed ones. Still the most affected countries, environmentally speaking, are those placed on a development trend. These countries are the subject of technological transfers. Most of the case, technologies that are too old and highly polluted to operate in high developed countries, but still financially efficient and able to give high productivity margins. These countries are also characterized by low environmental standards and lack of regulations.

As ecological footprint studies and assessments shown, less developed and developing countries seem to better maintain their environmental conditions stable and proper use their resources. Yet this is a false impression. The main reason for which this happens is the lack of technological potential to exploit the resources and to obtain economic progress by they own.

This is how Romania has a 2.71 ecological footprint and for example Switzerland, a developed high standards of living country scores 5.02. The translation of this scores is that the Romanian medium behavior and way of living demands 2.71 Earth planets to fulfill the needs of the population and in Switzerland, the population would need almost double to cover its common needs of living at the given standards.

It is also clear that the balance line between the economic development and the environmental sustainability is very thin and the two issues should be assessed with great care.

For instance the EU economic growth has doubled the ecological impact on the planet in the last 30 years. Although owns 7.7 % of the global population and 9.5 % of the planet's biocapacity, the EU is responsible for 16 % of global ecological footprint. Despite technological advances, environmental pressure has grown faster than the population of Europe, thereby creating a deficit of natural resources both for the world and for future generations. Except for few examples all EU countries face a shortage of natural resources.

In 1999, the average ecological footprint in all continents of the world population was 2.3 ha per person, a level 20 % higher natural production capacity of the Earth, with well defined differences by region. Average ecological footprint of Africans and Asians was 1.4 ha per person, the Europeans was 5 ha per person, while that of North Americans amounted to 9.6 ha per person. This imbalance is not surprising insofar as we know that one fifth of the world (North America, European Union and Japan holds 90 % of global consumption).

In future work we will test the validity of the Environmental Kuznets Curve for the case of Romania. The approach could encounter some difficulties given the conditions of the Romanian economy transformation during the last decades from a centralist to a free market one.

References

Aşici AA, Acar S (2016) Does income growth relocate ecological footprint? Ecol Indic 61 (2016):707–714

Baek J, Cho YS, Koo WW (2008) The environmental consequences of globalization: a country-specific time-series analysis, American Agricultural Economics Association Annual Meeting, Orlando, FL, July 27–29, 2008

Blasi E, Passeri N, Franco S, Galli A (2016) An ecological footprint approach to environmental–economic evaluation of farm results. Agric Syst 145(2016):76–82

Borucke M, Moore D, Cranston G, Gracey K, Iha K, Larson J, Lazarus E, Morales JC, Wackernagel M, Galli A (2013) Accounting for demand and supply of the biosphere's regenerative capacity: the National Footprint Accounts' underlying methodology and framework. Ecol Indic 24:518–533

Collier P, Venables AJ (2012) Greening Africa? Technologies, endowments and latercomer effect. Energy Econ 34(2012):575–584

Etzion D (2007) Research on organizations and the natural environment, 1992-present: a review. J Manag 33:637–664

Galli A, Giampietro M, Goldfinger S, Lazarus E, Lin D, Saltelli A, Wackernagel M, Muller F (2016) Questioning the ecological footprint. Ecol Indic 69(2016):224–232

Haung C-H, Cai D (1994) Constant returns endogenous growth with pollution control. Environ Resour Econ 4:383–400

Jamali-Zghal N, Amponsah NY, Lacarriere B, Le Corre O, Feidt M (2013) Carbon footprint and emergy combination for eco-environmental assessment of cleaner heat production. J Clean Prod 47:446–456

Jayanti RK, Gowda RMV (2014) Sustainability dilemmas in emerging economies. IIMB Manag Rev 26:130–142

Kathuria V (2007) Informal regulation of pollution in a developing country: evidence from India. Ecol Econ 63(2007):403–417

Lafferty W, Meadowcroft J (eds) (2000) Implementing sustainable development. Oxford University Press, Oxford

Margolis JD, Walsh JP (2003) Misery loves companies: rethinking social initiatives by business. Adm Sci Q 48(2):268–305

Mikulcic H, Cabezas H, Vujanovic M, Duic N (2016) Environmental assessment of different cement manufacturing processes based on Emergy and Ecological Footprint analysis. J Clean Prod. doi:10.1016/j.jclepro.2016.01.087

Nuţă AC (2011) A theoretical approach of fiscal and budgetary policies. Acta Univ Danub OEcon 7(6):91–98

O'Faircheallaigh C (2010) Public participation and environmental impact assessment: Purposes, implications and lessons for public policy making. Environ Impact Assess Rev 30:19–27

Vackár D (2012) Ecological Footprint, environmental performance and biodiversity: a cross-national comparison. Ecol Indic 16(2012):40–46

Wackernagel M, Rees WE (1996) Our ecological footprint: reducing human impact on the earth. New Society Publishers, Philadelphia, PA

Wackernagel M, Schulz NB, Deumling D, Callejas Linares A, Jenkins M, Kapos V, Monfreda C, Loh J, Myers N, Norgaard R, Randers J (2002) Tracking the ecological overshoot of the human economy. Proc Natl Acad Sci USA 99:9266–9271

Weber JL (2007) Implementation of land and ecosystem accounts at the European Environment Agency. Ecol Econ 61:695–707

***WWF, Living Planet Report 2012, www.panda.org

Authors Biography

Florian Marcel Nuţă holds a Ph.D. in accounting (2009), on a subject related to environmental protection accounting. He is currently Associate Professor at Danubius University from Galaţi, Faculty of Economic Sciences. Among the research interests we can mention green accounting, ecological footprint accounting, sustainable development, environmental responsibility, global performance, green taxation, etc. In 2010 he won a scholarship and from 2010 to 2013 developed postdoctoral research home and abroad in the field of environmental responsibility. He is a member of the European Association of Environmental and Resource Economists (EAERE). One of the most recent research is "An assessment upon the environmental policy in Romania" published in Economic Research-EkonomskaIstraživanja. (http://www.tandfonline.com/doi/full/10.1080/1331677X.2015.1083874).

Alina Cristina Nuţă holds a Ph.D. on Public finance (2011), a M.Sc. in Finance and Banking Management (1999) and 2 B.Sc in Finance and Insurance and Public Administration (1993) from Alexandru Ioan Cuza University, Romania. She is currently Associate Professor in the Department of Finance and Business Administration at Danubius University, Romania. She followed post doctoral research (2014–2015) in population ageing and socio-economic impact of it at national and European level. Mrs. Nuţă has participated in many research and development projects on the national and international scale. She is also a reviewer of several acclaimed journals, including International Business Research. Main research interests are: public policy, fiscal and social policy, ageing and migration, green taxation, social protection and sustainable development.

Transition to Green Economy: Green Procurement Implementation Strategies Experiences from Hungarian Public and Private Organizations

Orsolya Diófási-Kovács and László Valkó

Abstract

According to UNEP green economy reduces emissions and pollution, enhances energy and resource efficiency and prevents the loss of biodiversity and ecosystem services. With other words greening the economy is a key to achieving sustainable development. There are many strategies and many fields to take action and special enabling conditions will be needed in form of policies, subsidies, incentives and legal infrastructure. At national level a strategic step towards green economy can be the implementation of green public procurement. The aim of our paper is to explain the correlation between green procurement and green economy and give example on practical solutions and methods for implementing the concept of green procurement. In the recent years there was a lot of attention on green public procurement EU-wide, and many sensitization and awareness raising projects were funded by the Union. These have very different effects in each country. Realizing and analyzing this gap we found that there is a solution, and green procurement can be implemented in every country with the help of a toolkit specially designed for procurement development. We believe that these findings and solutions can give valuable information to scholars, practitioners and policy makers and will contribute to furthering green procurement and help achieving sustainable development.

O. Diófási-Kovács (✉) · L. Valkó
Department of Environmental Economics, Budapest University of Technology and Economics, Budapest, Hungary
e-mail: o.diofasi@survive.hu

L. Valkó
e-mail: valkol@eik.bme.hu

W. Leal Filho et al. (eds.), *Sustainable Economic Development*,
World Sustainability Series, DOI 10.1007/978-3-319-45081-0_14

Keywords
Green economy · Green public procurement · Green procurement · Sustainable consumption · Sustainable production

1 Introduction

The widely known and accepted definition of sustainability created by the Brundtland Commission in (1987) states, that sustainable development is "development that meets the needs of the present without compromising the ability of future generations to meet their own needs." (Brundtland Commission 1987). For achieving this goal, economic transition is necessary. Green economy, is a relatively new term (used since 2008) and has many definitions (UNDESA 2012), but we can sum up with the followings: it reduces emissions and pollution, enhances energy and resource efficiency and prevents the loss of biodiversity, ecosystem services and also takes social wellbeing into consideration. Green economy is not a state but a process of transformation and a constant dynamic progression (Danish Group 2012). With other words greening the economy is a key to achieving sustainable development. However, if we look at theory and practice, we can see that there is an implementation gap, with such knowledge and developed policy toolkit far better results should be achieved in terms of reducing environmental footprint of our economy and lives all together.

For this reason the aim of our paper is twofold: (1) to sum up the correlation between green procurement and green economy and (2) give detailed example on a practical solution and method for implementing the concept of green procurement successfully.

We find that it is highly important to get from theory to action and start effective, useful actions towards sustainability before it is too late. Procurement is quite a special field, but has great significance on the economy due to the great purchasing power.

This paper will report on and analyze green procurement practices in Hungary on the organizational level, which means that the presented case studies are from public authorities and private companies as well. The procurement cases were guided by the green procurement toolkit created by the authors. The evolution of the toolkit—the changes and developments done according to the needs of practitioners in different procurement cases (public procurement; below the threshold purchase done by a public authority; private purchases)—will be analyzed and a theory created which methods are the most successful in helping organizations implementing green criteria in their call for tenders.

For better understanding of the paper the relationship between cases and different versions of the toolkit is explained here: the development of the toolkit was a continuous, integrated process: we prepared one version of the toolkit, started to test it in practice and changed several things (the development/evolution), then at the end got the final toolkit.

Even though the case studies are from only one country, the results are significant for other countries too. Especially ones with similar economic and cultural situations, mainly in the Central Eastern European Region where—in our opinion—transition to green economy and sustainable consumption need additional support.

2 The Relationship Between Green Procurement and Green Economy

Sustainable procurement is a relatively new field of research (Nissinen et al. 2009), nonetheless attracts steadily growing interest from scholars and practitioners. Since 2000 the number of research articles has increased about tenfold (Johnsen et al. 2012).

There are many international examples of the implementation of sustainable procurement legislation and supporting tools from the USA (Swanson et al. 2005), South Africa (Bolton 2006), Canada and Japan too (Brammer and Walker 2011). On the EU level policy making green procurement has been a significant topic since 2000. With the help of public purchasing power it has direct effect on sustainable consumption and production patterns, furthers sustainable economic solutions so helps achieving sustainable development (Li and Geiser 2005; Ambec and Lanoie 2008; Diófási and Valkó 2014).

According to Johnsen et al. (2012) 'sustainable procurement can be defined as the pursuit of sustainable development objectives through the purchasing and supply processes'. Green Public Procurement (GPP) is defined in the European Union as'a process whereby public authorities seek to procure goods, services and works with a reduced environmental impact throughout their life cycle when compared to goods, services and works with the same primary function that would otherwise be procured' COM (2008: 400).

In other words green procurement is a possibility for everyone to buy green products and services on all three levels of economy: private, consumer level; green supply chains, corporate procurement; and public procurement. Green procurement has the following economic and social advantages on the macro level:

- Sets example, promotes sustainable solutions
- Furthers innovation, drives new technologies
- Through the high demand on eco efficient products, it results in decreasing prices and makes green solutions available for wider public

So green procurement has significant effects on the economy by creating demand for high quality, sustainable and innovative products and services.

However the concept only furthers green economy if it is implemented properly. This leads us to the next very practical part of the paper that describes solutions and methodology for the implementation of green procurement to different type of organizations.

2.1 Greening the Economy: Practical Solutions, Case Studies

2.1.1 Aims of the Green Procurement Hungary Project

The works of the authors started in 2010 within the Green Procurement Hungary Project (GPH project). This article gives a report on the results achieved until the end of 2014.

The method of the work had two main tasks: (1) collection of pilot projects and the implementation of green procurement (2) toolkit development and improvement according to the organizational needs and experiences of the pilot projects.

The following phases describe the cooperation with the volunteering organizations: (1) Contact: giving information about the project and green procurement, goal setting, development of project plan and schedule (2) Training of the employees, selection of product groups for green purchasing (3) Call for tenders, procurement preparations with the help of the toolkit and consultations (4) Evaluation of the procurement case, evaluation of the implementation process and the toolkit, drawing conclusions.

All the organizations worked closely together with the authors, were provided help and consultancy over the project period to implement green procurement practices. This raises the question about the need and effect of external support in green procurement practices. According to our experiences even with the most comprehensive training and toolkit practitioners asked for support and checking at least at the designing of the call for tenders. This can be due to the interdisciplinary nature of environmental aspects (Nissinen et al. 2009), the lack of experience of the procurers in green issues taking part in the project, and even the risk aversion especially in public procurement procedures.

Over the 4 year period since 2011–2014 eight public organizations and 27 private companies took part in the project, 120 people were trained and 19 pilots completed. There are three typical types of procurement processes depending on the type of organization: public procurement, tendering and ad hoc purchases. Public organizations fall within the scope of public procurement act, but below the threshold purchases can be done by normal tendering or on very small scale can also do ad hoc purchases. Private organizations have the choice of the latter two versions. On large scale they conduct tendering; on the small scale they have ad hoc purchases.

These different types of processes usually have different resources, personnel (know-how) and need a completely different approach in order to have successful implementation of environmental aspects in form of purchasing criteria into the procurement process. This was one of our most evident, but not yet discovered conclusions. It is more effective to differentiate between the processes than between organization types. For example it makes no difference in terms of training and supporting toolkit if public or private organization is going to conduct below threshold/normal tendering.

In Part 3 the contents and the development of the actual toolkit used will be described in detail.

3 Evolution of the Green Procurement Toolkit

This part of the article describes the most significant case studies that helped to further develop the toolkit, and lead us to significant results in terms of implementing green procurement successfully. The toolkit is meant to be easy-to-use and understand so that green procurement would not result a lot extra paperwork and new tasks for the officers. The new practice needs careful implementation and user friendliness because the procurement officers' commitment is crucial to the success of green procurement systems.

3.1 Version 1. Initial, Basic Version

The development of GPH project's toolkit began in 2010 after extensive literature review and the analysis of existing tools, projects (GPP Europe, Toolkit; Green Labels Purchase; Buy Smart; ICLEI-Procura+) and studies from 2000 to 2010 conducted about GPP (GPP Europe, Studies). The studies and toolkits mainly concentrated on Europe, and that gave a useful benchmark in a similar legal and economic environment. The authors had experience with German municipalities too, so these were also taken into consideration by the development.

The first version of the toolkit had three parts: (1) A **basic training** on green procurement with the following topics: principles of green procurement, economic and management aspects, life cycle costing, steps of green procurement—green criteria in the call for tenders, evaluation of offers, methods for monitoring the performance of green procurement. (2) The **performance sheets** contained only technical specifications in a very detailed form, based on ISO Type I ecolabel (UNEP web) criteria and EcoDesign Directives. (3) For calculation of the life cycle costs a **calculation tool** was developed. These tools were similar to the ones used in the BuySmart project. These tools are described in detail in one of the articles of the authors (Diófási and Valkó 2011a). This first toolkit's aim was to address the two biggest barriers of green procurement: lack of technical knowledge, and perception of financial burden. For this reason the emphasis was on technical specifications-performance sheets and calculation tools. The first 11 **product groups** were chosen with the intention to "find the low hanging fruits", so energy consuming products, which can achieve significant economic benefits for green procurement by the implementing organizations. These products were in the list of the most suitable products for greening (Bouwer et al. 2005).

3.2 Case Study 1. Public Purchasing Below the Threshold

With the above described toolkit we started working with a small municipality from the south west of Hungary. The municipality had all three types for procurement processes, but because of the product groups (IT equipment, household appliances, lighting) and financial situation, they chose to apply green criteria in below the threshold tendering and ad hoc purchases. After the training and introduction of the toolkit there were no purchases due to budgetary reasons for a few months. After there was a request for new product group: office paper. As we discovered by other small municipalities and companies paper, sanitary paper and cleaning products are always needed irrespective of current budget.

3.3 Version 2. Fixed and „Flexi" Parts of the Toolkit, for Public and Private Organizations

The conclusion of Case study 1 was that, there is a clear need for including three more product groups into the toolkit. This resulted in 14 **product groups** within the toolkit. Paper and cleaning products are the most popular ones for EU Ecolabel and all other leading European ISO Type I labels, they are even available on the Hungarian market in most cases with no price difference. For this reason the **training** material was completed with more information on ecolabels and their significance in tendering and evaluating the tenders. The training was designed to be less theoretic, contained more ready to use solutions, samples. The background and detailed description can be read in Diófási and Valkó (2011b). Our other conclusion was that a more straight forward approach is needed in terms of training and **green criteria** as well because by small municipalities sometimes the purchasing culture can be less developed, the slightest difficulty can disconcert the green procurement process. Although according to literature it isn't necessarily true that size of the organization ruins green procurement performance (Michelsen and de Boer (2009) analyzed this option in more detail). The **calculation tool** also seemed to be a bit too difficult in Case Study 1. The solution of this problem was that calculations were show as demonstrations within the training, but it was not anymore a compulsory part of the toolkit. To sum up the changes we divided the toolkit into fixed and „flexi" parts. The calculation tools and performance sheets (for 14 product groups already) were meant to stay the same for all types of organizations, and the training divided into public or private organization's needs.

3.4 Case Study 2. The First Green Public Procurement Process

The testing of Version 2. started at the end of 2012 when one of the biggest Hungarian public organizations joined the GPH project. They mainly conducted public purchases and had a team of professionals in terms of management and procurement law as well. Altogether eight procurement experts and legal experts were trained for green public procurement. This case brought the biggest change and step forward to the toolkit.

They demanded different product groups: smart phones, cars/taxi services, notebooks, furniture, paper and printing services. Because of public procurement law and processes the use of ecolabels was limited in terms of allowing different, but equal kind of proof for compliance. These professionals needed different kind of approach, requested implementable, straight to the point criteria, with exact and easy methods of evaluation.

In this case printing services had to be procured with environmental aspects. As the first comprehensive green public procurement tender in Hungary the risks had to be minimized. Market research was conducted with positive results. Green aspects were implemented into the performance clauses (packaging recycling), selection criteria (ISO 14001, EMAS or equivalent) and into the award criteria as well with the use of ecolabels. For minimizing the risk of having an invalid procurement process FSC, PEFC and EU Ecolabel, Blue Angel, Nordic Swan ecolabels or equivalent were requested for the paper used by printing services. These are of course not equivalent (ISO Type I and Type I-like label definition of UNEP), but as a first step procurers assumed that at least FSC or PEFC certified paper will be offered on the same price as not environmentally friendly paper. The financial calculations, cost-benefit analyses were made at the stage of market research with the organization's own tools. The calculation tool only acted as a demonstration for life cycle costs. Another assumption was that because of the high popularity and market availability of these labels none of the bidders will try to prove adequacy without certification.

3.5 Version 3a. Comprehensive Toolkit for Green Public Procurement

Because of the special needs realized in the above described pilot project some developments were made on the toolkit again. It was designed specifically for public procurement with already 18 available **product groups**. The special needs were fulfilled by changing the form, style and structure of the **green criteria**. The toolkit in this form was more similar to the GPP Europe Toolkit, with core and comprehensive criteria, in text form, strictly following the structure of tenders (subject, technical specifications, selection criteria, award criteria, performance clauses). In addition to this new form evaluation methods were also given. This way the green criteria and the evaluation methods became suitable for copy—pasting into the call for tenders one by one. The criteria was still based on ISO Type I ecolabels and the EU recommendations.

The **training** was completed with comprehensive explanations of the green criteria: environmental effects, the meaning of all criteria, along with proof that can be used for evaluating the bids. This case the participants get an overview of green procurement basics and become familiar with the product group specific, exact environmental criteria.

This 3a. version of the toolkit seems to be final, working very well since it's development. It was used in several public procurement processes (for example Case study 3) and in our opinion it seems to be the best solution so far for implementing green criteria into public procurement.

3.6 Case Study 3. Implementing Green Public Procurement with Toolkit Version 3a.

The third selected case study is about implementing environmental criteria into another major public authority's public procurement process with the help of toolkit version 3a. The cooperation started with training for their procurement experts, with emphasis on the selected product groups: long term car rental and facility management. In this case too, similar to Case study 2 a team of procurement and legal experts worked together to manage the call for tenders and evaluate the offers. In case of both product groups the selected and recommended green criteria had to be discussed with the technical experts, engineers of the organization. The interdisciplinary team worked together successfully and the call for tenders was published in Tenders Electronic Daily. There were several vehicle categories, but general criteria was complete tire management and the energy label of tires had to comply with the best two levels. Award criteria were fuel consumption, start and stop system, gear shift indicators, tire pressure monitoring systems, fuel consumption display. With these criteria the minimum of 20 % of fuel consumption and 128 232 kg CO_2 emissions will be avoided yearly compared to the old vehicle fleet.

3.7 Case Study 4. Public Purchasing Below the Threshold

This case study is an extract of two successful below the threshold tenders by two municipalities. One of them was about energy efficient street lighting the other about IT equipment. Both of them gave an interesting lesson to learn, something to consider in the future. In case of street lighting modernization the demand was basically raised by the market. The mayor received many offers and calculations about energy savings and the benefits of LED lighting on the streets from different companies. These external effects and personal commitment towards sustainability together formed the initial idea of changing the lighting system of the town. The call for tenders was compiled by the procurers of the municipality after the green procurement training provided. In this case no performance sheet was requested. In the call for tenders no ecolabel criteria was used, but energy consumption data and ROI calculations were in the award criteria. For this project no external funding was available but the relatively large savings and short return time gave the possibility for the town to invest from their own tight budget. 66 % cost savings and about 50 tons of CO_2 emissions were avoided. Awareness about new technologies, solutions and information about their market availability creates demand and is an important part of implementing green procurement.

3.8 Version 3b. Below the Threshold Purchases of Public Organizations and Private Tendering

The experiences from the above described case studies highlighted that below public procurement threshold purchases are slightly different from the normal public procurement processes. In this version of the toolkit the **training material** stayed the same as in Version 3a. with a bit more emphasis on ecolabels, because in case of under the threshold purchases they can be used without any restrictions. The structure of **green criteria** also stayed the same as in Version 3a., but as proof now ecolabels are requested and used, which makes the evaluation of the tenders a lot easier. After some more experiences with Version 3b. we found that this version of the toolkit can be used perfectly for private organizations' tendering needs because there isn't any significant difference between the two types of processes.

3.9 Case Study 5. Private Ad Hoc Purchases

The fourth case study is rather an extract of working with 23 private organizations in 2013 and the beginning of 2014. The participants of the Green Store Project (GSP 2013–2014) in Hungary were involved in implementing green procurement. The main goal of the Green Store Project is to promote environmental awareness among the owners of the stores, employees, suppliers and customers as well. By the responsible business development, the members learn about the principles of sustainability, and the possibilities of practical realization (GSP 2013–2014).

The participants of the Green Store Project are mostly SMEs, so they usually have ad hoc purchases and sometimes tendering by bigger investments. Ad hoc purchasing in this context is used for not planned, made-on-demand purchasing. Usually these purchases are done on small scale by the owner of the SME (or other responsible employee) and it is made of daily, small value products which have significant environmental impact. These are for example paper—and sanitary paper products, cleaning products, personal hygiene and other household goods. After working with the GPH toolkit (Version 3c.—to be described in Part 3.9) all participating stores succeeded in the implementation of green procurement: for their everyday purchases they use ecolabelled products and have access to version 3b of the toolkit for their higher value tendering. The results are annually audited within the Green Store Project.

3.10 Version 3.c Green Procurement for Ad Hoc Purchases

To implement green procurement by ad hoc purchases called for a slightly different approach again. The **training** was redesigned, included comprehensive sensitization about sustainable consumption and ecolabels. At the last part of the training the participants had a task: created a collection of product groups they daily purchase and checked if ecolabelled products are available in these groups. At these trainings

Table 1 The evolution of the green procurement toolkit

Version	Target group	Product groups	Description	Parts of the toolkit			
				Training	Performance sheets	Calculation tool	Evaluation methods
1	Public and private organizations	Electric oven, refrigerators, washing machines, drying machines, dishwashers, PCs, notebooks, monitors, printers, MFDs, in-door lighting	Initial version	Basic, about green procurement management	Data, in form of tables, useful for the technical specifications based on ecol abel criteria and Eco Design Directives	Detailed calculations for each product group	Partly present because of ecol abel criteria
2	Public and private organizations	Smart phones, taxi/courier services, paper and printing service, notebooks	Fixed parts are the performance sheets and calculators, "flexi" parts are personalized upon the request	Basic, about green procurement management + comprehensive about green criteria and ecolabels	Text, recommendations for all 5 stages of call for tenders/core and comprehensive criteria	For demonstration of LCC	Exact evaluation method for each criteria
3.a	Public organizations	Smart phones, taxi/courier services, long term car rental services, paper and printing service, notebooks, PCs, street lighting, facility management, cl eaning services	Procurement process based separation, not the type of organization matters: **public procurement**	GPP management, legal framework, comprehensive expl anation of environmental criteria	Text, recommendations for all 5 stages of call for tenders/core and comprehensive criteria/*e.g.: recycled paper*	For demonstration of LCC	Ecolabel or if ecol abel criteria used external help/*e.g.: EU Ecolabel or equivalent proof*

(continued)

Table 1 (continued)

Version	Target group	Product groups	Description	Parts of the toolkit			
				Training	Performance sheets	Calculation tool	Evaluation methods
3.b	Public and private organizations	Smart phones, taxi/courier services, long term car rental services, paper and printing service, notebooks, PCs, street l ighting, facility management, cl eaning products/services, sanitary paper products, in door l ighting	Procurement process based separation, not the type of organization matters: **below threshold or private tendering**	Green procurement management, ecolabels	Text, recommendations for all 5 stages of call for tenders/core and comprehensive criteria + availabl e ecolabels/*e.g.: EU Ecolabel certified paper*	For demonstration of LCC	Ecolabels or exact eval uation method if the product group does not have ecol abel criteria
3.c	Public and private organizations	Anything that has ecolabel criteria, and avail able product, but mainly small scale purchases: paper, cleaning or sanitary products	Procurement process based separation, not the type of organization matters: **ad hoc purchases**	Ecol abels, exact meanings-environmental cri teria, sustainable consumption	Mainl y a l ist of ecolabels and product groups, the possibl e differences highl ighted (ISO Type I or Type I-like labels)	For demonstration of LCC	Ecolabels

Source Own compilation

the importance of personal commitment was even more significant than in other cases. At some points we had to face skepticism. This in literature is called the behavior—impact gap problem by Csutora (2012), explaining that it can be questionable if these little steps for sustainable consumption add up and really further sustainability. The other objection was product availability. This had been analyzed by several studies even in Hungary in connection with ecolabelling (Baranyi 2011) and in practice we can say that ecolabelled products are available on the Hungarian market, but of course greater demand can facilitate greater range of green products. A **list of ecolabels** and information sources for ecolabelled products was made instead of green criteria for these type of purchases. This is quite similar to household shopping, so this version 3c. of the toolkit can be also used for personal trainings and awareness raising on sustainable consumption.

The similarities and differences of the toolkit versions are summarized in Table 1. It makes clear the evolution of methods and tools used for different type purchases in order to implement successful green procurement.

4 Conclusions

Table 1 contains the essence and conclusions of the case studies described in Part 3. and shows the development, evolution of the toolkit. From the initial version finally three types of toolkits emerged. According to the experiences of the authors the separation based on procurement types and modifying the support strategy accordingly is the most successful method for implementing green procurement.

There are two main findings (1–2) from the four years of this project, and a few others (3–6) that has to be taken into account anytime implementing green procurement.

(1) The procurement types are the most important categorization aspects, these influence the procedures, the resources needed and the support materials. (2) The use of ecolabels are crucial, but not unlimited in all types of procurement processes. According to GPP Europe (GPP Europe): 'Ecolabels are one of the key tools used to implement GPP—both to develop specifications or criteria and to verify the compliance of products and services with these standards. They can be used in a number of ways in procurement—although it is not allowed to require products or services to carry a particular ecolabel'—due to legislation (discrimination issues). To sum up ecolabels are vital for setting official and correct criteria, for describing what makes a product or service environmentally friendly and also help furthering innovation with green procurement (Diófási 2013). Nonetheless as it can be seen from Table 1. the practice shows that the use of ecolabel criteria can be limited, or at least has to be applied carefully in case of public procurement processes. This is because of the evaluation methods. If ecolabel criteria is used in the technical specifications ecolabel can be a proof for compliance, but any other equivalent proof also has to be accepted. In case any of the bidders decide to verify their compliance without an ecolabel very complex and interdisciplinary tasks can occur

by the evaluation. For this reason an external expert might have to be involved in the process, making it more complex. This aspect has to be considered by the preparing of the call for tenders. (3) The existence and ever growing popularity of ecolabels and their connection with green procurement suggests that producers, companies are able to take the challenge of fulfilling the ecolabel criteria if they find motivation, being green means competitive edge. (4) The question of an external expert guiding the whole process of green procurement or an interdisciplinary team working on criteria setting should help the implementation and having successful tenders. (5) As it was mentioned in the case studies (2 and 3) there is a high demand for risk avoidance: new methods (e.g. environmental criteria) cannot risk the success of the call for tenders in any aspect, neither in connection with product availability or price nor in the legal sense. These risks can also be avoided with professional guidance. (6) The question of how personal commitment and organizational culture affects green procurement performance (Tátrai 2013) is also vital. Resistance towards green procurement within the organization (at any level) can ruin the chances of successful implementation. In connection with this topic further research task of the authors is to follow these organizations (and people attending the trainings) to see what they implemented, and achieved with their knowledge and commitment.

In our opinion these findings have great significance for practitioners and policy makers as well (the cultural differences of each country and region of course has to be taken into account). Even on the European level by the development of funding opportunities and by the evaluation of project proposals these experiences are valuable. With careful coordination of the demand and supply side efforts green economic solutions can be reached.

References

Ambec S, Lanoie P (2008) Does it pay to be green? a systematic overview. Acad Manag Perspect 22(4):45–62

Baranyi R (2011) A környezetbarát termék védjegyezés "3win" kapcsolatrendszere, a hazai rendszer gazdasági elemzése, Doktori Értekezés, 2011. http://www.omikk.bme.hu/collections/phd/Gazdasag_es_Tarsadalomtudomanyi_Kar/2011/Peterne_Baranyi_Rita/ertekezes.pdf. Accessed 03 Feb 2012

Bolton P (2006) Government procurement as a policy tool in South Africa. J Pub Procure 2006(6): 193–217

Bouwer M, de Jong K, Jonk M, Berman T, Bersani R, Lusser H, Nissinen A, Parikka K and Szuppinger P (2005) Green public procurement in Europe 2005—Status overview. Virage Milieu and Management bv, Korte Spaarne 31, 2011 AJ Haarlem, the Netherlands. http://europa.eu.int/comm/environment/gpp/media.htm#state. Accessed 10 Mar 2011

Brammer S, Walker H (2011) Sustainable procurement in the public sector: an international comparative study. Int J Oper Prod Manag 31(4):452–476

Brundtland Commission (1987) Report of the World Commission on Environment and Development. United Nations

BuySmart Buy Smart—Green Procurement for Smart Purchasing, Project funded by Intelligent Energy Europe, www.buy-smart.info. Accessed 20 Feb 2010

BuySmart+, 2014 BuySmart + Green Procurement In Europe, http://www.buy-smart.info. Accessed 15 Sept 2014

COM (2008) 400 Communication from the Commission to the European Economic and Social Committee and the Committee of the Regions, Public procurement for a better environment

Csutora M (2012) One more awareness gap? the behaviour-impact gap problem. J Consum Policy 35:145–163

Danish Group (2012) The Danish 92 Group forum for sustainable development. Building an Equitable Green Economy

Diófási O, Valkó L (2011a) Green (Public) procurement in practice—methods and tools for the successful implementation. Reg Bus Stud 3(2):11–23

Diófási O, Valkó L (2011b) Training methods for purchasing officers to facilitate green procurement in Hungary. 2011. In: Koprivanac N, Hrvoje K, Loncaric B (eds) 3rd International symposium on environmental management conference proceedings: towards sustainable technologies. Zagreb, Croatia (ISBN: 978-953-6470-56-3)

Diófási O (2013) Green Public Procurement, the driver of innovation through product development and ecolabelling. In: Thomas B (szerk.) Proceedings 12th IAS-STS Annual Conference, Graz: pp. 1–15. Paper 2, Graz, Ausztria, 06–07 May 2013

GPP Europe European Union on Green Public Procurement, Official website: http://ec.europa.eu/environment/gpp/index_en.htm. Accessed 01 Oct 2015

GPP Europe, Studies European Union on Green Public Procurement, Official website, Studies, http://ec.europa.eu/environment/gpp/studies_en.htm. Accessed 01 Oct 2015

GPP Europe, Toolkit Official recommendations of the EU on GPP criteria, http://ec.europa.eu/environment/gpp/eu_gpp_criteria_en.htm, Accessed 01 Oct 2015

Green Labels Purchase Green Labels Purchase—making a greener procurement with energy labels project, http://ec.europa.eu/energy/intelligent/projects/en/projects/greenlabelspurchase. Accessed 01 Oct 2015

GSP (2013–2014) Green Store Program (2013–2014): http://zolduzlet.hu/green-store-competition. Accessed 26 Feb 2015

ICLEI-Procura + Procura + Sustainable Procurement Campaign, http://www.procuraplus.org, Accessed. 01 Oct 2015

Johnsen T, Miemczyk J, Spencer R, Walker H (2012) Sustainable procurement: past, present and future. Editor J Purch and Supp Manag 18(4):201–206

Li L, Geiser K (2005) Environmentally responsible public procurement (ERPP) and its implications for integrated product policy (IPP). J Clean Prod 13(7):705–715

Michelsen O, de Boer L (2009) Green procurement in Norway: A survey of practices at the municipal and county level. J Environ Manage 91:160–167

Nissinen A, Parikka-Alholaa K, Rita H (2009) Environmental criteria in the public purchases above the EU threshold values by three Nordic countries: 2003 and 2005. Ecol Econ 68:1838–1849

Swanson M, Weissman A, Davis G, Leet Socolof M, Davis K (2005) Developing priorities for greener state government purchasing: a California case study. J Clean Prod 13(7):669–677

Tátrai T (2013) Tátrai Tünde (2013): Közbeszerzés és Fejlődés 2013. Public Procurement and Development 2013, http://www.kozbeszkut.hu. Accessed 02 Nov 2015

UNDESA (2012) A guidebook to the Green Economy, 2012, Cameron Allen and Stuart Clouth, UN Division for Sustainable Development, Aug 2012

UNEP web UNEP—ecolabelling subsite. http://www.unep.org/resourceefficiency/Consumption/StandardsandLabels/Eco-labelling/tabid/101342/Default.aspx. Accessed 05 Dec 2014

Authors Biography

Orsolya Diófási-Kovács is Ph.D. student at Budapest University of Technology and Economics. She graduated at the Faculty of Economics and Social Sciences, Regional and Environmental Economics Program and has MSC in Engineering and Management. Now she is working in the field of environmental management consulting, implementing green procurement, supporting ecolabelling projects.

László Valkó professor at Budapest University of Technology and Economics, teaches at the Faculty of Economics and Social Sciences, Regional and Environmental Economics. His courses are Environmental Economics, Environmental Management, International Practices of Environmental Protection, Eco-Marketing.

Sustainable Economic Development: Green Economy and Green Growth. Analysing Economic Growth and Identifying Sensible Measures Addressing Socio and Environmental Concerns Whilst Promoting Green Growth

Paul Saliba

Abstract

The Paper reflects a critical, realistic and objective analysis of the current status of SED global-wise and locally from the perspective of an observant, keen on the wellbeing of our environment as common heritage of humankind without prejudice to economic sustainable development of countries and in full respect of the communities concerned. Conscious of the many inherent challenges and risks (environmental, political, economic and financial, social, cultural) that present themselves to governments striving to strike this balance, this Paper attempts to exploit opportunities to achieve positive outcomes in a win-win-scenario, through sensible and practical proposals and solutions. Reconciling the impacts and constraints of such development concurrently with environmental considerations and social sensitivity is the only way forward. The Paper consists of a series of proposals for concrete viable measures towards redressing the harmful economic and environmental deficits, taking into account the multifarious issues that these challenges would embrace, such as climate change, innovation, diversification of economies, policies and technologies which intrinsically or potentially enshrine green credentials, conservation of natural resources and ecosystems, financial burden of implementation, integration of social dimension in policy and decision-making processes (focusing on vulnerable people), taking on board pertinent stakeholders, and managing culture change.

P. Saliba (✉)
Valletta, Malta
e-mail: united50@hotmail.co.uk

W. Leal Filho et al. (eds.), *Sustainable Economic Development*,
World Sustainability Series, DOI 10.1007/978-3-319-45081-0_15

Keywords
Holistic sustainability · Local scenario · Environmental deficits · Climate change · Barriers, challenges and opportunities · Social sensitivity · Green energy · Green economy and green growth · Corporate social responsibility · Win-win scenario

1 Introduction

As an observant keen for the well-being of our environment, this Paper comes as a natural expression of personal deep concern for the prevailing deterioration of our environment and of a heartfelt desire to witness a marked uplifting of this common heritage for the generations to come. Inherent challenges and risks abound in the process of applying the brakes on and reversing the negative impacts brought about by the industrial revolution, decades of widespread neglect and degradation of ecosystems, over exploitation of natural resources and the wanton use of products that are harmful to biodiversity, to the holistic wellbeing of peoples and to the very air we breathe. The world has to come to its senses and whilst diversifying their economies to reflect today's challenges in a competitive world countries should invest more intensively on innovation and green economy to meet the global objective of sustainability in terms of economic development based on respect for the environment and social consolidation, focusing on support to the more vulnerable sectors of Society.

Malta too has sinned against nature over the past three decades in its drive to stimulate its economy and, notwithstanding its small size, in line with its international obligations is committed to contribute its share towards meeting this global objective whilst enhancing the quality of life of its own citizens.

The following literature is based on documented research ranging from local media reports, sundry studies, Malta Government portal, local independent bodies, and other sources.

On a broader base, my knowledge and competence in this field of sustainability is complemented with personal and enriching experiences accrued through the participation in various European and international conferences delving into the complexities of ESD and SED.

2 Analysis of the Current Status of SED: Risks, Challenges and Opportunities

The 2008–2012 global economic and financial crises, leading to the collapse of renowned banks and economies world-over, are still vivid in our minds. Millions of jobs—the livelihood of countless families—are still being lost. Europe is experiencing its fair share of this predicament. Official studies on the EU 28 reflect such a grim picture.

> It is estimated that more than 120 million persons (representing almost 25 %) were at risk of poverty or social exclusion in 2013 in the EU, whereas on the local front, 15.7 % of the population were in the risk of sustaining poverty and 24 % of social exclusion (Eurostat)

This poses a pertinent question warranting a solid answer. What is being done by governments to address this deficit?

Ultimately, it is the onus of governments to alleviate the sufferings of such vulnerable people. Empowering these vulnerable people, through various incentives and support services would in turn achieve productive results, rather than sidelining these sectors to depend more heavily or exclusively on social assistance. These could range from providing easier access to formal, non formal and informal education, upgrading of skills, exposure to information technology, participation in sports activities, health awareness and healthy lifestyles, giving citizens co-ownership (and hence pride of participation) of government projects or initiatives that affect them in their daily lives. *Digital Malta*: National Digital Strategy (2014–2020) and the plausible *Leap Programme* are practical tools towards attaining these objectives. *Digital Malta* outlines policy principles and actions on incorporating ICT for socio-economic development within different sectors of the economy and society towards overcoming the digital divide, with citizens and SMEs being the prime beneficiaries, for instance through the curtailment of administrative bureaucracy (Digital Malta). The *Leap Programme*, a 3-million euro project financed by the EU, aims to assist vulnerable sectors of society achieve enhanced social inclusion through "intense, specific and holistic training", whilst enabling them "to break the cycle of benefit dependence and poverty risk." (Maltatoday).

These stark challenges would be more pronounced in the case of employees in twilight of their careers and more so in the elderly retired classes where no effort ought to be spared to harness their inherent wealth of experience and expertise and capitalizing on their potential re-engagement in productive employment. Developing and consolidating an inclusive Learning Society is the key formula towards this end. Providing free or part-financed instruction and training in new and developing areas of economic activities, putting capable retired public officers and persons from the private sector on state and parastatal boards or committees are certainly initiatives that could boost this drive.

Compounding this unhappy situation is the incidence of workers engaged in precarious work, in part as a result of large influx of irregular immigrants across Europe—though in the case of Malta highly subdued these last couple of years—and whose plight is being exploited by unscrupulous employers offering degrading pay and working conditions.

Focusing on the local scenario, the Maltese Parliament on 26 January 2015 unanimously passed a bill allowing persons with disability to participate in boards and authorities and involve themselves in the lifelong learning commission. Today, persons with a disability and in employment still benefit from their special pension scheme, whilst employers are legally obliged to employ 2 % quota of workers with disability.

Defaulting companies would be subject to €2300 fine for every person with disability whom they should have employed in the first instance. The money collected from such fines goes towards a *National Fund for Integration of People with a Disability*.

Nonetheless, and within the broader picture, economic development with a social component but without due regard to environmental and health considerations represents half-baked sustainability.

Malta's vehicle density for instance takes its toll on the citizens' overall well-being. This translates into negative impact on the environment, the national economy, and health issues. A 2014 study commissioned by the European Commission Representation in Malta has found that Maltese drivers spend an average of 52 h in gridlock traffic every year.

> The loss of economic productivity cost €118 million, traffic accidents cost the country €84 million in disability benefits and loss of motorists' productivity, while the environmental impact of air pollution cost €14 m. Climate Change accounted for €47 million and Noise pollution €11 million, a total of €274 million, representing 4 % of GDP. On the other hand, shifting to public transport would net €132.5 million in savings by 2030 (Prof Maria Attard et al.)

Polluter Pays Principle (PPP) in Malta, for years mooted, is still at its conceptual phase. Its launching and strict application would indeed mark a milestone for our tiny Island where public perception persists in considering perpetrators still having their way with impunity.

Outside and beyond the Malta box, Europe is fast becoming host to a progressively aging population, a state of affairs posing fresh challenges and opportunities ahead.

Policy messages of a 2014 study by UNECE in conjunction with the European Commission highlight the way forward:

1. Greater policy efforts are required to fully mobilize the potential of active ageing in EU countries
2. Generating higher capacity and stronger enabling environments for active ageing will be a key prerequisite, particularly for Central and Eastern European Union countries

Addressing the issue of gender disparity in employment (leaning heavily on the male side), the study identifies that:

> it is particularly large in the two Mediterranean countries Malta (22.9 %) and Cyprus (18.1 %)" and confirms the policy priority in enriching older women's employment experiences of active ageing in many EU countries. The evidence also seeks ways to develop their full potential, not just to enhance their own well-being but also contribute towards improving the future sustainability of public welfare systems of the countries in which they live (UNECE)

Indeed, elderly people still have a lot to offer to society and the national economy. Critically crucial hence is the moral duty of governments to ensure their holistic wellbeing, empowerment and quality of life. A sensible measure would be to motivate idle ageing persons to engage in the labour market through viable, practical and tangible incentives such as providing subsidized instruction and training, as well as setting up more child care centres to relieve senior citizens from the burden-sharing of grandchildren in the absence of working parents.

The EU on its part is allocating a 30 billion euros injection to promote jobs in the member states through a series of ambitious strategic programmes and initiatives:

- promotion of more female active participation in the labour market to sustain health and other vital services
- promotion and stimulation of the private sector, boosting job creation and economic growth
- training of workers towards upgrading of skills
- shifting to a green economy

2.1 Climate Change and its Aftermath

A crucial yet often belittled or ignored phenomenon that impacts economic, social, health, and environmental sustainability is Climate Change.

Few people probably realize its full and direct impact on the well-being and lifestyle of large sections of communities, potentially provoking conflict and refugees from disasters. Displacement and dispersing of communities is a classic example of this phenomenon.

Climate Change, which ultimately leads to general degradation of ecosystem services and biodiversity, impacts also negatively on health (through malnutrition and disease), food security, agriculture, forestry, water resources, sea and coral systems…factors contributing to poverty risks and social exclusion. As professed by an Intergovernmental Panel on Climate Change there exist "linkages between Climate Change, Biodiversity Change, and Human Health" (COHAB 2 2008).

Were it not for man's lack of respect for the environment, today we would not be talking of greenhouse effects (WMO reporting that greenhouse gases had 'reached a record high in 2013'—2014 report, with The Climate Group claiming that globally deforestation was "the second biggest cause of greenhouse gas emissions, accounting for nearly 20 % of the total"), depletion of the ozone layer (some comforting news that this atmospheric shield is showing 'signs of recovery' comes from a UN 2014 study—BBC News), and man-instigated disasters. Man has through the years over-exploited, mismanaged or even abused natural resources for short-term gain.

Nature on the other hand seems to have absorbed all this unwarranted stress for ages…but now is coming round with a vengeance. Rise in global atmospheric temperatures, extreme stormy weather, thawing of glaciers, unpredictable avalanches and tsunamis, rise in the sea level, drought risk, encroaching desertification—all experienced these last years—are witness to this alarming phenomenon…and, lest we forget, the forced migration of alien and invasive animal and plant species to other 'foreign' territories threatening environmental contamination of their new-found habitats.

Rather belatedly, world leaders are waking up from their indifference and prolonged slumber and are attempting to take stock of this worsening scenario. *Green* is becoming synonymous with environmental friendliness and a catalyst agent for meaningful efforts to address environmental concerns and deficits.

The Lima Climate Summit (2014), a forerunner to the 2015 Paris Meeting earmarked to set specific targets on a global scale to meet the challenges of climate change, developed into sustained frustration from developing countries who perceive rich developed countries as non-committed towards supporting them in financing the highly ambitious project of cutting on greenhouse emissions and reduce their carbon footprint (through a "green climate fund", intended for the purpose of investing in clean energy technology) whilst coping with adaptation to climate change. Compounding this concern is the inherent fear that any pledges of support would eventually dissipate over time (BBC News).

3 Concrete Viable Measures to Promote Green Growth

On the local front, in the process of diversifying our economy to incorporate a greener component and meeting EU environmental obligations, successive governments have evaluated several alternative options, notably onshore and offshore wind farms and solar energy solutions, inspired by Malta's exuberance of these two natural sources. Ultimately, the choice of energy generated by wind farms (potentially at 40 % of the country's obligation to secure 10 % of its electricity needs through renewable sources by 2020) was discounted in favour of greater overall environmental protection, resulting in MEPA (Malta's hitherto autonomous authority concerned with development planning policies and environmental protection) to refuse the pertinent development application.

> The area surrounding the proposed wind farm is designated as a Special Protected Area and a Special Area of Conservation, where further human intervention, particularly in the form proposed, would have adverse impacts on the avifauna and marine ecology of the area, running counter to Structure Plan policy MCO 12.
>
> The proposal would therefore adversely affect the area, hinder its protection, and run counter to the rural conservation and ecological objectives of the Structure Plan (MEPA).

On a positive note, our governments have in these last years embarked on a robust investment in solar photo-voltaics solutions, co-financed by the EU. Government is also trying to lead by example through the installation of pv panels on its public buildings (notably state schools) and disused quarries. And latest news comes on the inauguration (4 May 2015) of the new iconic building housing Parliament "the Pride of Malta" (Renzo Piano) at the entrance of Valletta, built on designs by world-renowned architect Renzo Piano and representing an example of sustainability, exploiting undergound energy by means of 27 boreholes dug below sea level, thereby maximizing rain water and energy conservation, along with comprehensive insulation and renewable solar energy. Individual families, lured by government subsidy are following up, opting for solar pv and water heater solutions, roof thermal insulation and double glazing (NREAP/MRA), positively impacting on their energy bills and enhancing comfort. Some industries too are participating in lowering their carbon footprint through ongoing investment in energy conservation and best green practices.

And only very recently, a new power plant (running on HFO) was built in Delimara to produce additional power, increase production efficiency and reduce gaseous emissions. Government's additional commitment is to convert it into a gas–fired energy plant (through gas supplies from an LNG tanker berthed in Marsaxlokk harbour close to the power station—a highly controversial issue) to produce cleaner energy and further reduce emissions. A further boost in the right direction is the latest imported power procurement via an interconnector with neighbouring Sicily. Complementing this initiative was the 25 % reduction on electricity rates for households (since March 2014) and industry (since March 2015), thereby curtailing industrial production costs and enhancing competitiveness.

A positive outlook stems from the current steady restraint in the international price of crude oil that helps in no small measure in maintaining economic stability and keeping expenses for industry in check. Paradoxically, however, any sustained oil price depression could turn out to be counter-productive environmental-wise, potentially undermining incentives for consumers to restrain their energy consumption and opt for alternative clean sources.

An ISE Conference 2014 Paper on such energy conservation and efficiency measures concerns the construction industry, a driving force behind Malta's economic activities. Manifold are the benefits accrued in implementing these viable concrete measures:

- Contribution towards EU Targets and Directives Obligations (Energy Efficiency Directive 2006/32/EC; Renewable Energy Directive (2009/28/EC); Energy Performance of Buildings Directive (EPBD) of 2002 (2002/91/EC) and the EPBD recast of 2010 (2010/31/EC)
  Retrofitting dwellings will help Malta achieve the objectives set out in these 3 EC directives
- Reduction in CO2 Emissions and Pollution in an effort to combat Climate Change
  "Electricity generation accounts for about 64 % of all of Malta's greenhouse gas emissions in terms of CO2 equivalent...we would get a figure of 0.383 kgCO2/kWh as a rough estimate of CO2 emitted in Malta per kWh of electricity used by consumers by the end of 2015. Taking into account that 10 % of the energy consumed by Maltese households is used for space conditioning 60,000 MWh is taken as being used for air conditioning. The possible savings from retrofitting our dwellings by roof insulation is 45.5 % of 60 GWh, i.e. 27.3 GWh...the max. CO2 emissions that could be avoided was 27.0 kilotons with the former power plants, or 10.5 kilotons with the new Delimara gas power plants. Retrofitting would decrease emissions further since it also affects the amount of LPG gas used for heating purposes

Compared to total national greenhouse gas emissions of 3.0212 million tonnes of CO2 equivalent (2011), retrofitting can save up to 1.2 %, 0.89 % from electricity and 0.32 % from LPG consumption."

- Reduction in Energy Bills According to NREAP
  "for a 100 sqm typical roof, the saved energy would range from 470 to 2,200 kWh…a saving of €264.87 based on normal space conditioning and average saving of 45.5 % of corresponding electricity consumption."
- Higher Property Value
  "Buildings are a main asset to both individuals and the nation. Should investment in retrofitting increase the combined net monetary housing value even slightly it would render such investment worthwhile."
- Employment and local industry—with anticipated increase in job creation
  "An additional annual investment of €24 million would reflect an investment increase of 11 %….with 1326 jobs being created." (based on 2005 ECOFYS report estimating 1 additional job for every additional €35,000 turnover)."
- Social impact of policy measures for the building sector based on support to lowest-income households
  "Rising energy prices impact lowest-income households the most. Energy efficiency improvements would serve as a better means to combat fuel poverty. Energy efficiency policies have to be designed in a way that allows low income households to undertake the necessary investments or put the burden on stronger investors."
- Health and Thermal comfort Improvement
  "A thermal retrofit programme will definitely achieve more in comfort than in financial savings….insulation also lowers energy use, it generates environmental benefits (less air and water pollution, lower greenhouse gas emissions) and such benefits accrue to the wider community."

(Joseph Saliba A. and C.E. et al.)

Malta's first National Renewable Energy Action Plan (NREAP) to the European Commission (2010), outlines measures that Malta intends to adopt to achieve the national targets of 10 % renewable energy share of final energy consumption and 10 % share of renewable energy in transport, followed by the innovative introduction of energy from waste in 2011 (MRA). Further on the ground, tangible green initiatives are being implemented through legislation, designating protected areas for nature conservation covering *Natura 2000* sites and Special Areas of Conservation, waste segregation for RRR (supported by the provision of Civic Amenity Sites for bulky and hazardous household waste and recyclable materials), free CFL's for households, rebate on energy efficient domestic appliances (2006–2008), use of micro-wind turbines at public sites, smart metering on electricity consumption, building of new state colleges utilizing renewable energy sources for a quasi-autonomous energy requirement, public transport reform and promotion (introduction of energy-efficient buses running on cleaner fuels), grant schemes on RES and EE (industry-targeted), bio-fuel substitution (legislation addresses a percentage share of bio-fuels, incremental from 1.5 % in 2011 reaching 10 % by 2020), introduction of LPG as an alternative fuel for drivers, the promotion of bicycles and electric cars through incentives ranging from subsidies and installation of a network of solar powered charging points spread over the Island, scrappage schemes (grants on purchase of new Euro class IV or higher standard vehicles), and car-sharing benefiting from use of priority bus lanes (MRA).

Road network infrastructure however needs to be stepped up for better traffic management and to limit congestions. Designating more pedestrian zones, water transport across harbours and shores, and tele-working (reducing need of commuting to work) also need to be boosted.

At regional level, Gozo is being designated as an eco-island with authorities adopting measures in renewable energy resources and efficiency.

On water conservation, a praise-worthy effort is the launching of an MRA scheme targeted to further encourage the harvesting and use of rainwater in the domestic sector through financial support in repairs of wells and installation of second class water systems (of relevance is the reverse osmosis 3.2 % consumption of total electricity production for public water and waste water treatment for irrigation—timesofmalta.com). Complementing this effort would be the preservation and boosting of groundwater and guarding against poaching. Meanwhile, 3 sewage treatment plants are in operation for production of second-class water for irrigation and secondary purposes. This means that Malta is now fully compliant with the European Union's Urban Wastewater Treatment Directive (WSC, Malta).

Another laudable green cause is government's launching of a National Action Plan (NAP) for Green Public Procurement (implemented in 2011), through a series of National Guidelines governing public procurement, directing contracting authorities "to integrate green public procurement criteria, according to their respective targets" (Environment.gov.mt)…a win-win instrument conducive to the attainment of economic and environmental objectives.

On a global perspective, diversification of the world's economies to embrace innovation, green economy and an ingrained social conscience is the way forward…in other words, harmonizing economic growth with environmental and social objectives.

The risks and challenges associated with pollution, counterbalanced with opportunities for a greener economy and sustainable green growth are far-reaching and could not be underestimated.

> Air pollution alone, "the single biggest environment risk", costs European economies 1.6 US$ trillion a year in diseases and deaths, a 2010 study reveals, representing almost 10 % of the GDP of the whole European Union in 2013…curbing the health effects of air pollution pays dividends…"we not only save more lives but also achieve results that are worth astounding amounts of money" (WHO)

Investing more resourcefully in cleaner and renewable energy sources for industry, means of transport and households goes a long way towards this refreshing destination.

Seeking to achieve or boost energy's green credentials should be an overriding objective of all consumers. Whereas energy sources are plentiful it is crucial that these are clean and sustainable in the long-term. Sources with high Carbon content, such as fossil fuels (being the largest source of emissions of greenhouse carbon dioxide attributable to global warming) and CFCs (contributing towards depletion of the ozone layer, and potentially leading to harmful ultraviolet radiation) should be fervently discouraged. Governments investing in developing alternative sources of clean energy through research, infrastructure, co-operation with other countries advanced in green technology for support, in the long run are guaranteed of reaping rich dividends.

And yet, rationale demands that conservation of energy must be complementary to any initiatives for carbon footprint reduction. Indeed, green growth assumes enhanced dimension through this conservation.

Commendable policies should be those that focus on a holistic approach to sustainable energy and management of its waste, encompassing the full cycle of this resource: source, controlled production, transportation, conservation, waste treatment and disposal processes.

Governments harbour their own plausible concepts on green growth and how to achieve it through solid strategic policies and viable action plans. Some share common concrete measures, like investing in cleaner and renewable energy and making public transport services more accessible and environmentally-friendly, but other initiatives could be overlooked: like the offering of incentives to the automobile industries to invest in the production of vehicles with high green technology and efficiency.

Ultimately, the optimal way forward is the end result of converging initiatives and concrete practical measures…activities which are intrinsically or closely associated with green credentials based on the respect of our ecosystems and biodiversity. On top of the agenda, I would list: promotion of large scale sustainable forestation and farming projects; preservation of species habitats to assist in sustainable breeding and propagation; sustainable landscapes and seascapes; husbandry of rain water; clearing valleys of environmentally-alien development; preventing coastal over-development; treatment of polluted seas, rivers and lakes; investing in and promoting industries that produce green products and services; controlling with effective enforcement indiscriminate and unsustainable fishing, hunting, bird and animal trapping and poaching; controlling the production, stockpiling and use of fireworks; preserving public land and sea use and combating encroachment; preventing over-exploitation of **all** natural resources which impairs regeneration by nature; promotion of eco dwellings; a holistic approach to waste management; raising awareness and undertaking more research in environmental issues whilst sharing data, experiences and best practices.

Under the auspices of the UN, over 150 world leaders and major stake players meeting at the Special Summit on Sustainable Development in New York in September 2015 adopted the 2030 Agenda for Sustainable Development, as well as 17 "Global Goals", ranging from the fight against hunger and poverty, promotion of health and well-being, quality education and gender equality … to clean energy, making cities more sustainable, innovation, responsible production and consumption … to combating climate change, protecting oceans and forests ... to economic growth with a social conscience, and promoting peace, justice and good governance. "If the global community collectively is prepared to step up to the challenge of achieving the SDGs, then there's a chance of achieving sustainable development—and with it better prospects for people and our planet." (Helen Clark) (UNDP 2015).

Will the projected ambitious targets set at this Summit be achieved for the common good of mankind or would it result in irrelevant rhetoric? Only time will tell. With mutual goodwill, transparent negotiation processes, financial support to emerging economies and third world countries...and dogged perseverance light at the end of the tunnel would not be just a mirage or a distant milestone but a real and visible rainbow of solidarity.

On a state level, governments as powerful and influential prime movers, should act as role models and spearhead initiatives embracing and instigating corporate social responsibility standards aimed at safeguarding and promoting environmental and social wellbeing through all sectors and levels of participation—on environmental issues, be it infrastructural works, procurement and commissioning of products and services; whereas on social issues the promotion of gender equality, equal opportunities to all (irrespective of colour of skin, creed, or sexual orientation, physical ability or impairment), integration of family-friendly measures and just remuneration for services rendered. Other national major stakeholders, private industry, civil society, the Church and other institutions, NGOs, citizens would follow suit...all pooling resources and expertise at their disposal towards a common goal. Complementing this added-value asset would be the pursuance of good governance—the umbrella virtue—translated in transparency, accountability, promoting broad ownership of initiatives, schemes and changes, and engaging in social dialogue.

Implementing green economic measures could predictably and understandably meet quite a few barriers and incur risks, not only financial due to budgetary constraints but also socio-cultural ones.

In terms of financing such measures, governments have to manage their country's finances judiciously, identifying priority areas for expenditure, investing money on sensible projects to stimulate the economy and which yield return benefiting citizens at large, combating tax evasion and promoting incentives and implementing measures amenable to the upholding of the environment.

Confronting the challenge of cultural change might present a harder nut to crack. And here education plays a major role in instilling and cultivating a sense of national pride and developing a mindset in citizens for environmental concern and the attainment of goals for the common good.

4 Conclusion

Governments spend a substantial percentage of their national budgets on secondary health care that includes remedies for maladies attributable to air pollution potentially causing chronic or long-term respiratory problems and cancer. Hence, better protection of the environment makes real sense.

Leaders in power may choose to be the ostrich and dismiss the alarm bells and recommendations from scientists and environmental experts to redress environmental deficits. Furthermore, integrating social aspects in economic policies gives added value to efforts for such environmental redress.

Visionary governments would wisely focus their energies more on activities which intrinsically embrace green components based on conservation and rational use of natural resources and upholding ecosystems while safeguarding public health, empowering vulnerable people and promoting social inclusion based on equal rights and opportunities to all, irrespective of any known or perceived diversity.

In a nutshell: keen vision, strategic economic policies and planning followed by viable and practical action plans embracing green and socially-just components are the prevalent ingredients for a winning formula in entwining economic development and green growth in perfect harmony.

References

Attard M et al (2015) Institute for Climate Change and Sustainable Development, University of Malta. The External Costs of Passenger and Commercial Vehicles use in Malta—Study approved on 27 Feb 2015. http://ec.europa.eu/malta. Accessed on 13 April 2015

BBC News. 11 Sept 2014. Ozone layer showing 'signs' of recovery, UN says.—Harrabin R, environment analyst. www.bbc.com/news/science-environment-29152028. Accessed 11 April 15

BBC News. 14 December 2014. UN members agree deal at Lima climate talks t.www.bbc.com/news/science-environment-30468048

(COHAB 2), Galway Feb 2008—Official Conference Programme and Proceedings

Digital Malta. Government Launches Digital Malta: National Digital Strategy 2014–2020. www.digitmalta.gov.mt

Environment.gov.mt Accessed on 13 April 2015—National Green Public Procurement Guidelines Contracts Circular No. 5021/2011

Eurostat Press Office 168/2014—4.11.14 At risk of poverty or social exclusion in the EU28 eurostat-pressoffice@ec.europa.eu. Accessed 14 April 2015

Malta Today: Leap project to be extended for the next seven months. www.maltatoday.com.mt/news/national/53692/leap_project

MEPA, PA/01821/09. Off shore wind farm at Sikka l-Bajda, limits of Mellieha, including laying of power cables between turbines—26 February 2015. www.mepa.org.mt

MRA. Roof thermal insulation and Double Glazing 2012–2015. www.mra.org.mt/support-schemes/2012-roof-glazing-scheme/

MRA. Domestic Cisterns Restoration (wells) Scheme 2013–15. www.mra.org.mt/support-schemes/2013-wells/

MRA. Malta's Biennial Report on Policies and Measures and Projected Greenhouse Gas Emissions 2013. www.mra.org.mt/wp-content/uploads/…/Malta-PAMs-Report-2013-V1.5.pdf

NREAP www.ebb-eu.org/…/ActionPlanDirective2009…/national_renewable_energy

Saliba Joseph et al A. & C.E, MSc (Sustainable Energy). Consultant to Ministry for Energy & Health (Malta). Freelance Architect and Civil Engineer, Registered Energy Performance Certificate Assessor for Non-Dwelling Buildings. Extract from Paper on ISE. Policy Options for the Thermal Upgrade of Existing Maltese Housing Stock: An Environmental and Economic Approach

(THE) CLIMATE GROUP Financing the low carbon economy—19 Sept 2014. www.theclimategroup.org

TIMESOFMALTA.COM Mar 21, 2014. Less electricity being used in water production

UNECE/European Commission: AA1 2014 Active Ageing Index for 28 EU Countries. www1.unece.org/stat/platform/display/AAI/VI.+Documentation
WHO. 28 April 2015 www.euro.who.int/.../2015/air-pollution-costs-european-economies
World Meteorological Organization (WMO) Press Release No. 1002 'Carbon Dioxide Concentration Surges' Report dated 9 Sept 14. www.wmo.int. Accessed 14 April 2015
WSC. Wastewater Treatment and Collection. www.wsc.com.mt/content/wastewater-treatment-and-collection

Author Biography

Paul Saliba is currently a retired senior public officer who had assumed the role of a Green Leader, an integral component of his portfolio within the Ministry of Health, The Elderly and Community Services (Malta). (*The Green Initiative was launched on 1st January 2005 by the then Prime Minister Dr Lawrence Gonzi LL.D. The main aim of this Initiative was to create a green network within government in order to enable it to better meet its corporate responsibilities with regard to the environment through environmentally-friendly practices*). He has participated as representative of the Health Ministry (Malta) in several high profile international and European Conferences and Fairs as follows: *Developing Public Procurement Policies for Sustainable Development and Innovation*—Hamar, Norway (2006); 2nd European Conference and Fair: *Education for Sustainable Development*—Hamburg (2006) where he presented a power-point Paper entitled *Assuming the Green Leader's Role at Workplace—A Critical Analysis* and acted as a moderator in one of the Sessions at the same Conference; *Implementing Sustainable Public Sector Procurement* —London (2007); *COHAB 2* (Conference on Health and Biodiversity)—Galway (2008); *Working together on Education for Sustainable Development*—Bordeaux (2008); 3rd European Conference and Fair: *Education for Sustainable Development*—Hamburg (2009); CONCERTO Forum: *Energy and Climate Change Policy Solutions* ...studies on experiences of 58 European cities implementing the Concerto concept—Brussels (2010).

Understanding the Challenges Involved in Transitioning to a Low Carbon Economy in South Asia

Tapan Sarker

Abstract

This study stems from the hypothesis that there are gaps at local, national and regional levels concerning policy making and project formulation, related to climate change, sustainable development and human security. This chapter focuses on the South Asia region, with particular emphasis on India and Bangladesh, as these are two countries faced with high vulnerability to climate change. It is clear that the increased level of projected growth in the region needs to tread down a path towards a low carbon economy, in order to face up to the challenges of climate change. Inadequate regional cooperation, energy insecurity and weak governance are identified as key impediments to achieving a low carbon economy in South Asia. However, some progress has been made through the formation of climate change specific policy, establishment of forums for regional cooperation, and private sector initiatives, promoting sustainable practices. The step from policy and ideas to action needs to be taken. Barriers such as, political instability, lack of economic integration and limited natural resources need to be acknowledged and surmounted before any real progress towards sustainable development in the region can be achieved. The promotion of a low carbon economy is touted as a key step in the right direction. Emerging from this literature review is a series of examples and mechanisms that may help in the formulation of coordinated and integrated policy and mechanisms that would be needed to see real progress towards low carbon economy and an effective response to the challenges derived from climate change in South Asia.

T. Sarker (✉)
Griffith Centre for Sustainable Enterprise, Griffith Business School, Griffith University, 170 Kessels Road, Nathan, QLD 4111, Australia
e-mail: tapan.sarker@griffith.edu.au

W. Leal Filho et al. (eds.), *Sustainable Economic Development*, World Sustainability Series, DOI 10.1007/978-3-319-45081-0_16

Keywords
Climate change · Low carbon economy · South Asia

1 Introduction

Prior to the global financial crisis in late 2007, South Asian countries experienced moderate to high economic growth. For example, India experienced between 8–9 % growth and Bangladesh, 5–6 % growth (Hossain and Selvanathan 2011). Over the last quarter of the 20th century to present, overall living conditions have improved markedly in the region. It is widely acknowledged that globalisation has played a key role in these economic transitions by enhancing international trade and regional interconnectedness. Whilst there have been positive economic changes for the South Asian region in the early 21st century, there has also been an emergence of new challenges concerning regional cooperation, sustainable development, climate change and associated socio-economic impacts.

There is a limited portfolio of literature investigating the significance of regional cooperation strategies and their significance in providing capacity for sustainable development, low carbon economy and resilience and innovation, particularly within the context of South Asia and the impacts of climate change. This chapter aims to identify the key issues influencing regional integration and cooperation strategies concerning climate change and related human security challenges. These issues are contextualised from local, national and regional perspectives within developing regions. This paper is fundamentally a response to the lack of available empirical analysis regarding the implications of climate change and the relevance of cooperation and regional integration, towards the development of a low carbon economy in South Asia.

A key emphasis of this study is therefore on the process of transition to a low carbon economy in South Asia, with a particular focus on Bangladesh and India; the two most vulnerable countries in South Asia to climate change. This study stems from the hypothesis that, in South Asia, there are gaps at the local, national and regional levels concerning policy-making and project formulation, related to climate change, sustainable development and human security. Therefore, the key objective of this paper is to explore areas of cooperation at local, national, and regional levels that can aid in overcoming the challenges associated with the transition to a low carbon economy including examples of technology exchange, energy security and policy. This study aims to propose policies and frameworks geared towards enhancing regional integration and cooperation. South Asia is home to 24 % of the world's population and is faced with many issues related to continued rapid population growth, increased economic output and mounting natural resource pressures. These pressures combined with the region's geographical context make it highly vulnerable to the impacts of climate change. Projections suggest that these vulnerabilities will be exacerbated due to an increased likelihood of extreme weather events, depleted natural resources, and economic imbalances, as markets face the implications and multidimensional challenges associated with climate change. Therefore, in order to cope with these challenges, the South Asian economies and enterprises will have to focus on sustainable economic development by assessing the major risks, enhancing strengthening capacity, filling the knowledge gaps, and identifying adaptation investments at local, national and regional levels. Furthermore, regional integration has the potential to improve South Asia's chances of success when mitigating cross—border challenges by enhancing cooperative capacity and regional resilience. By exploring areas of cooperation at local, national and regional levels and addressing the challenges related to climate change in the South Asian region, this study aims to provide example policies and frameworks related to the mitigation of climate change and human security issues in South Asia.

An extensive review of literature has been conducted to identify the key factors that relate to climate change and human security issues in South Asia (focusing on India and Bangladesh). This review of scientific reports, policy papers and other relevant materials has provided a helpful means for identifying the gaps in addressing climate change and related threats at local, national and regional levels within the South Asian region. This literature review firstly provides a contextual outline of South Asia and climate change from multiple perspectives, including environmental impacts, socio-economic impacts, social impacts, existing climate change policy and policy directions. Secondly, existing policy and mechanisms for dealing with climate change are summarised. Having laid this groundwork, the next section provides an insight into the current South Asia regional cooperation scenario and its relevance to building resilience and fostering sustainable enterprise in the region. Finally, this review provides a list of challenges and changes including examples such as technology transfer, policy and energy security that need to occur before the propagation of a regional-scale low carbon economy can occur in South Asia, from government, institutional and private sector perspectives.

1.1 Aims and Objectives

The study has the following three aims:

- To identify the key issues influencing regional integration and cooperation strategies concerning climate change and related human security challenges;
- To explore areas of cooperation at state, national and regional levels that can aid in overcoming the challenges associated with the transition to a low carbon economy; and
- To propose mechanisms, policies and frameworks geared towards enhancing regional integration and cooperation for the cultivation of sustainable development.

2 Background: Impacts of Climate Change in the South Asia Region

2.1 Environmental Impacts

As outlined by Professor Nicholas Stern (2006), even a moderate rise in temperature could have serious implications for the environment of South Asia. Climatologist, Professor Mark New of the School of Geography and Environment at Oxford University, has indicated that over the past 30 years the east Himalayas have experienced a drop in snow and ice cover by around 30 %, and there is a real risk that these glaciers might diminish altogether within a matter of decades (Hossain and Selvanathan 2011). Bangladesh's reliance on these glaciers to fuel its major rivers and streams mean that the implications of diminished east Himalayan glaciers would be severe and result in drought, and therefore a lack of freshwater for irrigation and domestic uses. The Copenhagen Summit in 2009 has identified Bangladesh as one of the most vulnerable countries to the impacts of climate change (Hossain and Selvanathan 2011).

The extreme weather conditions of recent years in Bangladesh have been connected with climate change and the IPCC has made projections for this trend to continue and intensify into the next three decades. The implications of cyclones, seal level rise, sea water surge, flooding, drought, changes in rainfall patterns and temperature are far reaching and acutely impacts the human socio-economic systems operating within the affected regions of South Asia, in particular the coastal belt of the Bay of Bengal (Stern 2006).

2.2 Socio-economic Impacts

During the economic growth period in South Asia, prior to the 2007/2008 global financial crisis, overall living conditions improved, as is expected with economic development. Interestingly, during this period the South Asian region also maintained the highest overall population density and one of the lowest Gross National Incomes per capita (World Bank 2007).[1] The region has also displayed one of the highest child mortality, and maternal birth rates, correlating with high levels of poverty, as displayed in Table 1. Climate change has the potential to impact the social fabric of South Asia severely. Population displacement and climate induced mass migration are already occurring on a large scale in coastal Asia and Africa. The UN has stated that more people are displaced by climate induced phenomena than war, numbering in the tens of millions (UNFCCC 2008). Climate induced and exacerbated phenomena include drought, cyclones, storm surges, flooding, and

[1]For a comprehensive review see: World Bank (2004), 'World Development Indicators 2004'. During this period South Asia had a population density of more than 300 persons/sq. km and a GNI per capita of less than 1500 $US, compared with Europe EMU with a population density of less than 150 persons/sq. km and a HNI per capita of more than US$ 20,000.

Table 1 Indicators of poverty and mortality

Areas/regions	Share of people living on less than US $1.25 a day	Child mortality rate per 1000 live births	Maternal Mortality ratio per 100,000 live births-modelled estimates
East Asia and pacific	16.8 %	23	150
Europe and central Asia	3.7 %	19	45
Latin America and Caribbean	8.2 %	20	130
Middle East and North Africa	3.6 %	29	200
South Asia	40.3 %	58	500
Sub-Saharan Africa	50.9	86	900
Euro area	_	8	5

Source World Bank (2010)

Table 2 Demographic condition in South Asia

Country	Area (in '000 sq.km)	Population (million)			Population growth rate		Population ratio (rural/total)	
		1990	2005	2015	1990–2005	2000–2015	1990	2005
India	3287	849.5	1095	1248.5	1.4	1.0	74.5	71.3
Pakistan	796	108	156	190.5	2.4	2.0	69.4	65.1
Bangladesh	144	104	142	168	2.1	1.7	80.2	74.9
Nepal	147	19.1	27	32.7	2.3	1.9	91.1	84.2
Sri Lanka	66	17	20	21	1.0	0.7	82.8	84.9
Total	4440	1098	1440	1661	1.84	1.46	79.6	76.1

Source Hossain et al. (2010, p. 18), World Bank (2004, 2007)

dry-land salinity (USAID 2010). It is also now widely recognised that the South Asian region, in particular the coastal belt of the Bay of Bengal, will suffer severely in socio-economic terms from climate change due to its impacts on the local population's, predominantly agricultural livelihoods and the region's high population density; particularly the millions residing in the Bay of Bengal Delta, many of whom represent South Asia's poorest of the poor (Stern 2006).

Table 2 displays a comparison between land area, population, population growth rates and rural/urban ratios for each South Asian country (excluding the Maldives), reflecting a marginal migration away from rural areas to urban areas.

2.3 Climate Change Policy in South Asia

In the context of climate change, sustainable development policies and measures that have been implemented to accelerate development in emerging economies are uniquely powerful drivers in the move towards achieving global climate stability.

Policies that improve the productivity, efficiency, and ecological footprint of transport and energy infrastructure are being implemented by many developing countries. Following the Bali agreement of 2007, almost all South Asian countries have developed national climate change strategies (Hossain and Selvanathan 2011). Kelkar and Bhadwal (2007) indicate that while many 'South Asian public sector programmes are relevant for strengthening adaptive capacity to climate change, they do not as yet explicitly incorporate responses to the increased risks due to climate change'. The authors continue by explaining that, a comprehensive Preparedness, Adaptation and Mitigation Strategy (PAM) would require a long-term strategic plan which may not be entirely congruent with existing short-term political governance cycles. This incongruence, when combined with a number of other possible factors has resulted in a weak commitment to the formulation of a comprehensive PAM strategy. Kelkar and Bhadwal (2007) also suggest a number of other factors hindering the formulation of a PAM including:

- Both the extent and impacts of climate change are yet to be fully understood, mapped and properly costed;
- Inadequate decentralisation of public governance and low institutional capacity to link national with local and local with community, and vice versa, compromise a bottom-up planning process, vital to climate change planning;
- Inadequate resources and a lack of required technical know-how jeopardise development of suitable adaptive infrastructure; and
- Absence of meaningful regional cooperation—geo politics and the political mind-set of inter country rivalries continue to obstruct progress. Collective visioning of a regional framework is vital to address the hazards of climate change migrations, especially those that relate to cross-border migration.

2.3.1 Climate Change Policy: India in Focus

India's first National Action Plan on Climate Change (NAPCC) was released on June 30, 2008[2]; outlining existing and future policies and programs addressing climate mitigation and adaptation. The plan is arranged into eight core "national missions" through to 2017. The plan itself emphasises India's main priority, which is to maintain high economic growth to raise living standards. The plan also highlights the fact that although India is committed to reducing its overall carbon emissions, these measures would be more successful with assistance from developed countries. It also emphasises that India will at no point exceed the per capita green house gas emissions of developed countries, even as they pursue their development objectives (Pew Centre on Global Climate Change 2008). The eight 'National Missions' include:

[2]For a comprehensive review, refer to: *National Action Plan on Climate Change: Government of India*, Pew Centre on Global Climate Change (2008).

- National Solar Mission; which aims to promote the development and use of solar thermal technologies with the aim of making solar energy production more competitive. This mission also combines specific targets such as, increasing the production of photovoltaic cells to 1000 MW/year, with other objectives such as, the establishment of solar research facilities, promotion of international collaboration on technological development, increased government funding and international support;
- National Mission for Enhanced Energy Efficiency; aims to mandate specific energy consumption decreases in large energy consuming industries, with a system for companies to trade energy-savings certificates, as well as energy incentives for improved energy efficiency and increased financing for public-private partnerships that reduce energy consumption through demand side management programs in the construction and agricultural sectors;
- National Mission on Sustainable Habitat; promotes energy efficiency as a core component of urban planning;
- National Water Mission; sets a goal of 20 % improvement in water use efficiency through a range of methods including pricing, in light of the projected impacts of climate change on water supplies;
- National Mission for Sustaining the Himalayan Ecosystem; aims to conserve biodiversity, forest cover and other ecological values in the Himalayan region, where global warming is projected to cause glaciers to recede dramatically, thereby compromising much of India's current fresh water supplies and river systems;
- National Mission for a "Green India"; aims for the re-forestation of six million hectares of degraded forest lands and the expansion of India's overall forest cover from 23 to 33 % of India's territory;
- National Mission for Sustainable Agriculture; aims to support climate adaptation in agriculture through the development of climate-resilient crops, and agricultural practices as well as the expansion of weather insurance mechanisms; and
- National Mission on Strategic knowledge for Climate Change; aims to promote research through funding and increased international collaboration. Also encourages private sector initiatives to develop adaptation and mitigation technologies through venture capital funds.

The NAPCC also combines a number of individual and continuing initiatives such as the renewable energy initiative which states, under the Electricity Act 2003 and the National Tariff policy 2006, that the central and the state electricity regulatory commissions must purchase a certain percentage of grid-based power from renewable sources (Pew Centre on Global Climate Change 2008).

2.3.2 Climate Change Policy: Bangladesh in Focus

Currently there is no concrete climate change policy in Bangladesh that specifically aims at the projected risks associated with climate change in the region. The Government of Bangladesh has instead taken a policy stance that focuses on climate change vulnerability, with a vision to eradicate poverty and improve economic and social well-being through a 'pro-poor' climate resilient strategy. The strategy focuses on adaptation and disaster risk reduction mechanisms along with low carbon development goals, mitigation strategies and promotion of technology transfer in order to build capacity and meet projected climate change challenges into the future (25 years). This strategy is primarily based on the four building blocks of the Bali Action plan (UNFCCC 2008):

- Food Security: Bangladesh continues to lack sufficient food supplies despite the substantial quantities of rice, jute, potatoes etc. that it produces. A prime commitment of the Government of Bangladesh was to achieve self-sufficiency in food-grain production by 2013. Crop security is threatened by rising sea levels, changing precipitation patterns, increased regularity of flooding and drought (Jilani et al. 2011);
- Water Security: Fresh water supplies in Bangladesh are threatened by increased regularity and severity of drought, changes in precipitation patterns, salinisation of groundwater supplies and pollution of surface water supplies;
- Energy Security: The Government of Bangladesh presented its vision and policy statement in February 2000, aiming to eliminate the inequality between the urban and rural areas by upgrading energy supplies and producing 8500 MW by 2013; and
- Livelihood Security: Agriculture provides employment for approximately 48.1 % of the population. The government's aims and objectives are based around both improving the productivity of the agricultural sector whilst simultaneously increasing the contribution of other sectors including industry and service sectors.

In order to enhance funding in development and sustainability initiatives, the Government of Bangladesh has formulated a National Adaptation Program of Actions (NAPA). This was submitted to the UNFCCC in 2005 in order to gain official development assistance through the Least Developed Countries Fund (LDCF). More recently, the government has established two authorities as the National Clean Development Mechanism committee. These authorities have taken on four projects involving the waste and energy sectors.

In 2008, the government formulated the Bangladesh Climate Change Strategy and Action Plan (BCCSAP) 2008, revised in 2009. The BCCSAP identifies six priority themes as follows: Food security; Social protection and health; Comprehensive disaster management infrastructure; Research and knowledge management; Mitigation/low carbon development; Capacity building and Institutional strengthening. It has been suggested that although Bangladesh is acting to mitigate climate change vulnerability, the lack of a concrete national climate change policy may increase pressures on the government and institutions as the consequences of climate change are felt (UNFCCC 2008).

3 Background: Cooperation and Uncertainty in Transitioning to a Low Carbon Economy in South Asia

The concept of cooperation is important in the context of climate change, uncertainty, and transitioning to a 'green economy' in South Asia. Jhirad et al. (2011) outline that challenges induced by climate change can be shared between developed and developing nations, which foster an environment of learning in this arena. By aligning sustainable development goals and climate policy more closely with development interests, the global market place, key stakeholder and decision makers, private industry and the investment community may be more readily engaged. This also promotes wider international support for both climate change and development initiatives, and can thereby help to attract significantly greater investment. Atiken and Raihani (2011) outline the importance of incorporating mechanisms to foster cooperation whilst accounting for uncertainty and irrational responses that may inhibit collective action, concerning efforts to mitigate global climate change. There are many sources of uncertainty involved with the issue of climate change that may hinder collective action, such as an imperfect understanding of how increasing greenhouse gas stocks in the atmosphere will affect climatic patterns. Therefore, in the context of climate change, promoting cooperation may fundamentally rely on the ability of scientists and policy makers to convince stakeholders and investors of the high probability that negative consequences will arise if no collective action is taken on climate change, for example, if emission reduction targets are not met.

3.1 Regional Cooperation: South Asia Context

The South Asian region could particularly benefit from increased levels of regional integration and cooperation at all levels: state, intra-regional and international. Ahmed et al. (2010) provide an example of the impacts of climate change on Himalayan glaciers and water resources in South Asia, and the associated benefits of cooperation in the South Asia region. Glacial melt is already impacting water flow regimes and these impacts are likely to be to the detriment of downstream countries. The rivers and floodplains of Pakistan, India and Bangladesh rely on these glaciers for regular seasonal flow. Clearly issues such as this can be the catalyst for dialogue on regional water cooperation. Currently, cross-border cooperation between India, Bangladesh and Nepal may offer the only long-term solution regarding flood/drought mitigation, and could benefit over 400 million people (Ahmed et al. 2010).

In terms of intraregional trade, capital and ideas, South Asia is the least integrated region in the world. Factors contributing to this scenario include, poor infrastructural connectivity, cross border conflicts (primarily between India and Pakistan) and security concerns (World Bank 2007). Despite slow progress for regional integration, the South Asia region has experienced high economic growth over the past decade, and this is a trend set to continue. Projected economic growth spurred the unanimous signing of the South Asia Free Trade Agreement (SAFTA) on the 6th of January, 2004, which promotes free trade throughout South Asia; a step towards intraregional economic integration, and providing scope for further regional cooperation.[3] Despite this progress, the cost of conducting business remains high in the South Asia region and firms identify corruption and energy supply as the two biggest constraints (Ahmed et al. 2010).

4 Conditions Needed for the Propagation of Low Carbon Economy in South Asia

An environment conducive to the propagation of sustainable enterprise combines the traditional profit-seeking model with the need for development that not only respects environmental sustainability, but also human dignity and decent work (ILO 2007). Table 3 provides a summary of the conditions required for the propagation of SEE that are highly applicable to the South Asian regional context.

4.1 Transformations and Challenges: Green Economy and Sustainable Development in South Asia

There are many issues; social, economic and environmental that need to be addressed before any real progress towards a socially-equitable green economy can be made in developing regions such as South Asia. Table 4 provides a glimpse at the issues that need addressing, specific to South Asia, before any successful propagation of a 'green economy' can begin.

4.1.1 Challenges: Regional Cooperation and Energy Security in South Asia

In terms of regional cooperation and energy in South Asia, it must be highlighted that the region has energy surplus[4] (e.g. Bhutan, Nepal), and energy deficit (e.g. India, Bangladesh) countries. The development of greener, cross-border energy projects can enable countries to benefit from natural endowments, and improve capacity for increased grid efficiency, and secure supplies of fossil fuels and renewable energy. Cross border energy projects that focus on the development of renewable energy projects, particularly those involving hydro-electric generation and wind power, can reduce reliance on fossil fuels, and therefore limit

[3]The SAFTA agreement was made under the South Asia Agreement on Regional Cooperation (SAARC), which provides the overall regional forum for political level dialogue (Ahmad et al. 2006).

[4]In the 2007 fiscal year Bhutan's electricity export contributed to nearly 25 % of GDP and 60 % of government-revenues (ADB 2009).

Table 3 Conditions required for the propagation of a low carbon economy

Conditions for a conducive environment for low carbon economy	Role of Government in the promotion of low carbon economy
• Peace and political stability • Good governance • Social dialogue • Respect for universal human rights • Entrepreneurial culture • Sound and stable macroeconomic policy • Trade and sustainable economic integration • Enabling legal and regulatory environment • Rule of law and secure property rights • Fair competition • Access to financial services • Physical infrastructure • Information and communications technology • Education, training and lifelong learning • Social justice and social inclusion • Adequate social protection • Responsible stewardship of the environment	• Facilitating and participating in social dialogue • Labour law enforcement through efficient labour administration, including labour inspection • Encouragement of voluntary concept of corporate social responsibility • Promotion of socially and environmentally responsible public procurement, lending and investment • Promoting sectors in value chains • Flexibility and protection to manage change • Targeted programmes • Research and innovation • Access to information and business and financial services • Policy coordination and coherence • International policies • Production and consumption patterns • Supporting skills development
Enterprise-level principles for low carbon economy	Role of the social partners in the promotion of low carbon economy
• Social dialogue and good industrial relations • Human resource development • Conditions of work • Productivity, wages and shared benefits • Corporate social responsibility • Corporate governance	• Advocacy • Representation • Services • Implementation of policies and standards

Source Somavia (2007)

environmental harm. This is a particularly important factor for India, as their CO_2 emissions are set to grow substantially, as demonstrated by an overall 5 % increase in emissions between 2003 and 2008 (ADB 2011).

Using the example of hydropower development, it has been indicated that currently India exploits a mere 33 % of its hydropower capacity, and this is in large part due to inadequate management of ecological impacts and resettlement and rehabilitation issues. On top of this India's ability to exploit regional hydropower resources (in Nepal and Bhutan) has been limited due to an unwillingness of energy endowed countries to be tied so closely with Indian markets. In order to develop the capacity of hydropower development in India further, these two key issues would

Table 4 Issues that need addressing for the propagation of a 'Green Economy' in South Asia

Issues that need addressing for the propagation of a 'Green Economy' in South Asia	Relevance to Millennium Development Goals (MDG)	What can South Asian countries do?	Examples
Poverty alleviation	High	Remove trade barriers on agricultural products (UNEP 2009)	It is estimated that the removal of agricultural protectionism could reduce global poverty by as much as 8 % Developing countries can cater for the demand for sustainably produced products in developed countries by promoting practices such as organic production
Job creation and social equity	High	Nurture competitive small to medium sized enterprise (SME) (Somavia 2007)	Provide support such as, business pre-start-up, start-up and development assistance Promotion and development of enterprise based training Encourage entrepreneurship
Promotion of renewable energy and low carbon technologies	Med	Support the scaling up and diffusion of small scale off-grid technologies	Grameen Shakti, microfinance model, Bangladesh. This is a non-grid solution to clean energy for the poor. This model is particularly powerful as it is commercial in operation and microfinance driven (substitutes kerosene for photovoltaic electricity at the village level) (UNEP 2009)
Resource and energy efficiency	Med	1. Increase funding for energy infrastructure upgrades Apply energy efficiency standards	Improve energy transmission infrastructure to curb transmission and distribution losses Refurbish or decommission low- efficiency coal fired power-plants, and Adopt mandatory energy efficiency standards for household appliances, and new motor vehicles

(continued)

Table 4 (continued)

Issues that need addressing for the propagation of a 'Green Economy' in South Asia	Relevance to Millennium Development Goals (MDG)	What can South Asian countries do?	Examples
Sustainable Urbanism	Med	Encourage public transport sector	India's National Urban Transport Policy (NUTP), discourages the use of personal motorised vehicles in favour of public transport Jawaharlal Nehru National Urban Renewal Mission, has made funding for transport projects in cities conditional upon the proposals being in conformity with the NUTP (TERI 2010)
Climate change	High	Early Action	The major source of atmospheric carbon in South Asia is energy production Near future transfers of finance and technology are important to forestall energy intensive growth due to long lead times to install improved energy infrastructure (Hunt 2011)
Resilience	High	Adaptation in the agricultural sector	Drought tolerant crop cultivation Management of surface water for crop production Rainwater harvesting Community based pond management for supplementary irrigation (Yu et al. 2010)

need to be addressed. Climate change may also play a part in diminishing the capacity of hydropower, through accelerated glacial melt and changed precipitation patterns, Therefore, these projections would need to be accounted for in any expansion of the hydropower sector (World Bank 2007).

From a 'whole of Asia' perspective, long-term growth in the region will dependent on the efficiency at which natural resources are used (ADB 2011). The literature identifies energy security in South Asia as both a barrier to regional integration, and the development of a low carbon economy. Energy security may also be framed as providing the biggest opportunities for fostering regional cooperation and development of a renewable energy sector. Population growth, urbanization, industrialisation and economic growth have all contributed to an ever increasing demand for energy in South Asia.[5] Where this has been traditionally, and to varying degrees of success, managed by supply-increased production, the future presents a different scenario in which natural resource depletion has rendered these traditional, unsustainable supply-demand practices as inadequate. A region wide analysis reveals that South Asia's energy reserves are exceptionally low, despite it being the most populous region of Asia. Reserves are at 0.5 and 0.8 % of the world's oil and gas reserves respectively (Hossain and Sarker 2013). Energy security in South Asia provides an exemplary scenario, endowed with potential for catalysing regional integration beyond the sphere of commerce, and the development of a low carbon economy, through technological innovation and aptly directed policy-making (Hossain and Sarker 2013).

Presently, relative to global statistics, most regions of South Asia display low energy intensity levels. For example India sits at 20 % below world averages per capita emissions. However, as countries such as India become more energy intensive with high rates of growth, as outlined in Table 5, there exists numerous opportunities to pre-empt rapid carbon intensification and provide significant emission reduction opportunities such as, improving energy transmission infrastructure to curb transmission and distribution losses; refurbishing or decommissioning low-efficiency coal fired power-plants; and adopting mandatory energy efficiency standards for household appliances and new motor vehicles (ESMAP 2009).

In terms of carbon mitigation and the transition to a low carbon economy, an analysis of low-carbon scenarios for India reveals that renewable energy sources are a preferred choice when combined with either a 'Carbon tax' (CT) scenario or a 'sustainable society' (SS) scenario (Shukla et al. 2008).[6] In a CT scenario, faster penetration of renewable energy is enabled by reducing the relative price difference between renewable and fossil fuels. In the SS scenario, cooperation among stakeholders plays an important role in lowering transaction costs and increasing deployable potential of renewable energy production; these two aspects drive the penetration of renewable resources.

Environmental degradation, sustainable growth and energy security are three key challenges faced by South Asian countries, in particular India. This type of transition to low-carbon, renewable energy would not only reduce the region's carbon intensity, it would also be associated with environmental improvement, sustainable industry, and increased energy security for a rapidly expanding economy (ADB 2009). Using the example of India, as previously outlined, there are a number of

[5]See: ADB (2009, p. 39): "South Asia's primary energy demand will increase from 582.1 Mtoe in 2005 to 1264 Mtoe in 2030 at an annual rate of 3.2 %. South Asia is expected to account for 17.5 % of the TPED in Asia and the Pacific in 2030. In all the members in South Asia, per capita energy demand remains low, at below 2.0 Mtoeeven in 2030".

[6]See: Shukla et al. (2008, p. 160): Carbon Tax (CT) scenario; presumes a stringent carbon tax (\$10/t CO_2, during Kyoto Protocol period, rising to \$100/t CO_2in 2050. Sustainable Society scenario (SS) follows a distinct 'sustainability' rationale, similar to that of the IPCC SRES B1 global scenario (IPCC 2000).

Table 5 South Asian Commercial Energy Supply and Demand Figures and Projections

South Asia	Supply and demand figures and projections (MTOE)					
Primary energy demand	1990	2000	2005	2006	2015	2030
Total	345.0	496.0	582.1	611.3	753.4	1264.3
Coal	106.4	165.1	209.2	223.5	184.5	462.5
Oil	66.2	122.2	138.4	145.5	178.9	336.6
Natural gas	13.6	28.4	40.5	43.1	64.8	132.9
Hydro	6.7	7.1	9.6	10.7	13.5	23.8
Nuclear	1.6	4.4	4.5	4.8	20.0	56.3
Others	150.5	168.8	180.9	183.6	191.6	252.2

Source ADB (2009)

policies in the NAPCC directed at creating a low carbon society, however, these are very much in their infancy and do not, as yet, form the requirements or framework for an effective, long term PAM (Kelkar and Bhadwal 2007).

4.1.2 Socio-political Change for Regional Cooperation

Using energy security as an example, it is clear that there is an inclination towards regional cooperation, as demonstrated by the energy security plans devised in each South Asian country. Currently institutions such as the SAARC Energy Working Group, South Asia Forum for Infrastructure Regulation (SAFIR), the World Energy Council and others, have provided space for dialogue, however, adequate follow up to these discussions has not occurred. Political issues often become entangled with policy action, thereby slowing the progress. This is because of the entrenchment of the government as partners in all of these institutions and proposals getting mixed up with non-energy related political issues of separate countries. Discussions should involve emerging issues in the energy sector; establishing regional energy projects; facilitating training, testing and development of renewable technology applications; and exchanging information on new technologies such as, renewable energy, hydrogen fuel and gas hydrates. It is further suggested that, in order to support these ideas, a permanent body must be created in the form of a research institution as well as a high level think tank, backed by governments. This would enable in depth analysis of specific issues and proposals with the key purpose to insulate energy related issues from political interferences (Sankar et al. 2006).

4.1.3 Technological Change and Policy Considerations

The transition of developing countries to low-carbon growth can initially benefit from the experiences of relatively advanced regional economies. This has been termed a South-South cooperation framework. The next phase of cooperation may involve higher income economies, providing a more advanced level of technological diffusion (Rahman 2010).

Many low carbon technologies are owned by firms in developed nations. The transferral of these technologies to developing countries, and an understanding of the associated logistics, is becoming increasingly urgent. As outlined by Ockwell et al. (2008), there are six key considerations for the development of policy that is aimed at facilitating low carbon technology transfer to developing countries:

- Technology transfer needs to be seen as part of a broader process of sustained, low carbon technological capacity development in recipient countries.
- Because low carbon technologies are at different stages of development they may involve both vertical transfer[7] and horizontal transfer.[8]
- Less integrated/rigid technology transfer arrangements are more likely to involve knowledge exchange and diffusion through recipient country economies.
- Firms that build into their transfer process the strategic objective to obtain technological know-how and knowledge necessary for innovation during the transfer process, are more likely to develop capacity as a result.
- Access to intellectual property rights, whilst an important facilitator of technology transfer processes, must also be accompanied by consideration of both absorptive capacity and risk assessment of new technologies.
- National and international policy interventions can play a central role in achieving low carbon technology transfer.

It is evident that policy intervention can play a central role in the promotion of low-carbon technology transfer and overcoming energy supply shortages. For example, at the national level, domestic policies[9] that incentivise the use of low carbon, as opposed to conventional technologies, have a strong effect in surmounting cost barriers and developing markets for new low carbon technologies. National level policies can also be tailored to encourage innovation processes that are actively engaged with international research and development (R&D) collaborations. Within this sphere, it is also important to enable and foster an environment conducive to international business transactions (Ockwell et al. 2008).

4.1.4 CSR and Social-behavioural Change

Corporate Social Responsibility (CSR) can play an important role in changing the behaviour of employees, consumers and communities. Generally, companies can offer a more proactive choice of products and services to promote low carbon lifestyles and importantly, using the example of India, growth has been driven by corporatisation, not by government policy. The Indian Institute of Management (IIM) (2010) identifies several companies in India that have begun using practical environmental solutions to promote change.[10] One key example is TATA; India's leading private sector power producer, currently using 90–95 % of coal used in India and therefore emitting large amounts of carbon. In order to help offset these

[7]Transfer of technologies from the R&D stage through to commercialisation.

[8]Transfer of technologies from one geographical location to another.

[9]Such as, taxes, subsidies, and emissions trading schemes.

[10]*Proceedings of the Regional Workshop on Corporate Environmental Management: From Policy to Practice*, Indian Institute of Management (2010, p. 7).

carbon emissions, TATA has initiated an energy club involving its employees. The initiative encourages employees to calculate their own carbon footprints and work towards reducing them. The employees were also sensitised to the environmental issues through visits to power plants, and audio visual displays. Another example of corporate change, identified by IIM (2010), is Yes Bank, a private sector bank in India that promotes responsible banking via financial growth whilst enhancing environmental and social impacts through business initiatives that focus on agribusiness, social and rural banking, sustainable investment banking and micro-finance. Yes Bank also documents its carbon footprint and is India's first signatory to the Carbon Disclosure Project (CDP).[11]

From an idealistic perspective, the current global financial crisis has been presented in some literature[12] as an opportunity for change towards a more resilient economy base, by placing sustainability at the heart of policy and reform. This change can be viewed as a move away from the traditional paradigm of economic growth as an imperative. Apart from establishing resource and environmental limits on economic activity and re-designing economic models, it is suggested that there must be a change in the social logic of consumerism. Steps to achieving change in this sense, within South Asia may include tackling systematic inequality, measuring prosperity, strengthening human and social capital and ultimately reversing the culture of consumerism. It is suggested that decreasing the influence and goals of materialism can be achieved through investment in public amenities and spaces, strengthening communities and enhancing socio-cultural ties that can enrich life by providing opportunities for leisure and self development and ultimately move society away from materialism as fulfilment (Jackson 2011).

4.1.5 Institutional Change

As Somavia (2007) surmises, fostering sustainable enterprise has its basis in strengthening institutions and governance systems which have the capacity to nurture enterprises. Fundamentally strong and effective institutions provide the basis for strong and efficient markets. Somavia continues by adding that ensuring human, financial and natural resources are divided equitably and efficiently to promote innovation and enhanced productivity is essential.

As briefly mentioned within the technology and policy section of this paper, it is important to foster an environment conducive to international business. National policy can play a major role in this arena, especially in India, which is currently hampered by bureaucratic barriers to efficient business. Importantly, government led initiatives (information sharing, renewable technology, R&D) must engage private companies (Ockwell et al. 2008). Ockwell et al. (2008), outline that at the international level, initiatives can be targeted at overcoming barriers associated with

[11]CDP is an independent non-profit organisation that holds the largest database of primary corporate climate change information in the world. CDP collects and distributes information to investors, corporations and governments to motivate them to take action to prevent severe impacts on climate change.

[12]See: Jackson (2011, p. 162).

high costs of low-carbon technologies. These efforts are categorised as either direct financing (e.g. GEF,[13] CDM,[14] World Bank) or carbon pricing. Policies aimed at reducing the cost of business and improving competitiveness have already begun to be addressed by focussing on microeconomic policies. This attention has led to a set of objectives aimed at improving institutions and governance by expanding market-based allocation of resources; increasing transparency through improved disclosure and market discipline; regulation reform to minimise corruption; and introducing better checks and balances to improve governance (Ahmed et al. 2010).

4.1.6 Climate Change Adaptation and Mitigation

One of the key facets of climate change adaptation are the challenges faced at the design and implementation phase. As mentioned throughout this literature review, South Asia faces many impediments to implementing sustainable development policy; climate change adaptation falls into this category. Lack of knowledge, weak institutional capacity, corruption, and poor trans-boundary coordination are common themes in this paper, and directly impact on the development of effective climate change adaptation mechanisms. The foundations of a low carbon economy and ultimately sustainable development, if acted upon, can in itself act as long term climate change mitigation mechanisms. However, as frequently noted, many regions of South Asia, such as the Bay of Bengal Delta, are already feeling the direct consequences of climate change, and the short-to medium term objective must be adaptation. Options that promote low carbon economy, whilst improving adaptive capacity include, increased trans-boundary cooperation, promotion of economic development through diversified livelihoods, improved management of public services and maintenance of healthy ecosystems (USAID 2010).

5 Conclusions

Climate change poses a new set of challenges for the future development of South Asia. It is important to consider the impacts of climate change in South Asia due to the region's status as one of the world's fastest growing economic powers, containing the most densely populated region of the world that is highly vulnerable, and already experiencing changes to weather patterns, hydro-geographical flows, flood and drought patterns (Stern 2006). This paper is fundamentally a response to the lack of available empirical analysis regarding the implications of climate change and the relevance of cooperation and regional integration, towards the development of a low carbon economy in South Asia. A key emphasis of this study is therefore on the process of transition to a low carbon economy in South Asia, with a particular focus on Bangladesh and India; the two most vulnerable countries in South Asia to climate change. This review of scientific reports, policy papers and other

[13]Global Environment Facility.

[14]Clean Development Mechanism.

relevant materials has provided a helpful means for identifying the gaps in addressing climate change and related threats at local, national and regional levels within the South Asian region.

Gaps at local, national and regional levels, concerning policy-making and project formulation related to climate change and human security have been identified primarily as stemming from a lack of regional integration, energy security, corruption, weak institutional capacity and ineffective policy in terms of promoting private-sector action and cooperation with developed countries (Ahmed et al. 2010). Although there has been some progress in all of these areas, for example SAARC, there is a lack of coordinated and integrated policy and mechanisms that would be needed to see real progress towards SEE and an effective response to the challenges derived from climate change. However, further research is necessary to further fill the gap in the literature and expand the empirical work to other countries in South Asia looking beyond Bangladesh and India.

References

Ahmad QK, Warrick RA, Ericksen NJ, Mirza MQ (2006) The implications of climate change for Bangladesh: a synthesis. In: Warrick RA, Ahmad QK (eds) The implications of climate and sea-level change for Bangladesh. Springer, Dordrecht

Ahmed S, Kelegama S, Ghani E (2010) Promoting economic cooperation in South Asia: Beyond SAFTA, Sage Publications, New Delhi

Asian Development Bank (ADB) (2009) Energy outlook for Asia and the Pacific. ADB, Philippines

Asian Development Bank (ADB) (2011) Asia 2050: realizing the Asian Century. In: Kohil H, Sharma A, Sood A (eds). Sage Publications Pty Ltd

Atiken D, Raihani N (2011) Uncertainty, rationality and cooperation in the context of climate change. J Clim Change 108:47–55

Energy Sector Management Assistance Program (ESMAP) (2009) Low carbon growth country studies program: mitigating climate change through development. Energy Sector Management Assistance Program. http://sdwebx.worldbank.org/climateportalb/doc/ESMAP/KnowledgeProducts/Low_Carbon_Growth_Country_Studies_Getting_Started.pdf

Hossain M, Kathuria R, Islam I (2010) South Asian economic development, 2nd ed, Routledge

Hossain M, Selvanathan EA (eds) (2011) Climate change and growth in Asia. Edward Elgar Publishing, Cheltenham, Northampton

Hossain S, Sarker T (2013) Resource constraints and Asia's growth: role of integrated energy, water and food security policy. In: Hossain M, Sarker T, McIntosh M (eds) The Asian century, sustainable growth and climate change. Edward Elgar, Cheltenham

Hunt C (2011) Greenhouse gas abatement in Asia: imperatives, incentives and equity. In: Selvanathan A, Hossain M (eds) Climate change and growth in South Asia. Edward Elgar, Cheltenham, p 114

International Labour Organisation (ILO) (2007) The promotion of sustainable Enterprises, Report VI. http://www.ilo.org/wcmsp5/groups/public/@ed_emp/@emp_ent/documents/publication/wcms_093969.pdf

Indian Institute of Management (IIM) (2010) Regional workshop on corporate environmental management: from policy to practice. Summary of Proceedings, Bangalore, pp 1–15

Jackson T (2011) Societal transformations for a sustainable economy. Nat Resour Forum 35:155–164

Jhirad DJ, Nitze WA, Gollwitzer, L (2011) Oil in the 21st century: responding to key drivers of change. In: The oil era: emerging challenges. Emirates Center for Strategic Studies and Research, Abu Dhabi. ISBN 9789948144304

Jilani T, Gomi K, Yuzuku M (2011) Low carbon society development towards 2025 in Bangladesh. Department of Urban and Environmental Engineering, Graduate School of Engineering, Kyoto University, Kyoto

Kelkar U, Bhadwal S (2007) South Asian regional study on climate change impacts and adaptation: implications for human development. Human Development Report 2007/2008: UNDP Stern Report on the Economics of Climate Change, 2006

Ockwell D, Watson J, Mackerron G, Pal Prosanto, Yamin F (2008) Key policy considerations for facilitating low carbon technology transfer to developing countries. Energy Policy 36:4104–4115

Pew Center on Global Climate Change (2008) National Action plan on climate change. Government of India

Rahman SH (2010) Cooperation to promote low-carbon economic development: pathways and vision. Asia Pacific Finance and Development Center (AFDC) 2010 Biennial Forum Fiscal and Financial Policies for Low-carbon Economic Development Shanghai National Accounting Institute, Shanghai, P.R. China, 26 November 2010

Sankar T, Awardee P, Raza H, Barkat A, Wijayatunga P, Acharya M, Raina D (2006) Regional energy security for South Asia: regional report. USAID South Asia Initiative for Energy (SARI/Energy). USAID, pp 1–66. http://www.sari-energy.org

Shukla P, Dhar S, Mahapatra D (2008) Low-carbon scenarios for India. Climate Policy 1752–7457

Somavia J (2007) Conclusions concerning the promotion of sustainable enterprises. In: International labour conference, June. International Labour Office, Geneva

Stern N (2006) The stern review: economics of climate change. http://mudancasclimaticas.cptec.inpe.br/~rmclima/pdfs/destaques/sternreview_report_complete.pdf

The Energy and Resources Institute TERI (2010) India's energy security: new opportunities for a sustainable future. TERI, June 2009, p 16, New Delhi

The World Bank (2004) World development indicators 2004. http://data.worldbank.org/data-catalog/world-development-indicators/wdi-2004

The World Bank (2007) South Asia: growth and regional integration. In: Ahmed S, Ghani E (eds) Washington DC

The World Bank (2010) World development indicators 2010. http://data.worldbank.org/data-catalog/world-development-indicators/wdi-2010

UNFCCC (United Nations Framework Convention on Climate Change) (2008) Report of the conference of the parties on its thirteenth session, held in Bali from 3 to 15 December 2007. United Nations Framework Convention on Climate Change, Bonn

UNEP (2009) Global green new deal: policy brief. United Nations Environment Program, Nairobi, Kenya.

USAID (United States Agency for International Development) (2010) Final report: findings and recommendations. Asia-Pacific Climate Change Adaptation Assessment, Washington, DC

Yu W, Alam M, Hassan M, Khan AS, Ruane AC, Rosenzweig C, Major DC, Thurlow T (2010) Climate change risks and food security in Bangladesh. Earthscan, London, pp 144

Explanatory Factors of Social Responsibility Disclosure on Portuguese Municipalities' Websites

Verónica Paula Lima Ribeiro, Sónia Maria da Silva Monteiro and Ana Maria de Abreu e Moura

Abstract

Purpose Municipalities can play a key role in supporting the development of green growth, by linking environmental, social and economics goals, in a sustainability strategy at the local level. At the same time, sustainability reporting allows for a deeper understanding of the local green policies. The aim of this paper is to analyse the disclosure of social responsibility (SR) information by Portuguese municipalities and to identify their determining factors. **Design/Methodology/Approach** A content analysis was performed on webpages from 60 sampled municipalities, and an information disclosure index was created (based on generic, economic, social and environmental information). For each information category, a disclosure sub-index was also developed. To analyse the relationship between the developed information disclosure indices and potential determinant factors, several tests were conducted using previously formulated hypothesis, via univariate and bivariate analyses. **Findings** Descriptive statistics indicate average levels of SR disclosure. The Total Disclosure Index (TDI) value was 0.46. The Economic Information sub-category exhibits the highest value (0.66), followed by the Social and Environmental information categories (0.61 and 0.36, respectively). Univariate and bivariate analyses show that certain variables can explain degrees of information disclosure. The Size (SIZE2) (measured by the number of inhabitants), Education Level (EDUC) and

V.P.L. Ribeiro (✉) · S.M. da Silva Monteiro · A.M. de Abreu e Moura
School of Management - Polytechnic Institute of Cávado and Ave,
Campus do IPCA, 4750-810 Arcozelo Barcelos, Portugal
e-mail: veronica@ipca.pt

S.M. da Silva Monteiro
e-mail: smonteiro@ipca.pt

A.M. de Abreu e Moura
e-mail: amoura@ipca.pt

W. Leal Filho et al. (eds.), *Sustainable Economic Development*,
World Sustainability Series, DOI 10.1007/978-3-319-45081-0_17

Tax Burden (TAX) variables positively influenced all indices, while the Unemployment Rate (UNEMP) variable does not influence the indices. Political Competition (POLCOM) only influenced EnDI, and the remaining variables influenced at least four of the five indices studied. **Originality/Value** This study adds to the international research on SR in public sector by providing empirical data from a country, Portugal, where empirical evidence is still relatively limited.

Keywords
Social responsibility · Disclosure · Municipalities · Internet

1 Introduction

The green economy, in the context of sustainable development, was one of the themes for the United Nations Conference on Sustainable Development held in Rio de Janeiro in 2012 (or Rio + 20). The green economy as been seen an important tool for sustainable development, because that is inclusive and can drive economic growth, employment, and poverty eradication, whilst maintaining the healthy functioning of the Earth's ecosystems (United Nations Department of Economic and Social Affairs—UNDESA 2012b).

To UNDESA (2012a) the various definitions of green growth and green economy are generally consistent, both having sustainable development as their basic objective and being a means to reconcile the economic and environmental pillars, without ignoring social aspects.

The green economy "designates a set of activities associated with research, technologies and industries, which are directly geared to improve environmental outcomes, reduce pollution, conserve energy and protect natural resources (…). The concept of sustainable development emphasises the broader perspective of sustainability encompassing economic, social and environmental dimensions. Sustainable development provides an important context for green growth" (Organisation For Economic Co-Operation And Development—OECD 2012: 24). The concept of green growth "is closely related to that of sustainable development, but with more emphasis on growth and on mitigating climate change, which is widely perceived as the major long-term challenge to sustainability" (Bowen 2012: 11).

According to the European Environment Agency—EEA (2011: 5), "the green growth isn't just a preferable approach to economic development. On the contrary, in the long term it's the only way to sustain economic growth".

The organizations have been submitted to different pressures in order to achieve a balance between economic growth and environmental preservation, and the concept of environmental management has been included in the general management system of the organization.

Although the public sector has different and unique characteristics from the private sector, Rondinelli and Berry (2000) emphasize the need for greater cooperation of central and local governments with the private sector. They argue that to overcome the "gap" between public policies regarding the environment requires the adoption by both sectors of a new philosophy that emphasizes eco-efficiency and recognizes that economic growth and environmental quality are two mutually interdependent goals.

Several authors, such us Ball and Grubnic (2007) have defended that public sector organizations should play an important role in promoting sustainable development. On the one hand, as policy-makers, governments can drive the environmental agenda by translating long-term environmental priorities into concrete policies and programs; communicating, implementing and monitoring such environmental policies and programs and controlling compliance of environmental regulations (Emilsson and Hjelm 2004; Moon 2004; Ball and Grubnic 2007). On the other hand, as providers and/or producers of public services and utilities, public sector organizations carry out activities that have a significant impact on the environment (GRI 2005). Therefore, public entities, such us local authorities, are called by Taylor et al. (1994) as "silent destroyers" because they are considered "clean" businesses that, supposedly, do not originate products or apparently pollutant wastes. Therefore, environmental issues must also be integrated into public entities at organizational level in order to reduce the environmental impacts associated with their own operations.

Sustainable development provides an important context for green growth. However, has not been conceived as a replacement for sustainable development, but can be considered a subset of it. "It is narrower in scope, entailing an operational policy agenda that can help achieve concrete, measurable progress at the interface between the economy and the environment" (OECD 2012: 24).

The public sector plays a key role in supporting the development of green growth, by linking environmental goals to broader social equity goals, in a sustainable economic development strategy, at the local level (OECD 2012: 75).

Amongst the different government levels, local authorities (municipalities) have been considered key actors for implementing sustainability (Ball 2005). From a practical standpoint, municipalities operate closest to the people and can place green economy policies in a tangible context (Federation of Canadian Municipalities—FCM 2011). These policies include "articulating an economic and sustainability vision for their areas, diffusing sustainability ideas and practices, managing local regulatory, planning and approval instruments, designing green infrastructure, promoting new industry and skill opportunities, and working with industry, households and energy authorities to promote critical mass in low-carbon energy use" (OECD 2012: 89).

The transition to a green economy poses many challenges for local authorities, particularly in ensuring the development of economic activity in cleaner sectors and providing incentives to develop green technologies (OECD 2012). During the last decades many local entities have undergone a number of initiatives to become environmentally friendly by reducing resources consumption and wastage and promoting polities and actions aimed at achieving the sustainable development within their geographical area (Emilsson and Hjelm 2004).

When moving towards green growth, reporting and accounting frameworks are necessary to determine what and how to measure, and how to report information about the progress made to achieve the goals of green growth.

Discussion on the importance of promoting public access to information on environmental and social issues is a familiar part of the sustainable development agenda. Following the trend in the private sector, public entities have also undertaken a range of sustainability reporting practices with the aim of disclosing information on their social and environmental impacts and the initiatives carried out to manage them.

Reporting on sustainability by municipalities allows for a deeper understanding of the concerns and strategies that guide decision-making as well as messages that are conveyed to citizens.

Regarding the publication of SR information, we underline the Supplement for Public Agencies (Suplemento para o Sector Público—SSPA) issued by the Global Reporting Initiative (GRI) in 2005 (GRI 2005), which contains recommendations on sustainability evaluation based on performance indicators grouped into economic, environmental and social categories.

Stakeholders increasingly follow the actions of public entities and analyse whether these entities behave in economically, socially and environmentally responsible ways, thus displaying sustainable behaviours. The Internet acts as an important channel for the disclosure of these practices. Currently, most public entities create websites as means of conveying information to potential stakeholders.

Sustainability reporting by public entities is still far from the level of development reached in the private sector. The extent to which a public entity develops sustainability reporting practices can be associated with several factors (such us, size, political competition, etc.). However, they can differ from those of private companies and, therefore, require greater consideration (CPASR 2005).

The majority of empirical studies on SR disclosure in public sector been headed by Anglo-Saxon countries, mainly the UK and Australia. To our knowledge, no empirical studies on social responsibility (SR) information disclosure by Portuguese municipalities currently exist. There is currently only one empirical study that focuses on SR disclosure by municipalities, and it focuses on one component of SR: the environment (Ribeiro 2007a, b; Ribeiro and Guzmán 2008a, 2011).

With the aim of covering the scarcity of studies in this area, this paper seeks to analyse the disclosure of social responsibility information by Portuguese municipalities and to identify their determining factors.

The rest of the paper is organized as follows: the next section briefly presents the literature review and the construction of the hypotheses on the impact of several factors on the level of SR information disclosed by local public entities. The research design, including research methodology, sample selection and variables, is described in Sect. 2. In Sect. 3, we present and discuss our results including both the descriptive and univariate/bivariate analyses. The final section summarizes the main conclusions of the study with a discussion of its limitations and implications for future research.

2 Literature Review and Hypotheses

According to our international review of existing literature, most studies related to SR or sustainability information disclosure focus on the private sector. However, according to Lodhia (2010), the public sector provides a theoretically useful context for researching topics related to sustainability. On this issue, Lewis (2008) states that there has been an increase in the number of studies conducted on sustainability reporting.

Thus, an increased research focus on the public sector has been observed, and of the three components of SR (economic, social and environmental), the environmental component has predominated as the focus of analysis for researchers (Ball 2005; Burrit and Welch 1997; Frost and Seamer 2002; McElroy et al. 2005; Ribeiro 2007a, b; Ribeiro and Guzmán 2008a, b; Ribeiro and Guzmán 2011; Sciulli 2009).

A geographical review of existing studies shows that Australia predominates as the geographical area in which the majority of studies on public sector SR disclosure have been conducted (Farneti and Guthrie 2009; Guthrie and Farneti 2008; Lodhia 2010; Mack and Power 2006; Mucciarone 2012; Sciulli 2011; Williams 2011, 2012; Williams et al. 2009). Studies have also examined public entities in Malaysia (Joseph and Taplin 2011; Joseph 2010; Pilcher et al. 2008), local and/or regional governments in Spain (Navarro et al. 2010a, b, 2011a), universities in the United States (Sanchez et al. 2011), universities in Spain (Moneva and Martin 2012, and public entities in Italy (Farneti and Siboni 2008; Marcuccio and Steccolini 2005, 2009). Other authors have conducted comparative analyses of public entities in different countries, e.g., the United Kingdom and Ireland (Navarro et al. 2011b); EU-15 states (Martins 2011); Australia, China, New Zealand and the United Arab Emirates (Tort 2010); and Australia, New Zealand, Europe, North America and Asia (CPASR 2005).

Some studies concerning SR information disclosure have focused on published annual reports (Guthrie and Farneti 2008; Moneva and Martin 2012; Mucciarone 2012; Sciulli 2011) and/or sustainability reports (CPASR 2005; Farneti and Siboni 2008; Guthrie and Farneti 2008; Mack and Power 2006; Marcuccio and Steccolini 2005, 2009; Sanchez et al. 2011; Tort 2010), which are both used for disclosing information. These studies identify a diverse range of reporting practices and, in addition to annual reports, several other forms of disclosure.

Online information disclosure has played a significant role in the public sector, as the approach increases engagement between public entities and citizens. Hence, several studies have applied methodologies for analysing organisational webpages (Joseph 2010; Joseph and Taplin 2011; Martins 2011; Moneva and Martin 2012; Navarro et al. 2010a, b, 2011a, b; Sanchez et al. 2011; Pilcher et al. 2008).

The degree of SR information disclosure by municipalities is generally low, and information disclosed on webpages have been found to be irregularly provided, nonhomogeneous, diverse and predominantly descriptive.

The determinant factors examined are diverse, and empirical evidence demonstrates a number of diverging results. In local public entities, emphasis is placed on the size of the entity and on the implementation of Agenda 21 as possible determinant factors influencing the degree of disclosure (Joseph 2010; Joseph and Taplin 2011), although Pilcher et al. (2008) concluded that only size has an impact, while LA21 application does not. Among other factors that have been considered by various authors, we stress the finding that municipality characterisation as urban (Williams 2011; Joseph 2010) is a determinant factor of disclosure levels, while political competition, political party affiliation and municipal expenditure volume do not affect the degree of disclosure (Navarro et al. 2010a).

3 Hypotheses Development

This study aims to analyse the influence of certain factors on SR information disclosure practices by Portuguese municipalities. The following discussion relates to the construction of the hypotheses on the impact of some variables on the level of SR disclosure.

H1: The degree of SR information disclosure is greater in larger municipalities.

According to Mucciarone (2012), size affects an organisation's ability to collect, retain and use information for reporting on organisational performance. Using this same reasoning, Navarro et al. (2010a) reported that larger municipalities typically employ more skilled workers, which can foster the application of SR practices. Moreover, the size of a municipality determines the size and diversified nature of interest groups as well as access to available resources for disclosing information (Navarro et al. 2010b).

According to Joseph (2010), larger municipalities are subjected to higher degrees of scrutiny by stakeholders than smaller municipalities. Therefore, larger municipalities are more likely to institutionalise their activities and programs to legitimate themselves in agreement with societal values and norms.

Likewise, Joseph and Taplin (2011) find that the disclosure of information by large organisations can be motivated by the higher degree of political visibility to which they are subjected, resulting in greater coercive pressures to disclose information.

H2: The degree of SR information disclosure is higher in coastal municipalities.

Ribeiro and Guzmán (2011) argue that the degree of economic development in a given region influences the degree of economic development of entities within it. If entities are located in more developed regions, these entities will tend to have access to more financial and human resources that can be leveraged to more intensively exercise practices of social and environmental responsibility and SR information disclosure.

In Portugal, municipalities are grouped into Municipal Associations, largely based on geographic location. These associations promote professional training seminars and courses and allow for the sharing of experiences, knowledge and various forms of information. In this sense, professionals practicing in each of these municipalities may regularly meet, discuss and share similar topics related to their work, forming networks of regulatory pressure and leading to a higher degree of information disclosure. Mimetic isomorphism may therefore underlie this process because if one municipality discloses information, other municipalities may follow suit due to the sharing of experiences.

H3: The degree of SR information disclosure is higher in urban municipalities.

According to Joseph (2010), urban municipalities may differ from their rural counterparts with respect to SR information disclosure. Higher standards of living and more frequent tendencies for residents to express opinions in urban regions may indirectly affect sustainability reports. In rural municipalities, more emphasis is placed on meeting basic needs, such as access to water, electricity and road maintenance.

McElroy et al. (2005) found that this variable affects environmental information disclosure by local Swedish public entities. Williams (2011) also classified local Australian governments as either rural or urban according to the *Australian Classification of Local Governments* and found that there may be a significant difference in levels of sustainability information disclosure between rural and urban governmental entities.

H4: The degree of SR information disclosure is higher in municipalities with environmental/SR certification.

Archel Domench and Lizarraga Dallo (2001) report that entities with ISO 14001 or EMAS (European Eco-Management and Audit Scheme) environmental certification engage in higher degrees of environmental disclosure because these regulations impose requirements on disclosure; i.e., they obligate entities to distribute environmental information and statements. Institutional Theory can thus explain a higher degree of SR information disclosure to the extent that certification (either ISO14000 or ISO26000 and NP4469-1) involves being subjected to pressures (coercive and normative) to comply with guidelines and standards, resulting in disclosure requirements.

Thus, while to our knowledge, this variable has not been the subject of empirical study at the public sector level, we believe that environmental and/or SR certification enhances SR information disclosure for Portuguese municipalities.

H5: The degree of SR information disclosure is higher in municipalities that have implemented Local Agenda 21.

Ribeiro (2007a) states that Local Agenda 21 (LA21)[1] implementation, in addition to encouraging transparency in environmental information disclosure by municipalities, presents advantages at the strategic, social, economic and financial levels. According to Joseph (2010), greater commitment to LA21 implementation

[1]Local Agenda 21 is a participatory and multi-sectoral action program that seeks local authority and various local player involvement in sustainable development (arising from economic, social and environmental spheres).

by municipalities strongly influences the breadth of information provided in sustainability reports.

In Portugal, no specific legislation exists that requires local authorities to design and implement LA21. Central Government level programs and structures that support such implementation do not exist either. Regional municipal associations have been the major drivers and facilitators of LA21 processes in Portugal insofar as they provide information, technical and financial support (through community funds) to local authorities. These associations act as "transfer institutions" that allow for the creation of inter-municipal interaction platforms (Sousa 2009).

In this sense, normative and mimetic isomorphisms that underlie Institutional Theory can provide a satisfactory explanation for higher degrees of SR information disclosure in municipalities that implement LA21. In fact, municipal associations provide specific guidelines that demonstrate a movement towards professionalisation and a strong possibility that certain municipalities will "imitate" others in the process of LA21 strategic development and implementation.

H6: The greater the level of political competition present, the greater the degree of SR information disclosure.

In democratic regimes, political competition is often played out as a form of partisan struggle that aims at either access to or preservation of political power. Political parties seek to obtain an absolute majority during elections by electing councillors (in the case of municipalities) with the aim of outnumbering opposition parties and to obtain political strength. This predictably gives rise to governmental stability.

However, the existence of a strong opposition necessitates responsible management from the ruling party and public awareness of behaviour that deviate from the electoral program. Assuming that politicians wish to remain in power for several terms, there will be an interest in voluntarily demonstrating a commitment to efficient management (Cárcaba García and García-García 2010) and socially responsible behaviour. This commitment will often be carried out through information disclosure. Hence, (coercive) pressures from the opposition may propel municipality leaders to disclose more information as a means of legitimising themselves before the citizens who elected them, thus demonstrating that they can offer more benefits and successes than their alleged competitors.

In this sense, it is expected that in municipalities with less political competition, i.e., municipalities in which the winning political party won with an absolute majority (more than 50.1 % of the votes), there is less pressure from oppositional parties/political groups to disclose SR information. This explanation is proposed by Navarro et al. (2010a, 2011a), who found a predictable relationship between political competition and sustainability information disclosure levels in Spanish regional and local governments.

Serrano-Cinca et al. (2009) and Jorge et al. (2011) also examined the variable of political competition when analysing possible factors influencing financial information disclosure and information transparency on municipal websites.

H7: Higher educational levels in a municipality correlate with higher degrees of SR information disclosure.

According to Navarro et al. (2011a), higher stakeholder education levels can increase stakeholder demands for SR while raising awareness of municipality commitments in this area, thus meeting citizen demands.

For Jorge et al. (2011), education precludes ICTs use, and a lack of skills in using the Internet, especially related to education, lies at the heart of what is called the "digital divide". Qualified citizens with higher levels of education/training feel more comfortable using technological tools and are thus more inclined to demand information from public authorities (Jorge et al. 2011).

Thus, because higher education levels initially correlate with greater demands and scrutiny of leaders by citizens, information disclosure by municipalities may preserve and improve municipality legitimacy by meeting stakeholder expectations (residents and others).

H8: Higher unemployment rates correlate with a greater degree of SR information disclosure.

According to Navarro et al. (2011a), a large number of unemployed persons implies the existence of greater social needs. Hence, such municipalities are expected to develop projects and actions in this area. In such cases, greater pressures will be placed on municipalities to disclose social information. Thus, higher degrees of SR information disclosure may increase municipality reputations as entities that are socially accountable to citizens.

H9: A larger inactive population (aged below 19 and above 65 years of age) correlates with higher degrees of SR information disclosure.

The proportion of the population under 19 and over 65 years of age typically constitutes the inactive population that requires the most public spending. Municipalities contribute to the activities and investments of Private Institutions of Social Solidarity that directly support this population, largely through financial support.

Navarro et al. (2011a) argued that the size of this population in relation to the entire population can influence sustainability information disclosure, specifically the disclosure of social information due to the demands of these age groups for this type of information.

Following the same reasoning as that demonstrated in Hypothesis 8, municipalities with more residents belonging these age groups (in principle, citizens with more social needs) will disclose more information to demonstrate their concerns for the social context in which they operate, thus legitimating themselves before society.

H10: The degree of SR information disclosure is greater in municipalities with greater tax burdens.

Tax burden can be defined as the sum of direct and indirect taxes per capita for each municipality. According to Navarro et al. (2011a), an increase in the tax burden may be attributable to less concern for economic sustainability by municipalities due to a preoccupation with the current period that limits the predisposition to disclose economic information. We disagree with these authors. In our opinion, a greater tax burden should correlate with a greater propensity to disclose information, as citizens and opposition parties can demand information on municipality tax dollar spending. Additionally, motivations described by Legitimacy Theory may also be present, such as disclosing information to demonstrate that taxes are being allocated to projects of municipal interest, thus increasing the municipality's legitimacy in the eyes of society.

Though consensus on the effect of this variable in terms of information disclosure does not exist, we formulate the hypothesis expecting a positive correlation for this association.

H11: Higher environmental municipality expenses correlate with a higher degree of SR information disclosure.

Municipalities, according to their duties and responsibilities, play an important role in environmental management, and expenditures dedicated to environmental management and protection by municipalities is important compared to central government entities (Queirós 2002).

Navarro et al. (2011a) argue that higher levels of environmental expenditure reflect a greater municipal sensitivity to environmental issues, which may promote the disclosure of such information. Legitimacy Theory may explain this positive relationship between environmental expenditures and information disclosure because municipalities may wish to demonstrate that these expenses are fair and legitimate and that they contribute to the development of the municipality.

4 Research Design

4.1 Sample Characterisation

In selecting our sample, we considered the global ranking system published in the Financial Yearbook of Portuguese Municipalities of 2010 (Carvalho et al. 2012), which ranks Portuguese municipalities based on the efficient use of financial resources. The global ranking list is based on 15 indicators defined above that best assess the financial, economic, asset and budgetary management of municipalities. Based on this ranking, the 60 top Portuguese municipalities include 10 of large dimension, 20 of medium dimension and 30 of small dimension, constituting our

Table 1 Summary of sample characteristics

Main characteristics	% Sample
Size	
Small (≤ 20,000 inhabitants)	50.00
Medium (20,000 < inhabitants ≤ 100,000)	33.33
Large (>100,000 inhabitants)	16.67
Location	
Coast	51.67
Inland	48.33
Characterisation	
Rural	31.67
Urban	68.33
Political party	
"Partido socialista" (PS)	50.00
With environmental certification	5.00
With Local Agenda 21 certification	45.00

sample.[2] This classification thus allows us to evaluate all types of Portuguese municipalities rather than only the largest (a criterion typically used in other studies, e.g., Marcuccio and Steccolini 2005; Navarro et al. 2010a, b) and Navarro et al. 2011b). Table 1 summarises the main characteristics of the sample.

4.2 Methodology

According to Gandía and Archidona (2008), the Internet has partly caused public administrations to develop methods for improving information access and levels of responsibility. Gallego-Alvaréz et al. (2010) concluded that Internet use has reduced barriers between citizens and public administrations.

Several studies conducted in Spain have focused on public sector information disclosure online (Navarro et al. 2010a, b, 2011a, b; Pina et al. 2007, 2009; Rodríguez Bolivar et al. 2006, 2007) because municipality websites are considered to provide an adequate means for studying published information on SR.

In conducting our study, we collected SR information published on sampled municipality websites from July 23, 2012 to September 16, 2012.

Information sources include municipality website content, which includes all reports and other documents (minutes, resolutions, notices, flyers, posters, news, etc.) published on these sites as well as information listed on other related sites that are linked to the municipality websites examined.

[2]The structure of the sample is similar to that of municipalities at the national level; i.e., in Portugal in 2010, of the existing 308 municipalities, 179 are considered of small dimension (58.1 %), 106 of medium dimension (34.4 %) and 23 of large dimension (7.5 %). We can thus conclude that this is a representative sample of the population.

In collecting data, we applied a content analysis approach, as it is the most commonly used technique in studies that evaluate social and environmental information disclosure, both on websites and through other communication channels (Joseph 2010; Martins 2011; Moneva and Martin 2012; Navarro et al. 2010a, b, 2011a, b; Sanchez et al. 2011).

As Ribeiro and Guzmán (2011) state, this technique involves coding qualitative information into pre-defined categories. In our study, a list of 46 items divided into four categories or blocks of information was used[3]:

Category 1—Generic Information	10 items
Category 2—Economic Information	9 items
Category 3—Social Information	13 items
Category 4—Environmental Information	14 items

Appendix 1 presents items related to the information covered in each category. This classification scheme is based on the approach used by Navarro et al. (2010a, b, 2011a, b) as well as on GRI Guidelines (2006) that were adjusted to specific characteristics of the public sector.

We analysed municipality website content and documents disclosed therein to determine the presence or absence of information on our category list. We assigned a weight of 0 or 1 to each item according to the following criteria:

0—The item is not disclosed;
1—The item is disclosed.

Because this method indicates presence or absence, 1 is the maximum value that can be assigned to each item for each municipality, i.e., when webpages promote information on an item several times, only one instance is considered.

This method does not detect the quality or quantity of information disclosed but only recognises whether an item is disclosed or not. However, as Navarro et al. (2010a, b) argue, this scoring system allows one to measure levels of information disclosure in an objective and transparent manner, hence explaining why the method has been applied in several empirical studies similar to the present study (Navarro et al. 2010a, b, 2011a, b; Pina et al. 2007; Rodriguez Bolivar et al. 2006, 2007; Ribeiro 2007a, b).

[3]A final item on autonomous SR disclosure was included to assess whether municipalities in this study formally release sustainability or SR reports. After finding a positive response for this item, four more information-related variables were studied: the site location, the period covered, the frequency of compilation and GRI guideline adherence.

4.3 Variables

Total Disclosure Index (TDI): To analyse factors that influence the level of SR information disclosure by the sampled municipalities, we developed a Total Disclosure Index (TDI), that includes all information items mentioned above (presented in Appendix 1). For each information category, a disclosure sub-index that integrates items in each category was developed.

The use of indices for measuring information disclosure levels on this topic has been widely applied in previous studies (Gandía and Archidona 2008; Jorge et al. 2011; Joseph 2010; Joseph and Taplin 2011; Moneva and Martin 2012; Mucciarone 2012; Navarro et al. 2010a, b, 2011b; Ribeiro 2007a, b; Sciulli 2009; Beck et al. 2010).

The disclosure index is a quantitative variable that represents the degree of disclosure. The index is calculated by dividing the total score obtained for each category by the maximum number of items that comprise it (described below), with the value of each index ranging between 0 and 1.

$$\mathbf{IDx}_j = \sum_{\mathbf{n}=1}^{\mathbf{i}} \mathbf{i}_n/\mathbf{i}$$

where:

- $\mathbf{IDx}_j$ Information Disclosure Index (Category X)[4] of municipality j
- $\mathbf{i}_n$ General item n under analysis. A dichotomous variable (dummy) with a value of 1 if the municipality publishes item n and a value of 0 if the municipality does not disclose item n
- **I** Maximum number of information items in category X

Dutta and Lawson (2009) note that the three dimensions of SR (economic, social and environmental) should be balanced so that the three dimensions hold the same degree of importance. Thus, in constructing the Total Disclosure Index, we used the method followed by Rodriguez Bolivar et al. (2006, 2007) and Navarro et al. (2010a, b, 2011b), which allows for an equitable balance among the categories to equalise the importance of each category.

Thus, the Total Information Disclosure Index for SR (TDI) is calculated with a weight of 25 % for each of information sub-indices (GDI, EcDI, SDI, EnDI) as shown below:

$$\mathbf{TDI}_j = \mathbf{25\ \%} \times \mathbf{GDI}_j + \mathbf{25\ \%\ x\, EcDI}_j + \mathbf{25\ \%} \times \mathbf{SDI}_j + \mathbf{25\ \%} \times \mathbf{EnDI}_j$$

[4]Where GDI = Generic Information Disclosure Index, EcDI = Economic Information Disclosure Index, SDI = Social Information Disclosure Index, EnDI = Environmental Information Disclosure Index.

Size: The size (or dimension) of a municipality can be measured using several parameters: the number of inhabitants, the number of voters, the size of the geographical area, the size of the annual budget or the number of employees. Size based on the number of inhabitants is typically employed in studies related to municipalities (Carvalho et al. 2012; McElroy et al. 2005). We apply the municipality classification scheme defined in the Financial Yearbook of Portuguese Municipalities of 2010, which defines three municipality types based on the number of inhabitants: the small, medium and large dimension.

Location (LOC): Regarding geographical location, we follow studies conducted by Ribeiro (2007a, b) and Guerra (2011) by classifying municipalities into the seven geographic regions that comprise the Nomenclature of Territorial Units for Statistics (NUTS) II. Most counties examined in this study (85 %) are located in the Northern, Central and Alentejo regions.

As dispersion between the sampled municipalities complicates the statistical analysis, we follow Ribeiro (2007a) by grouping municipalities by coastal or interior location within continental Portugal and exclude municipalities located in the Archipelagos. For this classification, we follow criteria outlined by Alegria (2004), which distinguishes coastal NUTS III regions from those regions in which the "main" municipality (the most populous in 1991) is located less than 60 km from Lisbon or Porto (approximately 30 min away via transport on a major road). Remaining regions that do meet these criteria are classified as inland regions.

Characterisation (CHARAC): In classifying a municipality as rural or urban, we adopt the classification scheme applied by the Rural Development Program (Programa de Desenvolvimento Rural—PRODER) and Rural Development Program of the Autonomous Region of the Azores (Programa de Desenvolvimento Rural da Região Autónoma dos Açores—PRORURAL), which define rural areas according to the Organisation for Economic Co-operation and Development (OECD) classification, and adapt this scheme to the Portuguese reality (SRAF 2007). This classification is assigned for each *freguesia* (akin to a civil parish), with each defined as either urban or rural. The "rural" classification is assigned to areas meeting certain criteria, including a population density lower than 150 inhabitants per km^2, economic disadvantage and at least 10 % of the population employed in the agriculture and forestry sectors.

Our sample consists of municipalities that we classify as rural or urban depending on the classifications of the majority of *freguesias* that compose them. Thus, if at least 50 % of *freguesias* that compose a municipality are classified as rural, the municipality is considered rural.

Certification (CERT): This variable assumes a value of 1 if the municipality has environmental and/or SR certification and 0 otherwise.

Local Agenda 21 (LA21): This variable assumes a value of 1 if the municipality has implemented Local Agenda 21 and 0 otherwise.

Political Competition (POLCOM): To identify the existence of political competition, we examine the percentage of votes obtained in most recent elections and translate this figure into the number of councillors elected from each party. Thus, if a political party that heads a municipal executive enjoys an absolute majority (more than 50.1 % of the votes), it is expected to feel less pressure from the opposition, resulting in less political competition; if the party won <50.1 % of the votes, then there is likely more political pressure from the opposition.

Educational Level (EDUC): This variable corresponds in our study to the percentage of the municipal population that has completed a higher education degree (i.e., university and polytechnic education) in 2010.

Unemployment Rate (UNEMP): This variable corresponds to the percentage of unemployed individuals in the municipality in 2010.

Inactive Population (INAPOP): This variable corresponds to the percentage of people aged ≤ 19 and ≥ 65 years in 2010.[5]

Tax Burden (TAX): This variable corresponds to the amount of taxes per capita in 2010

Environmental Expenditure (ENVEXP): This variable corresponds to the proportion of environmental expenditure[6] in the total expenditure of the municipality in 2010.

5 Analysis and Discussion of Results

5.1 Descriptive Analysis

Appendix 1 presents a detailed list of descriptive statistics related to the number of entities disclosing information on each of the four pre-defined categories.

Regarding the level of SR disclosure measured by the disclosure index, Table 2 summarises statistical results individually obtained for each sub-index (GDI, EcDI, SDI, EnDI) and the TDI value based on all municipalities included in the sample.

An analysis of the table shows that in general terms, SR information disclosure reaches close to 0.5 (TDI = 0.4613), demonstrating that municipalities exhibit a moderate degree of disclosure. Compared to Navarro et al. (2010a, b) and Joseph

[5]Using the PORDATA Database, we collected for each municipality the number of residents whose ages in 2010 were less than or equal to 19 years of age and more than or equal to 65 years of age. Subsequently, by dividing the number of residents into these age groups based on the total population of each municipality, we obtained a percentage. The percentage of the population falling within these intervals relative to the total population constitutes the INAPOP quantitative variable.

[6]Namely, the extent of municipality expenditure on environmental management and protection in 2010.

Table 2 Disclosure indices: descriptive statistics

Índex	Average	Minimum	Maximum	Standard deviation
GDI	0.2150	0	1	0.11323
EcDI	0.6593	0	1	0.14733
SDI	0.6103	0	1	0.12006
EnDI	0.3607	0	1	0.19481
TDI	0.4613	0	1	0.10881

(2010), who obtained, respectively, values of 0.4042 and 0.286, this value is higher than average, though it is lower than the value obtained by Navarro et al. (2011b) (0.634).

Regarding sub-indices, the index concerning economic information disclosure exhibits the highest value (EcDI = 0.6593), followed by social information (SDI = 0.6103). The index concerning environmental information disclosure (EnDI) reaches a value of 0.3607, indicating that municipalities included in the sample reported only slightly more than one third of the items included in this category. Finally, the index on generic information exhibits the lowest value (GDI = 0.2150). This result echoes results from the study by Navarro et al. (2011a) on regional Spanish governments, in which the generic information disclosure index also generated the lowest value (0.53) while the SDI and EcDI indices reached the highest values (0.66 and 0.61, respectively). Navarro et al. (2010a, 2011b) obtained different results, with the social information index exhibiting a greater degree of disclosure (0.52 and 0.7, respectively) and the environmental information index showing the lowest disclosure degree (0.24 and 0.5, respectively).

5.2 Univariate and Bivariate Analysis

At this stage of the study, we conducted a univariate analysis to identify differences between the GDI, EcDI, SDI, EnDI and TDI indices based on sample characteristics. Similarly, we analysed bivariate relationships between each of the sub-indices and quantitative variables.[7] Tables 3 and 4 show the results obtained.

Table 3 shows that for significance levels of 5 % or less, several associations between various disclosure sub-indices and analysed characteristics exist. In summary, we conclude the following on average:

- urban municipalities and those that implement Local Agenda 21 tend to present higher degrees of generic information disclosure;

[7]First, the normality of variable distributions was analyzed via the Shapiro-Wilk test. Test results show that most of the variables do not follow a roughly normal distribution. Thus, the relationship between indices and the remaining variables was calculated using Spearman's correlation coefficient (nonparametric test). However, because UNEMP and TDI variables exhibit a roughly normal distribution, the relationship between these two variables was assessed using Pearson's correlation coefficient (parametric test).

Table 3 Univariate analysis of disclosure indices based on sample characteristics

	n	GDI		EcDI		SDI		EnDI		TDI	
		Medium rank	Tests	Medium rank	Tests	Medium rank	Tests	Medium rank	Tests	Medium rank	Tests
LOC											
Inland	29	31.31	*M-W* 426.00	25.64	*M-W* 308.50*	23.69	*M-W* 252.00**	23.69	*M-W* 281.50*	0.42 ± 0.10	*T-test* 2.92*
Coastl	31	29.74		35.05		36.87		36.87		0.50 ± 0.10	
CHARAT											
Rural	41	27.49	*M-W* 266.00*	24.48	*M-W* 183.50***	28.02	*M-W* 288.0	24.07	*M-W* 126.00***	0.42 ± 0.10	*T-test* 4.78***
Urban	19	37.00		41.34		35.84		44.37		0.50 ± 0.08	
SIZE1											
Large	10	37.70	*K-W* 4.64	42.90	*K-W* 8.85*	0.70 ± 0.09	*ANOVA* 7.65***	48.80	*K-W* 22.93*	0.57 ± 0.07	*ANOVA* 19.61***
Medium	20	33.30		32.10		0.64 ± 0.08		36.08		0.50 ± 0.09	
Small	30	32.10		25.30		0.56 ± 0.13		20.68		0.40 ± 0.09	
LA21											
No	33	23.92	*M-W* 228.50*	27.62	*M-W* 350.50	0.56 ± 0.13	*T-test* 2.57*	25.94	*M-W* 295.00*	0.42 ± 0.10	*T-test* 2.57*
Yes	27	38.54		34.02		0.56 ± 0.13		36.07		0.51 ± 0.10	
CERT											
No	57	29.62	*M-W* 35.5	29.39	*M-W* 22.0*	29.30	*M-W* 17.0*	29.45	*M-W* 25.5*	29.21	*M-W* 12.0**
Yes	3	47.17		51.67		53.33		50.50		55.00	
POLCOM											
No	33	27.48	*M-W* 346.00	28.47	*M-W* 378.50	26.56	*M-W* 315.50*	0.34 ± 0.20	*T-test* 0.73	0.44 ± 0.11	*T-test* 1.90
Yes	27	34.19		32.98		35.31		0.38 ± 0.19		0.49 ± 0.11	

* $p < 0.05$; ** $p < 0.01$; *** $p < 0.001$; *T-teste* (Teste *Levene*, $p > 0.05$)

Table 4 Bivariate analysis: correlation coefficients

Quantitative independent variables	Index				
	GDI	EcDI	SDI	EnDI	TDI
Size 2	0.26*	0.46***	0.44***	0.62***	0.65***
UNEMP	0.05	0.21	0.06	−0.01	0.13
INAPOP	−0.04	−0.30*	−0.26*	−0.40**	−0.38**
TAX	0.45***	0.43***	0.55***	0.61***	0.70***
EDUC	0.29*	0.40***	0.32*	0.54***	0.56***
ENVEXP	0.28*	0.24	0.32*	0.38**	0.38**

* $p < 0.05$; ** $p < 0.01$; *** $p < 0.001$

- coastal, urban and large municipalities and those with certification tend to present higher degrees of economic information disclosure;
- coastal and large municipalities, those that implement Local Agenda 21, those possessing certification and those that exhibit political competition demonstrate higher degrees of social information disclosure; and
- coastal, urban and large municipalities, those that implement Local Agenda 21 and those with certification exhibit higher degrees of environmental information disclosure. Overall, these municipalities also demonstrate higher degrees of SR information disclosure.

Table 4 shows that significant correlations were found between disclosure indices and independent quantitative variables. These results show that for significance levels below 5 %, SIZE2 (measured by the number of inhabitants), TAX (measured by the tax burden) and EDUC (measured by the percentage of graduates) variables are positively and significantly associated with all indices. The INAPOP variable exhibits a negative and significant relationship with EcDI, SDI, EnDI and TDI. The ENVEXP variable exhibits a positive and significant association with GDI, SDI, EnDI and TDI. UNEMP has no significant relationship with any of the indices.

The table below summarises the (positive or negative) association of each of the possible explanatory factors (independent variables) on the information disclosure index, emphasising situations in which a given explanatory factor has a significant influence either in the Univariate Analysis (UA) and Bivariate Analysis (BA).

Table 5 indicates that all analysed factors that were found to be statistically significant show a positive correlation with the disclosure index with the exception of INAPOP (population aged ≤ 19 and ≥ 65 years), which shows a negative association.

Regarding the TDI, we observe a significant and positive association with all variables with the exception of Political Competition (POLCOM) and Unemployment (UNEMP), and hence, we cannot validate Hypotheses 6 and 8. In this regard, it is important to note the following:

Table 5 Summary of determinant factors influencing the degree of SR information disclosure

	Variables	GDI	EcDI	SDI	EnDI	TDI
H1	Size1		+	+	+	+
	Size2	+	+	+	+	+
H2	LOC		+	+	+	+
H3	CHARAT	+	+		+	+
H4	CERT		+	+	+	+
H5	AL21	+		+	+	+
H6	POLCOM			+		
H7	EDUC	+	+	+	+	+
H8	UNEMP					
H9	INAPOP		−	−	−	−
H10	TAX	+	+	+	+	+
H11	ENVEXP	+		+	+	+

+, i positive influence; −, negative influence

- Regarding the Size (SIZE) variable, the results suggest that, on average, larger municipalities disclose more information and exhibit higher degrees of disclosure. Our results are consistent with most of studies that find a positive and significant association between municipality size and SR information disclosure levels for general (Joseph 2010; Joseph and Taplin 2011; Mucciarone 2012; Pilcher et al. 2008), economic and financial information in particular (Cárcaba García and García-García 2010; Jorge et al. 2011; Martani and Lestiani 2012; Serrano-Cinca et al. 2009; Yu 2010);
- Concerning the coastal and inland variable (LOC), our results are not consistent with Ribeiro (2007a) and Ribeiro and Guzmán (2011). These authors focused on environmental information disclosure, include this variable, and did not find a significant association.
- Regarding the Characterisation (CHARAC) variable for urban/rural classification, the results agree with those obtained by McElroy et al. (2005) with respect to environmental information but contradict those obtained by Joseph (2010) regarding sustainability information disclosure.
- Regarding Certification (CERT), we are not aware of studies focusing on the public sector that examine this variable. However, those focusing on the private sector (e.g., Monteiro and Aibar Guzmán 2010; Barros and Monteiro 2011; Fuente García and González Ayala 2010) have found a positive relationship between environmental certification and environmental information disclosure levels.
- The positive and significant influence of LA21 on SR information disclosure is consistent with previous studies by Joseph (2010) and Joseph and Taplin (2011). Pilcher et al. (2008), despite finding a tendency for municipalities that implement LA21 to disclose more information, did not find a statistically significant influence.
- Concerning the Environmental Expenditure (ENVEXP) and Tax Burden (TAX) variables, the results oppose those found by Navarro et al. (2011a), who

did not find any significant influence of this variable on information disclosure levels. However, regarding the Educational Level (EDUC) variable the results are similar to these authors.

The results also show that the inactive population (INAPOP) negatively and significantly influences the TDI, contradicting the predicted direction (positive). Thus, it is not possible to accept Hypothesis 9. This result does not coincide with results obtained by Navarro et al. (2011a), who found a significant and positive association with social information disclosure. Despite our initial expectations, which are perfectly admissible, this negative association may be attributable to the fact that individuals aged over 65 are less likely to use ICT while those under the age of 19, although likely to resort to ICT if necessary, will likely not view accessing their municipality's website as a priority.

6 Conclusions, Limitations and Suggestions for Future Research

Although several authors (CPASR 2005; Guthrie et al. 2010) have highlighted the need to understand public entities' current sustainability reporting practices, little research has been conducted in this field (Ball and Grubnic 2007; Guthrie and Farneti 2008; Farneti and Guthrie 2009).

The disclosure of SR information using internet has attracted the attention of researchers (Joseph 2010; Joseph and Taplin 2011; Martins 2011; Moneva and Martin 2012; Navarro et al. 2010a, b, 2011a, b; Pilcher et al. 2008), and this study responds to recent calls for more research on this subject.

This paper is based on empirical research directed to both the study of SR reporting of Portuguese municipalitties and the identification of the factors that explain the extent to which these entities disclose SR information. Thus, we conducted a content analysis of websites operated by 60 municipalities and constructed an SR Information Disclosure Index (TDI) that was subdivided into four sub-indices: Generic Information (GDI), Economic Information (EcDI), Social Information (SDI) and Environmental Information (EnDI).

Overall, the results show average disclosure levels across most indices. The calculated Total Disclosure of SR Information Index (TDI) value was 0.46. For each information category or sub-index, the highest value was found in the EcDI category (0.66), followed by SDI (0.61) and EnDI (0.36). GDI presented the lowest value (0.22).

Univariate and bivariate analyses show that certain variables can explain degrees of information disclosure. The Size (SIZE2) (measured by the number of inhabitants), Education Level (EDUC) and Tax Burden (TAX) variables positively influenced all indices, while the Unemployment Rate (UNEMP) variable does not influence the indices. Political Competition (POLCOM) only influenced EnDI, and the remaining variables influenced at least four of the five indices studied.

Concerning the TDI results show a significant and positive association with all variables with the exception of Political Competition (POLCOM) and Unemployment (UNEMP), and hence, we cannot validate Hypotheses 6 and 8. Similarly, Hypothesis 9, regarding the inactive population (INAPOP), is not validated because despite exhibiting a significant association with TDI, the variable presents the opposite sign to what was expected.

In interpreting the findings, one must take into account underlying limitations of this study. With respect to the sample, only 60 municipalities were analysed from the 308 existing municipalities in Portugal. Another limitation relates to the subjectivity of the content analysis technique, which involves making value judgments throughout the data analysis period and in making decisions on whether to include certain variables based on pre-defined items. Hence, the results should be interpreted with caution.

Despite the afore-mentioned limitations, we believe that our results can be considered as a starting point for future investigations. In this sense, future research projects should expand our sample and conduct a longitudinal study that investigates changes in the type and volume of disclosure. We can also identity what factors have a significant influence on SR disclosure level, through a multivariate analysis. Another relevant direction would involve personal interviews as a research method in order to deepen our understanding of the factors that motivate municipalities to disclose SR information.

Appendix 1: Information items by category: descriptive statistics

Category 1: Generic Information						
	Strategy and analysis:	**Yes**	**% Yes**	**No**	**% No**
1.	Is there a separate Corporate social responsibility (CSR) section included on the website?	2	3,33%	58	96,67%
2.	In the general message from the President, and is there any reference to sustainability (CSR) and its importance to the municipality and its strategy?	1	1,67%	59	98,33%
3.	Is there a clear and exclusive statement about sustainability and its importance to the municipality and its strategy?	4	6,67%	56	93,33%
4.	Does this statement include priorities, strategies, objectives and goals?	3	5,00%	57	95,00%
5.	Does this statement include entity events, successes and failures?	0	0,00%	60	100,0%
6.	Is a reference made to the implementation of a strategic plan for sustainable development?	43	71,67%	17	28,33%

(continued)

(continued)

Category 1: Generic Information					
	Governance, commitment and interest group participation				
7.	Is the governance structure of the municipality presented?	50	83,33%	10	16,67%
8.	Are clearly defined social, economic and environmental sections listed on the website?	9	15,00%	51	85,00%
9.	Is reference made to the existence of a workers council or employee representatives?	1	1,67%	59	98,33%
10.	Are interest groups (stakeholders) of the organisation identified?	16	26,67%	44	73,33%
Category 2: Economic Information		**Yes**	**% Yes**	**No**	**% No**
1.	Is information on Budget expenses disclosed?	58	96,67%	2	3,33%
2.	Is information on Budget revenues of the State Budget disclosed?	58	96,67%	2	3,33%
3.	Is information on transfers of revenue from other entities (other than SB), such as community funds, disclosed?	58	96,67%	2	3,33%
4.	Are tariff/tax values for services rendered by the municipality published?	53	88,33%	7	11,67%
5.	Is information on the average payment period disclosed?	18	30,00%	42	70,00%
6.	Is information on the purchasing policy, which takes into account sustainability issues (environmental or social), disclosed?	7	11,67%	53	88,33%
7.	Is information on current or future bids for the supply of goods or services disclosed?	44	73,33%	16	26,67%
8.	Is information on key economic data (GDP growth forecast, employment, unemployment rate, inflation and interest rates) provided?	5	8,33%	55	91,67%
9.	Are annual accounts (accounting) published online?	55	91,67%	5	8,33%
Category 3: Social Information		**Yes**	**% Yes**	**No**	**% No**
1.	Is information on the development of social programs/measures disclosed?	60	100,00%	0	0,00%
2.	Are expenditures on social programs disclosed?	50	83,33%	10	16,67%
3.	Is information on expenditures on local suppliers reported?	0	0,00%	60	100,0%
4.	Are indicators of effectiveness and efficiency in the social field presented?	29	48,33%	31	51,67%
5.	Is information on services provided by the municipality made public?	48	80,00%	12	20,00%
6.	Are public employment offers advertised?	54	90,00%	6	10,00%

(continued)

(continued)

Category 3: Social Information		Yes	% Yes	No	% No
7.	Is information on subsidies allocated/to be allocated for business/associations (recreational, cultural, or otherwise)/NGOs/students (scholarships/awards) disclosed?	50	83,33%	10	16,67%
8.	Is financial data on allocated grants/subsidies disclosed?	47	78,33%	13	21,67%
9.	Is information on workers (total number, training, age, etc.) disclosed?	57	95,00%	3	5,00%
10.	Is information on worker training activities (on any topic) held/to be held disclosed?	24	40,00%	36	60,00%
11.	Is information on community CSR training activities/workshops held/to be held disclosed?	11	18,33%	49	81,67%
12.	Is information on awards or distinctions obtained in the area of CSR disclosed?	1	1,67%	59	98,33%
13.	Is reference made to a plan for preventing risks of corruption and related offenses?	45	75,00%	15	25,00%
Category 4: Environmental Information		**Yes**	**% Yes**	**No**	**% No**
1.	Does the webpage disclose initiatives undertaken to mitigate environmental impacts of municipal products and services?	55	91,67%	5	8,33%
2.	Does the webpage reveal the extent to which this impact has been mitigated?	6	10,00%	54	90,00%
3.	Does the webpage reveal energy consumption figures?	11	18,33%	49	81,67%
4.	Does the webpage disclose actions/initiatives designed to increase energy savings?	37	61,67%	23	38,33%
5.	Are reductions in energy consumption as a result of such initiatives revealed?	6	10,00%	54	90,00%
6.	Does the webpage show the percentage and total volume of recycled and reused water generated by the municipality?	0	0,00%	60	100,00%
7.	Does the webpage provide information on discharge and/or destinations of city sewage?	22	36,67%	38	63,33%
8.	Does the webpage reveal information on air emissions?	14	23,33%	46	76,67%
9.	Does the webpage disclose actions/initiatives executed to reduce emissions?	30	50,00%	30	50,00%
10.	Does the webpage disclose the amount and types of environmental expenditures?	37	61,67%	23	38,33%
11.	Does the webpage disclose the amount and types of environmental investments?	50	83,33%	10	16,67%

(continued)

(continued)

Category 4: Environmental Information		Yes	% Yes	No	% No
12.	Does the webpage disclose information on environmental policies/statements?	20	33,33%	40	66,67%
13.	Does the webpage disclose information on environmental management systems?	6	10,00%	54	90,00%
14.	Is information on prizes obtained on environmental issues provided?	9	15,00%	51	85,00%
	Autonomous information	**Yes**	**% Yes**	**No**	**% No**
	Is an online CSR/sustainability report disclosed? (yes, no)	1	1,67%	49	81,67%
	If yes:				
	Where is it located? municipality documentation / in the CSR "tab"	Local agenda 21 "tab"			
	What is the period covered by the report?	2005 to 2008			
	How frequently is the report produced (annually, biannually, etc.)?	1 report (2009)			
	Are GRI guidelines followed? (yes, no)	yes			

References

Alegria E (2004) Litoral/interior: clivagem com sentido?—Variações populacionais inter-censitárias nas últimas duas décadas. II Congresso Português de Demografia—Demografia e População: os Novos Desafios, Lisboa, 27 a 29 de Setembro

Archel Domench PP, Lizarraga Dallo F (2001) Algunos determinantes de la información medioambiental divulgada por las empresas españolas cotizadas. Revista de Contabilidad 4 (7):139–153

Ball A (2005) Environmental accounting and change in UK local government. Account Audit Account J 18(3):346–373

Ball A, Grubnic S (2007) Sustainability accounting and accountability in the public sector. In: Unerman J, Bebbington J, O'Dwyer B (eds) Sustainability accounting and accountability. Routledge, London, capítulo 13, pp 243–265

Barros C, Monteiro S (2011) O relato ambiental obrigatório e seus fatores explicativos: o caso português do setor metalomecânico e metalurgia de base. XVI Congreso AECA. Granada. España. 21–23 Septiembre. Obra digital—ISBN: 978-84-96648-53-1

Beck A, Campbell D, Shrives P (2010) Content analysis in environmental reporting research: enrichment and rehearsal of the method in a British–German context. Br Account Rev 42 (3):207–222

Bowen A (2012) Green growth: what does it mean? Environmental Scientist. December 6–11

Burrit R, Welch S (1997) Australian commonwealth entities: an analysis of their environmental disclosures. ABACUS, 12(3):293–322

Cárcaba García A, García-García J (2010) Determinants of online reporting of accounting information by spanish local government authorities. Local Gov Stud 36(5):679–695

Carvalho J, Fernandes MJ, Camões P, Jorge S (2012) Anuário Financeiro dos Municípios Portugueses 2010. Edições OTOC

Center for Public Agency Sustainability Reporting (Cpasr) (2005) Research report—sustainability reporting: international uptake, forms and practice, November. http://www.publicagencyreporting.org

Dutta SK, Lawson RA (2009) Aligning performance evaluation and reward system with corporate sustainability goals. Cost Manag 23(6):15–23

Emilsson S, Hjelm O (2004) Different Approaches to standardizes environmental management systems in local authorities—two case studies in Gothenburg and Newcastle. Corp Soc Responsib Environ Manag 11(1):48–60

European Environment Agency (EEA) (2011) Green economy. http://www.eea.europa.eu/themes/economy/intro

Farneti F, Siboni B (2008) Sustainability reporting in Italian local governments. What they do not report? In: 7th Australasian conference for social and environmental accounting research—7 a 9 December 2008, Adelaide, South Australia

Farneti F, Guthrie J (2009) Sustainability reporting by Australian public sector organisations: why they report. Account Forum 33:89–98

Federation of Canadian Municipalities (FCM) (2011) Building Canada's green economy: the municipal role. www.fcm.ca

Frost G, Seamer M (2002) Adoption of environmental reporting and management practices: an analysis of New South Wales public sector entities. Financ Account Manag 18(2):103–127

Fuente García JL, González Ayala FR (2010). La Divulgación de Información Medioambiental en España y México: Un estudio aplicado a las empresas de la bolsa de Madrid y de la bolsa Mexicana de valores. Cuadernos de Trabajo. Facultad De Economía. Universidad Autónoma De San Luis Potosí. México

Gallego-Alvaréz I, Rodriguez-Dominguez L, García-Sanchez MI (2010) Are determining factors of municipal E-government common to a worldwide municipal view? An intra-country comparison. Gov Inf Q 27:423–430

Gandía MC, Archidona JL (2008) Determinants of web site information by Spanish city councils. Online Inf Rev 32(1):35–57

Global Reporting Initiative (GRI) (2005) Sector supplement for public agencies, pilot version 1, incorporating an abridged version of the 2002 Sustainability Reporting Guidelines, March. Global Reporting Initiative, Amsterdam

Global Reporting Initiative (GRI) (2006) Sustainability reporting guidelines version 3, October. http://www.globalreporting.org. Accessed 13 August 2012

Guerra J (2011) Municípios, Participação e Sustentabilidade—Dinâmicas Locais de Imperativos Globais, Tese de Doutoramento. Universidade de Lisboa, Instituto de Ciências Sociais, Lisbon

Guthrie J, Ball A, Farneti F (2010) Advancing sustainable management of public and not for profit organizations. Public Manag Rev 12(4):449–459

Guthrie J, Farneti F (2008) GRI sustainability reporting by Australian public sector organizations. Public Money Manag 361–266

Jorge S, Sá PM, Pattaro A, Lourenço R (2011) Local government financial transparency in Portugal and Italy: a comparative exploratoty study on its determinants. In: 13th Biennial CIGAR conference, bridging public sector and non-profit sector accounting, 9–10 June 2011, Ghent, Belgium

Joseph C (2010) Sustainability reporting on malaysian local authority websites. Doctoral thesis, Curtin University of Technology

Joseph C, Taplin R (2011) The measurement of sustainability disclosure: abundance versus occurrence. Account Forum 35:19–31

Lewis T (2008) Debate: public sector sustainability reporting—implications for accountants. Public Money Manag 28(6):329–331

Lodhia SK (2010) Sustainability reporting by the Australian commonwealth public sector: an application of new institutional theory. In: Proceedings of the 9th Australasian conference on social and environment accounting research (CSEAR) 2010, 5–7 December, Albury Wodonga, Australia

Mack J, Power C (2006) The influences on Queensland and local government to adopt triple bottom line reporting. Research Paper no. 2006-001, School of Accounting, Curtin University of Technology

Marcuccio M, Steccolini I (2009) Patterns of voluntary extended performance reporting in Italian local authorities. Int J Public Sector Manag 22(2):146–167

Marcuccio M, Steccolini L (2005) Social and environmental reporting in local authorities. A new Italian fashion? Public Manag Rev 7(2):155–176

Martani D, Lestiani A (2012) Disclosure in local government financial statements: the Case of Indonesia. Global Rev Account Finance 3(1):67–84

Martins M (2011) Comunicação da Sustentabilidade nas páginas web dos municípios da Europa dos 15. Tese de Mestrado. Faculdade de Economia, Universidade do Porto, Porto

McElroy J, Bisman J, Mathews R (2005) Environmental accounting in NSW local government. In: 7th Asian-Pacific conference on international accounting issues, 20–23 November, Wellington, New Zeland

Moneva J, Martin E (2012) Universidad y Desarrollo sostenible: Análisis de la rendición de cuentas de las universidades públicas desde un enfoque de responsabilidade social. Revista Iberoamericana de Contabilidad de Gestión X(19), Enero_Junio 2012

Monteiro S, Aibar Guzmán B (2010) Determinants of environmental disclosure in the annual reports of large companies operating in Portugal. Corpor Soc Responsib Environ Manag 17 (4):185–204

Moon J (2004) Government as a driver of corporate social responsibility. Research Paper Series No. 20-2004. International Center for Corporate Social Responsibility. Nottingham University. United Kingdom

Mucciarone M (2012) Factors affecting the GRI disclosures in the annual reports of Australian state government departments. Working Paper no. 3. February 2012, Murdoch Business Scholl, Perth, Australia

Navarro A, Alcaraz FJ, Ortiz D (2010a) La divulgación de información sobre responsabilidad corporativa en administraciones públicas: un estúdio empírico en goviernos locales. Revista Española de Contabilidad, RC-SAC 13:2, 285–314

Navarro A, Ortiz D, Alcaraz FJ (2010b) Desafios de la contabilidade pública ante las prácticas de responsabilidade social corporativa. ASEPUC, pp 1–28

Navarro A, Ortiz D, Alcaraz FJ, Zafra JL (2011a) La divulgación de información sobre sostenibilidad en los gobiernos regionales y sus factores influyentes: el caso de España. XVI Congreso AECA, pp 1–24

Navarro A, Ruiz M, Ríos A, Tirado P (2011b) Responsabilidad social y administración pública local: un análisis del grado de divulgación de información en Reino Unido e Irlanda). In: XVI Congreso AECA "Nuevo modelo económico: Empresa, Mercados y Culturas", Granada, Espanha

Organisation for Economic Co-operation and Development (OECD) (2012) Enabling local green growth: addressing climate change effects on employment and local development. www.oecd.org/cfe/leed/49387595.pdf

Pilcher R, Ross T, Joseph C (2008) Sustainability reporting on local authority websites within an institutional theory framework. In: The 7th Australiasian conference for social and environmental accounting research, edited by Roger Burrit, pp 510–531, South Australia: Center for Accounting Governance and Sustainability

Pina V, Torres L, Acerete B (2007) Are ICTs promoting governements accountability?: A comparative analysis of e-governance developments in 19 OECD countries. Crit Perspect Account 18(5):583–602

Pina V, Torres L, Martini C (2009) Accrual Accounting in EU local governments: one method, several aprproaches. Eur Account Rev 18(4):765–807

Queirós M (2002) O ambiente nas políticas públicas em Portugal. Finisterra: Revista portuguesa de geografia 37(73):33–59. ISSN 0430-5027

Ribeiro V (2007a) La gestión medioambiental en las Entidades Públicas Locales Portuguesas. Una Perspectiva Contable. Tese de Doutoramento, Universidade de Santiago de Compostela, Espanha

Ribeiro V (2007b) A divulgação de informação ambiental nas autarquias locais portuguesas. In: XVII Jornadas Hispano-Lusas de Gestión científica—"Conocimiento, Innovación y Empreendedores. Camino al futuro, Logroño, pp 1024–1040, ISBN 978-84-690-3573-3

Ribeiro V, Guzmán C (2008a) Factores determinantes de la divulgacion de informacion medioambiental en las autarquias portuguesas. In: XIII Encuentro de la Asociación Española de Contabilidad y Administración de Empresas, 25–26 Setembro, Aveiro

Ribeiro V, Guzmán C (2008b) Factores explicativos de la extensión de la divulgación medioambiental en las entidades locales portugueses. Conferência GECAMB, 16–17 Outubro, Leiria

Ribeiro V, Guzmán C (2011) La Divulgación de Información Medioambiental en las Autarquias Portuguesas: Caracterización y Factores explicativos. Revista de Contabilidade e Gestão—Revista Científica da Ordem dos Técnicos Oficiais de Contas, Novembro

Rodríguez Bolivar MP, Caba Pérez MC, López Hernandez AM (2007) E-Government and public financial reporting: the case of Spanish regional governments. Am Rev Public Adm 37(2):142–177

Rodríguez Bolivar MP, Caba Pérez MC, López Hernandez AM (2006) Cultural contexts and governmental digital reporting. Int Rev Adm Sci 72(2):269–290

Rondinelli DA, Berry MA (2000) Corporate environmental management and public policy. Am Behav Sci 44(2):168–187

Sanchez R, Muñoz L, Bolívar M, Díaz A (2011) La divulgation online de responsabilidad social universitária. Un estúdio comparativo en las universidades públicas y privadas estaduonidenses. XVI Congresso AECA, Granada, 21 a 23 Setembro

Sciulli N (2009) Sustainability reporting by local councils in coastal regions: an Australian study. Asian J Finance Account 1(1:E1):76–86. ISSN 1946-052X 2009

Sciulli N (2011) Influences on sustainability reporting within local government. Int Rev Bus Res Pap 7(2):282–291

Serrano-Cinca C, Rueda-Tomás M, Portillo-Tarragona P (2009) Factors influencing e-disclosure in local public administrations. Environ Plann C Gov Policy 27(2):355–378

Sousa MJ (2009) Agenda 21 Local. O Caso de Estudo do Norte Alentejano. Dissertação de Mestrado. Faculdade de Ciências. Universidade de Lisboa, Lisboa

Taylor B, Hutchinson C, Pollack S, Tapper R (1994) The environmental management handbook. Pitman Publishing

Tort LE (2010) GRI reporting in government agencies. Global Reporting Initiative (GRI), Amsterdam

United Nations Department of Economic and Social Affairs (UNDESA) (2012a) A guidebook to the green economy, issue 1: green economy, green growth, and low-carbon development—history, definitions and a guide to recent publications, prepared by Cameron Allen and Stuart Clouth, UN Division Sustainable Development, August. https://sustainabledevelopment.un.org/

United Nations Department of Economic and Social Affairs (UNDESA) UNDESA (2012b) A guidebook to the green economy issue 3: exploring green economy policies and international experience with national strategies, prepared by Cameron Allen, UN Division Sustainable Development, November. https://sustainabledevelopment.un.org/

Williams B (2011) Are local government authorities in Australia reporting on sustainability? Doctoral Thesis, University of Tasmania, Australia

Williams B (2012) GRI reporting guidelines in the local government sector—are they being utilized? In: Paper to present in 10th Australasian conference on social and environmental accounting research, 5–7 December 2012, Launceston EJ

Williams B, Wilmshurst T, Clift R (2009) Sustainability reporting in local government in Australia —a preliminary analysis. In: Proceedings of the 8th Australasian conference on social and environment accounting research (CSEAR), 6–8 December. Christchurch, New Zealand EJ

Yu H (2010) On the determinants of internet-based disclosure of government financial information. IEEEXplore Digital Library

Development of Green Economy in Belarus—New Possibilities

Siarhei Zenchanka

Abstract

XXI century is a century of the green economy. The Republic of Belarus seeks to reform its economy and reduce its dependence on energy supplies. Last two years new international and national initiatives were taken to further development of the green economy and turning it into the real economy. The purpose of this article is to present new Belarusian initiatives directed on development of the green economy. The methodology is based on analysis of legislative norms, scientific reports and articles and international and national projects realizing in the Republic of Belarus. This paper considers some new project in the field of green economy realizing in the Republic of Belarus together with international organizations. Besides that paper describes legal acts directed on development of green economy in the Republic of Belarus.

Keywords

Green economy · Green growth · Green industry · Renewable energy

1 Introduction

The Concept of green economy has several decades in his history. The term "green economy" was suggested by Pearce et al. (1989) and received wide distribution in XXI century.

S. Zenchanka (✉)
Plekhanov Russian University of Economics, Moscow, Russia
e-mail: SZenchenko@mesi.ru

W. Leal Filho et al. (eds.), *Sustainable Economic Development*,
World Sustainability Series, DOI 10.1007/978-3-319-45081-0_18

In 2008 the United Nations "Green Economy Initiative" was launched in collaboration with a wide range of partners for motivation governments to invest in green economies. "Green Economy Initiative" consists of several components to provide the analysis and policy support for investing in green sectors and for greening environmentally unfriendly sectors (http://www.unep.org/greeneconomy/AboutGEI/WhatisGEI/tabid/29784/Default.aspx):

- Producing a Green Economy Report and related research materials, which will analyze the macroeconomic, sustainability, and poverty reduction implications of green investment in a range of sectors from renewable energy to sustainable agriculture and providing guidance on policies that can catalyze increased investment in these sectors;
- Providing advisory services on ways to move towards a green economy in specific countries;
- Engaging a wide range of research, non-governmental organizations, business and UN partners in implementing the Green Economy Initiative.

UNEP (2011) defined a green economy as "*one that results in improved human well-being and social equity, while significantly reducing environmental risks and ecological scarcities. In its simplest expression, a green economy can be thought as one which is low carbon, resource efficient and socially inclusive*".

At the same time ICC (2011) defined the term "Green Economy" as part of the sustainable development concept in which business and industry work together for providing economic growth, environmental responsibility and social development.

The "Green Economy" concept is developed in parallel with such initiative as "Green Industry Platform". "Green Industry Platform" is a platform to combine business, government and civil society to promote the green industry agenda. United Nations Industry Development Organization (UNIDO) presented a Green Industry Policy (UNIDO 2011):

> Green Industry promotes sustainable patterns of production and consumption, i.e. patterns that are resource and energy efficient, low-carbon and low waste, non-polluting and safe, and which produce products that are responsibly managed throughout their lifecycle. The Green Industry agenda covers the greening of industries, under which all industries continuously improve their resource productivity and environmental performance. It also aims to create green industries, that deliver environmental goods and services in an industrial manner, including, for example, waste management and recycling services, renewable energy technologies, and environmental analytical and advisory services.

The Outcome document of the Rio+20 Conference "The future we want" (UN 2012) considers "*green economy in the context of sustainable development and poverty eradication as one of the important tools available for achieving sustainable development and that it could provide options for policy making but should not be a rigid set of rules*".

It is clear that all these initiatives consolidate economic, environmental and social priorities of the sustainable development and are directed to solve its problems. Ten Brink et al. (2012) pointed that "*the green economy concept is more than merely "greening" economic sectors, it is a means of achieving the sustainable development imperatives*:

- Improving human well-being: securing better healthcare, education and job security;
- Increasing social equity: ending persistent poverty and ensuring social, economic and financial inclusion;
- Reducing environmental risks: addressing climate change, ocean acidification, the release of hazardous chemicals and pollutants, and excessive or mismanaged waste; and
- Reducing ecological scarcities: securing access to fresh water, natural resources and improving soil fertility".

As it was noted in the report of the ICC (2011), "*a Green Economy requires a holistic approach to decision-making. It integrates and balances policies with respect to environmental, social and economic priorities by considering the intended and unintended consequences of interlinked policies that may result in synergies or barriers and promote or hinder economy-wide, greener growth*".

The National Report (2012) "Sustainable development of the Republic of Belarus based on "green" economy principles" considered achievements in sustainable development and progress in the development of a green economy and identified the directions and principles of the transition of the Republic of Belarus to a "green" economy. These principles include economic policy and instruments, education for "green" economy and science and innovation.

This article considers last international and Belarusian initiatives in the development of the green economy concept and presents some international projects in the Republic of Belarus in this area.

2 Methods and Research Instruments

This investigation is based on an analysis of legislative norms, scientific reports and articles and international and national project realizing in the Republic of Belarus. New international initiatives such as "Agenda 2030", "Circular Economy" and "Inclusive Green Economy" have been analyzed too.

3 Development of a "Green Economy" Concept

In recent years the concept of the green economy has been further developed and turned into the real economy. The green economy is a priority objective of the 7th Environmental Action Program (EAP) (Program 2014). First of all, this program is directed to protect, conserve and enhance European natural capital. Secondly, it establishes goals in developing low-carbon, resource efficient and green economy. It is supposed that the global market for eco-industries will double over the next 10 years; the European renewable energy sector will generate more than 400,000 new jobs by 2020, etc.

The 2030 Agenda (UN 2015) for sustainable development suggests "Plan of action for people, planet and prosperity" for next 15 years. In this agenda 17 goals and 169 targets have been established. Sustainable Development Goals are considered as the starting point for any green economy policy process. Jackson and Victor (2013) consider an investment as the single most important element in the green economy: it embodies the relationship between the present and the future.

The Report (2014) "Resource-efficient green economy and EU policies" analyzed the transition to a resource-efficient green economy in Europe. The authors considered four main factors influencing on achieving green economy goals—eco-innovations, green knowledge, fiscal reforms and financial resources. Considering eco-innovation the authors showed that there is a shift of the European economy to the service type, but service is not such innovative as manufacturing and although "*some types of technological innovation lead to an increased use of resources, green or eco-innovation can be a powerful lever for achieving resource efficiency*". As a result, they pointed that "*there is a policy push in Europe to foster a shift to a green economy*".

UNEP developed manual for policymakers on how to conduct a target-driven Green Economy Policy Assessment (GEPA) with next activities (UNEP 2014):

1. Establishing priority sustainable development targets based on the overall development plans of countries;
2. Estimating the amount of investments required to achieve the targets;
3. Identifying the policies or policy reforms that are essential for enabling the required investments;
4. Assessing the impacts of the required investments as well as the enabling policies using a range of economic, social and environmental indicators and comparing the results with the business-as-usual scenario; and
5. Presenting the assessment results to inform the making of specific decisions.

Transition to green economy demands a development of a corresponding policy. From the perspectives of UNEP (2014), green economy as a tool (as opposed to a particular state of an economy) focuses on mobilizing a more efficient allocation of resources through society's investments to achieve sustainable development. Hence, any policy that is able to mobilize and shift investments to attain specific sustainable development targets can be considered as a "green economy" policy.

Green economy assessment is a necessary step in developing the "green economy" policy and achieving sustainable development. Towards the end of the assessment, analysts need to effectively present the analytic results to policymakers and then to citizens. This will allow investments in an economy greening.

In its "A summary for leaders" UNEP (2015) considers an economy that is circular and green starting from "chains of challenge and the challenges of change" in the process of a transaction to the "Inclusive Green economy". The design principles for an "Inclusive Green Economy" include:

- Centrality of Jobs and the Economy;
- Focus on Public Wealth;
- Investment in Ecological Infrastructure;
- Operationalizing the Precautionary Principle;
- Innovation for Sustainability;
- Natural Resource Conservation;
- Human Resource Development;

- Building Institutions;
- Long-term versus Short-term.

The authors of this report suppose that the realization of these principles will ease the pathways towards achieving the sustainable development goals.

Published in 2015, Synthesis Report (EEA 2015) considered different aspects of a green economy. In accordance with the 7th Environment Action Program it analyzed 4 pillars of green economy—implementation, integration, information and investments. The implementation is directed on protecting, conserving and enhancing of the natural capital and increasing the resource efficiency; the needs for integrated management of natural capital and integrated approach to production-consumption systems were underlined. Besides that "*the wider perspective of the green economy provides a framework for the integration of current policies*". The report discusses the importance of the information on the way to a "circular green economy", what role the knowledge plays in the process of decision-making. Investment in research and innovation has an important role in economy greening.

On the base of discussions with a group of 25 stakeholders from science, business, policy and civil society four clusters of innovations were defined:

1. Collaborative consumption focuses on the ways that consumers can obtain products or services more effectively and resource-efficiently. This may involve fundamentally changing the ways that consumer demands are met, including shifting from individual decisions to organizational or collective demand.
2. Prosumerism reduces the distinction between producer and consumer and can be seen as a particular type of collaborative consumption. An example is distributed energy production systems, enabled by technological innovations such as smart metering and smart grids.
3. Social innovation entails developing new concepts, strategies and organizational forms to better meet societal needs. Both examples above are examples of social innovation, with prosumerism a social innovation that is enabled in part by technological innovation. Social innovation is a problem solving approach that carries a strong potential for generating new social relationships, and is perhaps the most crucial element required for fostering sustainability transitions.
4. Eco-innovation and eco-design go further than technological innovation, to incorporate environmental considerations by either reducing the environmental impact of products or production processes, or incorporating environmental concerns into the product design and life cycle. Harvesting energy from food waste, multi-trophic agriculture, and retrofit building insulation from recycled paper products are just a few examples of eco-innovation and design.

More and more attention in Europe and around the world is paid to the "circular economy" which converts the production-consumption process into one loop of the circular economy (Report 2015). This approach supposes a "growth within" which is focused on getting much more value from the existing stock of products and materials. The European Commission adopted "Circular Economy Package" which consists of an "EU Action Plan for the Circular Economy" (Communication 2015) and Annex (2015) to the Action plan. As an advantages of the circular economy the report pointed at benefits for the environment, boost competitiveness and resilience. This approach is largely the same as the green economy.

It is clear that the transition to a green economy is a long-term process. It requires a multi-dimensional approach directed on solving of sustainable development goals.

The Commission has approved an investment package of €264.8 million to support Europe's environment, nature and green growth.

4 New Belarusian Initiatives and Projects

In PIE (2015) the analysis of the key areas of "green economy" and legislative acts of the Republic of Belarus was presented till the middle of 2014. These results showed that in the field of green economy the Republic of Belarus seeks to act in accordance with international agreements. The legal acts of the Republic of Belarus refer to different aspects of the green economy, such as an energy saving, a drinking water, a waste management, etc. These initiatives in Belarusian legislation were directed on its convergence with European one.

National Agency of Investment and Privatization (NAIP 2015) denoted that the Republic of Belarus has at its disposal a significant resource potential for development of renewable energy sources. This potential in Belarus accounts for around 80 million tons of oil equivalent, which notably exceeds the Belarus's total energy consumption.

Consider new initiatives realized in the Republic of Belarus in the green economy.

4.1 International Cooperation

In 2014 three Belarusian institutions joined to "INNOVER-EAST" Project (2014a) "Building a more effective pathway leading from research to innovation through cooperation between the European Union and Eastern Partnership countries in the field of energy efficiency (2014–2017)". It is supposed that new skills and competences in innovation services at relevant organizations in partner countries will be developed and project results will be contributed in learning of business on energy efficiency.

The UNDP/GEF project "Landscape approach to management of peatlands aiming at multiple ecological benefits" was launched in March 2013 and is set to run until September 2017. The United Nations Development Program and the Global Environment Facility are the partners of this project. The national executing agency of the project is the Ministry of Natural Resources and Environment Protection of the Republic of Belarus. The objective of the project is to promote a landscape approach to a management of a peatland to conserve biodiversity, enhance carbon stocks, and secure multiple ecosystem services with the demonstration in the Poozerie landscape. The project proposes an integrated approach to decision-making on peatland's use that takes into account ecological as well as economic criteria, and considers carbon benefits that may be derived from participation in the voluntary and compliance markets, in addition to biodiversity, land degradation and SFM benefits. A National Strategy for Weatlands Management including a scheme for peatlands management will be developed as a consensus policy document and demonstrations of the restoration and sustainable use of peatlands will take place in a number of sites ranging from protected areas, to agricultural and forested peatlands. The existing MRV protocol for emission reductions from peatlands will be extended to agriculture and forestry's biotopes (Project 2013).

The World Bank supported "Belarus Biomass District Heating Project (2014–2019)". The project has three components. The first component, district heating energy efficiency, includes the following energy efficiency investments in the selected district heating systems: (a) modernization and/or construction of heat substations by installing individual building level heat substations with temperature controls; and (b) reconstruction and/or construction of district heating networks and upgrading of peak load gas boilers. The second component, biomass heat generation, includes investment in base-load biomass boilers, biomass-based small combined heat and power (CHP) plants, and in some towns also wood chipping equipment. The third component, technical assistance, supposes financing the capacity building for the participating district heating utilities and implementation support to the project management unit (PMU), including the following: (a) improvement of existing social accountability mechanisms; (b) support for a shift to energy-content-based biomass pricing; and (c) other project implementation support (Project 2014b).

Project "Clima-East: Conservation and sustainable management of peatlands in Belarus to minimize carbon emissions and help ecosystems to adapt to climate change" started in 2014 and will finish in 2018. The project aims to address the most critical problems of peatland conservation and management in Belarus, seeking to demonstrate innovative approaches to ecosystem-based climate mitigation and adaptation at peatlands (Project 2014c).

Project "Removing Barriers to Wind Power Development in Belarus (2014–2019)" is funded by UNEP. This project proposes to establish a financially viable Private company funded by an investment grant to facilitate investment into wind energy in Belarus and establishing market based empirical precedents. By the end of the project, it is expected that the project will enable the development of sound Feed-in-Tariff and procedures to directly foster >25 MW of installed wind capacity during the 5-year program and the generation of over 1 million MW-h of renewable energy over the project (Project 2014d).

In 2015 the project "Supporting the Transition to a Green Economy in the Republic of Belarus (2015–2017)" financed by the European Union started. The project is aimed on assistance to the Republic of Belarus in the formation of economic growth based on "green" principles, including environmentally sustainable and cost-effective using of natural resources, promotion of environmentally sustainable production and consumption, the creation of "green" jobs, and greater environmental sustainability of the target groups. As a project results the public knowledge of the principles and ideas of "green" economy should be increased, pilot initiatives in areas of waste management, water management, biodiversity conservation, eco-tourism, the creation of environmental information centers, etc. should be realized. Much attention is paid to the development of effective project partnership of business, NGO, industry and local authorities. At the present time an independent organization conducts potential evaluation of organizations-winners (Project 2015a).

Most of these projects are aimed at implementing the individual components of the green economy, such as climate change, biodiversity, energy efficiency, etc. Some of them aimed at achieving the social benefits, i.e. creating the green jobs, improving the social accountability.

4.2 National Projects

In 2015 the project of cooperation between the Ministry of Natural Resources and Environmental Protection and the Belarusian Orthodox Church (BOC) on environmental issues for 2016–2020 should be prepared (Project 2015b). This project supposes specific environmental activities and most important of them will be the events dedicated to the 30th anniversary of the disaster at the Chernobyl Nuclear Power Plant.

In accordance with "State program of collecting (harvesting) and processing of secondary raw materials in the Republic of Belarus for 2009–2015" (Program 2009) the first enterprise on recycling the oil waste started its work in the middle of 2015.

4.3 Legislation

The Decree (2015) of the President of the Republic of Belarus of May 18, 2015 No 209 "On using of renewable energy sources", the Resolution (2015a) of Council of Ministry of the Republic of Belarus of August 6, 2015 No 662 "On the establishment and distribution of quotas to create installations for renewable energy" and the Resolution (2015b) of the Ministry of Economy "On tariffs for electric energy produced from renewables…" of August 7, 2015 No 45 were accepted. These acts change the rules on renewable energy regulation—all investment projects in renewable energy may be realized only within range of quota.

4.4 Events

Annually for 20 years the Belarussian Energy and Ecology Forums are held in Minsk. During the Forum the state of the fuel and energy complex of the republic and environment conditions are analyzed, the tactics and strategy of development in various sectors of the economy are discussed, current problems and solutions are considered. These discussions help in accelerating the implementation of the principles of sustainable development in different sectors of the economy.

Five International agreements between Belarusian and foreign companies from Russia, Moldova, Germany and Finland were signed in the energy area during the XIX Forum in 2014 (Forum 2014).

XX Belarussian Energy and Ecology Forum (2015) was held in Minsk on 13–16 October 2015. In included specialized fair "EnergyExpo" and Energy and Ecological congress. Seminar (2015) "Sustainable City" was a part of this forum. Problems of sustainable development of cities over the world and the attraction of renewable energy sources for well-being were discussed. It was noted that the development of renewable energy sources in the Republic of Belarus lags behind European level.

5 Conclusion

The realization of the described initiatives and projects will help the Republic of Belarus on its path towards a green economy and sustainable development. The development of legal norms is an essential step towards a green economy.

Despite some progress in the development of the concept of "green economy", it should be noted that a small part of Belarusian organizations are implementing their own "green" technologies. According to the National Statistical Committee (Statistical compilation 2015) 9.4 % of organizations reduce energy consumption per unit of production, 4.6 % of organizations carry recycling of industrial waste, water or materials, 6.2 % of organizations reduce energy consumption or energy losses, 1.9 % of organizations improve the possibility of products recycling after use and 4.8 % of organizations cut air pollution, land and water resources, noise reduction.

These data show possible areas of further implementation of green economics principles in the Republic of Belarus.

New Paris Agreement on climate change, adopted 12th December 2015, (Agreement 2015) sets for States the objective of such reducing the emissions of GHG to prevent "*the increase in the global average temperature to well below 2 °C above pre-industrial levels and to pursue efforts to limit the temperature increase to 1.5 °C above pre-industrial levels, recognizing that this would significantly reduce the risks and impacts of climate change*".

The goals of "The national strategy for the sustainable social and economic development of Belarus for the period till 2030" (Strategy 2030), approved in 2015, are directed on decreasing of GHG emission, energy saving and increasing energy efficiency of the economy.

References

Agreement (2015) Adoption of the Paris Agreement. United Nations. FCCC/CP/2015/L.9/Rev.1
Annex (2015) to the Communication from the Commission to the European Parliament, the Council, the European Economic and Social Committee and the Committee of the Regions. "Closing the loop—An EU action plan for the Circular Economy". European Commission. Brussels, COM(2015) 614. http://ec.europa.eu/priorities/jobs-growth-investment/circular-economy/docs/annex-communication-action-plan-for-circular-economy_en.pdf. Accessed 5 Dec 2015
Communication (2015) from the Commission to the European Parliament, the Council, the European Economic and Social Committee and the Committee of the Regions. "Closing the loop—An EU action plan for the Circular Economy". European Commission. Brussels, COM (2015) 614/2. http://ec.europa.eu/priorities/jobs-growth-investment/circular-economy/docs/communication-action-plan-for-circular-economy_en.pdf. Accessed 05 Dec 2015
Statistical compilation (2015) Science and innovation activity in the republic of Belarus. Minsk (in Russian)
Decree (2015) Of the President of the Republic of Belarus of May 18, 2015 No 209 "On using of renewable energy sources". http://cis-legislation.com/document.fwx?rgn=75698. Accessed 16 Nov 2015 (in Russian)
EEA (2015) The European environment—State and Outlook 2015: synthesis report. European Environment Agency, Copenhagen
Forum (2014) Report on the XIX Belarussian Energetically and Environmental Forums. Minsk (in Russian)
Forum (2015) Seminar "Sustainable City". http://www.tc.by/exhibitions/energyexpo2015/news/2983.html. Accessed 16 Nov 2015 (in Russian)
Green Economy Initiative. http://www.unep.org/greeneconomy/AboutGEI/WhatisGEI/tabid/29784/Default.aspx. Accessed 16 Nov 2015
Green Industry Platform. http://www.greenindustryplatform.org. Accessed 16 Nov 2015
ICC (2011) Ten conditions for a transition toward a "Green Economy". Document No. 213-18/7–8 December 2011. www.iccwbo.org/…/ICC-Ten-Conditions-for-a-transition-to. Accessed 12 April 2016
Jackson T, Victor PA (2013) Green economy at community scale. Metcalf Foundation
NAIP (2015) Renewable energy in Belarus. http://www.tender.investinbelarus.by/print/en/invest/opportunities/directions/renewable-energy-and-new-materials. Accessed 16 Nov 2015
Pearce D, Markandya A, Barbier EB (1989) Blueprint for a green economy. Earthscan Publication, London
PIE (2015) Zenchanka, S., Korshiuk, E. The 'green economy' concept in Belarus: today and tomorrow. Progr Ind Ecol Int J 9(1):33–45
Program (2009) A state program of collecting (harvesting) and processing of secondary raw materials in the Republic of Belarus for 2009–2015. http://www.pravoby.info/docum09/part00/akt00928/page3.htm. Accessed 25 Nov 2015
Program (2014) General union environment action programme to 2020. Living well, within the limits of our planet. Publications Office of the European Union, Luxembourg
Project (2013) The landscape approach to management of peatlands aiming at multiple ecological benefits. http://un.by/f/file/00082884-en.pdf. Accessed 25 Nov 2015 (in Russian)
Project (2014a) INNOVER-EAST. http://www.innovereast.eu/. Accessed 16 April 2016 (in Russian)
Project (2014b) Belarus biomass district heating project (2014–2019). http://www.worldbank.org/projects/P146194?lang=en. Accessed 25 Nov 2015 (in Russian)
Project (2014c) Clima-east: conservation and sustainable management of peatlands in Belarus to minimize carbon emissions and help ecosystems to adapt to climate change. http://www.by.undp.org/content/belarus/en/home/operations/projects/environment_and_energy/Clima-East.html. Accessed 25 Nov 2015 (in Russian)

Project (2014d) Removing barriers to wind power development in Belarus (2014–2019). http://www.by.undp.org/content/belarus/en/home/operations/projects/environment_and_energy/00087557/. Accessed 25 Nov 2015 (in Russian)
Project (2015a) Supporting the transition to a green economy in the Republic of Belarus. http://www.greenlogic.by. Accessed 16 Nov 2015
Project (2015b) Project of cooperation program between the ministry of the natural resources and Belarus orthodox church from 2016 to 2020. http://ecoinfo.bas-net.by/ecology-belarus/Nomera/2015/news_belarus_23.pdf. Accessed 16 Nov 2015 (in Russian)
Renewable Energy in Belarus http://www.tender.investinbelarus.by/print/en/invest/opportunities/directions/renewable-energy-and-new-materials. Accessed 16 Nov 2015 (in Russian)
Report (2012) Sustainable development of the Republic of Belarus on "Green" Economy Principles. National report. Minsk. Scientific-Research Economic Institute of the Ministry of Economy of Belarus
Report (2014) Resource-efficient green economy and EU policies. EEA Report No 2. Luxembourg: Publications Office of the European Union
Report (2015) Growth within: a circular economic vision for a competitive Europe. Report by the Ellen MacArthur Foundation, the McKinsey Centre for Business and Environment and the Stiftungsfonds für Umweltökonomie und Nachhaltigkeit (SUN)
Resolution (2015a) of Council of Ministry of the Republic of Belarus on August 6, 2015, No 662 "On the establishment and distribution of quotas to create installations for renewable energy". http://cis-legislation.com/document.fwx?rgn=78720. Accessed 16 Nov 2015
Resolution (2015b) of the Ministry of Economy "On tariffs for electric energy produced from renewables…" of August 7, 2015 No 45. http://pravo.by/main.aspx?guid=3871&p0=W21530189. Accessed 16 Nov 2015 (in Russian)
Seminar (2015) "Sustainable City" http://www.tc.by/exhibitions/energyexpo2015/news/2983.html. Accessed 16 Nov 2015 (in Russian)
Strategy (2030) The national strategy for the sustainable social and economic development of Belarus for the period till 2030. Econ Bull Econ Res Inst 4(214):6–99 (in Russian)
Ten Brink P, Mazza L, Badura T, Kettunen M, Withana S (2012) Nature and its role in the transition to a green economy. Executive summary
UN (2012) The future we want: outcome document adopted at Rio+20. United Nation. A/RES/66/288
UN (2015) Transforming our world: the 2030 agenda for sustainable development. United Nations A/RES/70/1
UNEP (2011) Towards a green economy: pathways to sustainable development and poverty eradication—a synthesis for policy makers. www.unep.org/greeneconomy. Accessed 16 Nov 2015
UNEP (2014) A guidance manual for green economy policy assessment
UNEP (2015) Uncovering pathways towards an inclusive green economy: a summary for leaders. United Nations Environment Program
UNIDO (2011) UNIDO green industry policies for supporting green industry

Author Biography

Siarhei Zenchanka is Professor of Economics Chair of Minsk Branch of Plekhanov Russian University of Economics, and has a Ph.D. in Applied Science. His research focus is in the areas of sustainable development, education for sustainable development. He is an auditor in environmental management.

MIX
Papier aus verantwortungsvollen Quellen
Paper from responsible sources
FSC® C105338
www.fsc.org

If you have any concerns about our products,
you can contact us on
ProductSafety@springernature.com

In case Publisher is established outside the EU,
the EU authorized representative is:
Springer Nature Customer Service Center GmbH
Europaplatz 3, 69115 Heidelberg, Germany

Printed by Libri Plureos GmbH
in Hamburg, Germany